Mineral and Energy Resources
Occurrence, Exploitation, and Environmental Impact

DOUGLAS G. BROOKINS
University of New Mexico

Merrill Publishing Company
A Bell & Howell Information Company
Columbus   Toronto   London   Melbourne

For my daughters,
  Laura and Rachel

Cover Photo: George Elich

Published by Merrill Publishing Company
A Bell & Howell Information Company
Columbus, Ohio 43216

This book was set in Meridien.

Executive Editor: Stephen Helba
Production Editor: Julie A. Higgins
Art Coordinator: Gilda Edwards
Cover Designer: Brian Deep

Library of Congress Catalog Card Number: 89–62128
International Standard Book Number: 0–675–20755–X
Printed in the United States of America
1 2 3 4 5 6 7 8 9—94 93 92 91 90

Preface

The world community is today cognizant of our precarious situation with regard to both energy and mineral resources. Sophisticated techniques to find elusive mineral deposits and fossil fuels are being developed at an ever increasing pace, although the return seems to be dwindling. How to successfully exploit new discoveries of minerals or energy sources is also given wide attention. Finally, an awareness of the environmental impacts of the exploitation of these resources is widespread, in some countries more so than in others.

The point is well taken that many of our mineral resources, both worldwide as well as in the United States, are in short supply. The United States, which consumes roughly thirty percent of the world total for many substances (iron, copper, fertilizer chemicals, etc.), is becoming more and more reliant on imports as increasing costs and other factors cause shutdowns of many domestic operations. In the future, the per capita consumption of many of these commodities will increase in developing nations to the point that this "endless" supply of imports will be severely taxed. This text takes a careful look at mineral resources, with emphasis on United States supplies, but it also appraises the world resources picture. Environmental concerns about mineral resources exploitation are very real, and both necessary and unnecessary environmental restrictions have been placed on the United States industries involved with mineral resources exploitation.

There are several missions in this book. First, I wish to cover the adequacy of supply of mineral and energy resources, and data are included that allow projections for twenty or more years in most cases. For example, water resources are dwindling, and the Greenhouse Effect will make the United States situation far worse. Soils are being eroded senselessly, and wetlands and other lands are being contaminated on a wide scale. Reasons for our decline in water resources are explored, and the finger of blame points clearly to man's activities.

Second, I wish to cover the environmental impact of mineral and energy resource exploitation, as well as the environmental aspects of chemical and other wastes. There is no "free ride" in today's energy picture. All energy options—coal, nuclear, oil, gas, hydropower, solar, geothermal, and future sources—have associated risks that must all be assessed in terms of environmental impact and potentially adverse public health effects. In this assessment, care is taken to explore aspects of milling and other front end parts of the energy chains, as well as a hard look at disposal of wastes.

The third mission of this book is the most important and the most obvious: education. It is imperative that the world become educated on the issues of mineral and energy exploitation and how to achieve minimal environmental impact and minimal adverse impact on public health. The world faces very hard decisions in the near future on minerals, energy, water supply and use, population control, food availability, and disposal of materials from all aspects of human use. It is hoped that this book will present the reader with enough factual information to help overcome this situation.

Many individuals have contributed their expertise to the development of this book. I particularly want to thank the reviewers for their valuable feedback: Arthur H. Brownlow, Boston University; Peter L. Calengas, Western Illinois University; Peter Crowley, Amherst College; Gregory Eicholz, Georgia Institute of Technology; C. M. Lesher, University of Alabama; Nancy Lindsley-Griffin, University of Nebraska (Lincoln); Paul D. Nelson, St. Louis Community College (Forest Park); Donald B. Potter, Hamilton College; and William C. Rense, Shippensburg University.

Finally, I wish to thank Ms. Judith Metelits for her painstaking efforts in editing and typing the several drafts of this manuscript. Her hard work, conscientious attention to detail, and word processing expertise have made completion of this book in my lifetime possible.

Contents

v

1
Introduction

The world community is dependent on earth and energy resources, and in order for the world to meet its impending demand for goods and energy, there must be an awareness of how and where earth resources occur and how best to use all energy resources. The assumed goals of every nation are to become industrialized and to promote a healthy society. Although food is abundant throughout the world, it is, like the earth and energy resources, distributed unevenly. So, too, are water resources.

Population growth may be the single most important factor leading to long-term environmental degradation on this planet. At the time of this writing, the world population is steadily creeping, and sometimes almost leaping, toward the 6 billion mark. Further, the world population has been doubling every 33 years or so. Interestingly, due to the herculean efforts of the People's Republic of China, with over one-fifth of the world's people, there is some indication that this doubling period of 33 years may be lengthening. Let us certainly hope so, because in some undeveloped countries the population doubling period is 20 years. The U.S. National Academy of Science has estimated that this planet can produce food to meet the needs of just over 30 billion people—at a starvation level, and assuming worldwide food distribution. This 30 billion figure could be reached roughly half to two-thirds of the way through the twenty-first century, within our children's lifetimes.

There are other factors to consider. Chemicals for the agriculture industry and metals for farm equipment will experience a doubling within the next 10 years or so, but supplies of both are finite and not overly large. So, too, are fossil-fuel energy resources. It is predicted that the world supply of petroleum products will peak early in the next century.

Though this perspective is rather gloomy, all is not as bleak as indicated. In the chapters to follow, resources of water, metals, nonmetals, and energy will be discussed. It will be pointed out that in many instances supplies of various earth resources may be greater than determined just a few years ago, and that scientific data have revealed intelligent ways to meet our energy demands as well.

However, the response of many nations is an attempt to exploit resources rapidly and often inefficiently. Commonly this goes hand in hand with a blatant disregard for the environmental consequences of poorly harnessing resources and energy.

Therefore, the chapters to follow also focus on the environmental aspects of earth resources and energy sources. In them I discuss mining, milling and other preparatory techniques, then waste disposal, water pollution, and atmospheric pollution. The final chapter focuses entirely on environmental issues. Because this book is not intended to be a text for economic geology, nor is it, an exhaustive look at energy or the environment, additional readings are listed at the end of each chapter.

Any resource from any part of the animal, vegetable, or mineral kingdoms can be called an earth resource. In this book, however, we are concerned with only those commodities that are normally discussed together as *mineral resources.* Wind energy and solar energy are included in this book as well, although they are not "mineral" resources. It is to all of these resources that we owe our existence.

Before proceeding further, a note about units of measurement. There is no worldwide agreement on units for reporting different types of earth resources or environmental problems. Where possible, I have used the International System (S.I.) of units, but the reader will note many instances where measurements are reported in U.S. Customary units. I have felt this necessary so that when reference is made to the original source, the units will be consistent. Further, many metals are reported in ounces, pounds, short tons, long tons, metric tons—even, in the case of mercury, flasks (where one flask equals 76 pounds). Conversion tables are included in the Appendix to help the reader convert one set of units to another.

RENEWABLE AND NONRENEWABLE RESOURCES

Mineral resources may be classified as renewable or nonrenewable. A *renewable* resource, after use, will eventually be available again. For some resources, such as groundwater, the lag time between initial use and renewal may be so long that for practical purposes the resource becomes nonrenewable. *Nonrenewable* resources include those commodities that, once used, are never again available. For exaple, burning fossil fuels and splitting atoms remove these substances from use by future generations.

A *reserve* is that part of a resource for which we have precise quality and quantity information (for example, at least a 90 percent probability that a certain amount of rock will yield a specific tonnage of metal). For mineral commodities, such information is best obtained by drilling, assaying the drill core or cuttings, and combining the drill information with geologic data. Depending upon the degree of confidence we have in our estimates, reserves can be classified further as

FIGURE 1–1   Distinguishing between reserves and resources. *Reserves* are economically recoverable materials from identified deposits, as distinguished from *resources* of the same material from identified but subeconomic or hypothetical deposits. (Source: From Douglas G. Brookins, *Earth Resources, Energy, and the Environment,* Merrill, 1981.)

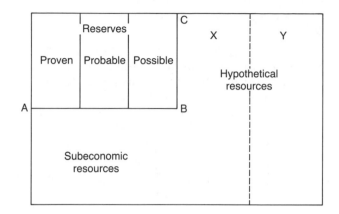

proven, probable, or *possible.* These three terms range from a very high level of confidence to a relatively uncertain degree of confidence, but all are still based on firm criteria.

As shown in Figures 1–1 and 1–2, we distinguish reserves from resources. Some resources will be added to or subtracted from reserves as economic conditions change. Other resources are classified as hypothetical, or undiscovered. The assumption is that additions to reserves are more likely from known mining areas and less likely from similar areas of little or no mineral exploitation. In Figure 1–1 the line A–B, representing the limit of profitability, can fluctuate up or down. It merely serves to separate the proven-probable-possible reserves from lower-grade *subeconomic resources.* Should the price of the material rise, the line will move down. For *hypothetical resources,* two categories can be defined: undiscovered ore from known areas of ore occurrence, (X on Figure 1–1), and undiscovered ore from areas where some geologic criteria indicate ore may exist but from which no deposits are known (Y on Figure 1–1). Both hypothetical resources are subject to the same economic considerations as the reserves, as represented by line A–B. Only when sufficient drilling or other techniques have indicated the

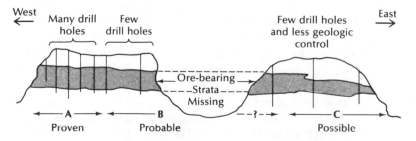

FIGURE 1–2   Classification of reserves. This cross section of ore-bearing strata (from west to east) illustrates proven-probable-possible ore reserves. (Source: From Douglas G. Brookins, *Earth Resources, Energy, and the Environment,* Merrill, 1981.)

presence of some ore can hypothetical resources be properly classified as a reserve.

In Figure 1–2, area A, with its many drill holes (the vertical lines) from which the rock has been assayed, is an area of proven reserves. In area B there are fewer drill holes and thus less certainty, but the grade (tenor) of the ore and its thickness are about the same as for area A, and it is classified as a probable reserve. Further east (area C) the ore horizon thins and the tenor of the ore drops, too. These factors, coupled with lack of drill holes, would dictate that the ore reserves be classified as possible. If the tenor of the ore did not decrease as shown but the thickness of the unit did decrease, then the ore classified as possible might be more properly classified as subeconomic in that too much barren rock would have to be removed to recover the ore profitably.

MINERALS

Rocks are generally composed of one or more minerals. *Minerals* are naturally occurring inorganic chemical compounds with crystalline structures. Of the 92 naturally occurring elements, nine (oxygen, silicon, aluminum, iron, calcium, magnesium, sodium, potassium, and titanium) make up over 90 percent of all the rocks of the earth's crust. Oxygen and silicon together account for almost 75 percent by weight (47.5 percent and 27.5 percent, respectively). Oxygen bonded in rocks and minerals accounts for 94 percent of their volume.

Of these nine elements, none is an economically recoverable resource in the common rock-forming minerals. This is because all nine elements are found in rocks and minerals as ions, which are bonded together in different ways. *Ions* are merely atoms that have gained or lost electrons. An atom that loses one or more electrons becomes a positive charged ion, called a *cation* because it will be attracted to the cathode in an electrolytic cell. Atoms that gain electrons take on a negative charge and are called *anions* because they would be attracted to the positively charged anode in the same cell. Of the nine elements listed above, only oxygen forms an anion (O^{2-}) while the others form cations (Si^{4+}, Al^{3+}, Fe^{2+}, Fe^{3+}, Mg^{2+}, Ca^{2+}, Na^+, K^+, and Ti^{4+}).

Oxygen and silicon commonly combine to form the silica tetrahedron (SiO_4^{4-}), a central ion of Si^{4+} surrounded by four O^{2-} ions. This structure is possible because the radius for the Si^{4+} ion is only about one-fourth that of each O^{2-} ion. The net charge on the silica tetrahedron is -4, as the four positive charges of the cation only partially offset the eight negative charges of the anions. This negative charge is usually balanced by other elements surrounding the tetrahedron in natural compounds.

A few major rock-forming minerals constitute about 90 percent of the rocks found in the earth's crust. In Table 1–1 they are written in simplest form as compounds. At first glance, this list of minerals looks complex. When we remember that there are more than 2500 minerals found in nature, then the handful listed in Table 1–1 are impressive because they are so much more abundant than all the others.

TABLE 1–1   Major rock-forming minerals.

Mineral	Formula
Olivine	$(Mg,Fe)_2SiO_4$
Pyroxene	
Enstatite	$MgSiO_3$
Diopside	$CaMgSi_2O_6$
Augite	$(Ca,Na)\ (Mg,Fe^{2+},Fe^{3+},Al)\ (Al,Si)_2O_6$
Amphibole	
Hornblende	$(Ca,Na)_{2-3}(Mg,Fe^{2+},Fe^{3+},Al)_5(Al,Si)_8O_{22}(OH)_2$
Potassium feldspars	$KAlSi_3O_8$
(sanidine, microcline, orthoclase)	
Plagioclase—a mixture of	
Albite	$NaAlSi_3O_8$
Anorthite	$CaAl_2Si_2O_8$
Micas	
Biotite	$K(Mg,Fe)_3AlSi_3O_{10}(OH)_2$
Muscovite	$K(Al_2)AlSi_3O_{10}(OH)_2$
Quartz	SiO_2

It is noteworthy that titanium (as Ti^{4+}) has not specifically been included in any of the minerals listed so far. This is because titanium is usually hidden in silicate minerals such as augite and hornblende, where due to its similar size (ionic radius) it can substitute for Fe^{3+} or, less commonly, for Fe^{2+} or Mg^{2+}. Owing to its large charge and size relative to Si^{4+} and Al^{3+}, titanium commonly forms accessory minerals where it is a dominant ion—ilmenite ($FeTiO_3$) and sphene ($CaTiSiO_5$), for example. Other important accessory minerals are listed in Table 1–2.

Only in the mineral olivine is the net charge of -4 balanced by a simple charge of $+4$. In most other minerals corners of the silica tetrahedron are shared with other tetrahedra to form chains (pyroxenes and amphiboles), sheets (micas and clay minerals), and three-dimensional stacks or frameworks (feldspars and quartz). All minerals with silica tetrahedra are known as silicates. If silicon is absent and only oxygen is present as an anion, then the mineral is called an oxide. Hematite (Fe_2O_3) and magnetite (Fe_3O_4) are oxides.

When carbon or phosphorus combines with oxygen to form an anion such as carbonate ion (CO_3^{2-}) or phosphate ion (PO_4^{3-}), then carbonate minerals such as calcite ($CaCO_3$, the main mineral in limestone) or dolomite ($CaMg(CO_3)_2$) occur as well as phosphate minerals such as apatite, $Ca_5(PO_4)_3(OH,F)$. Sulfur usu-

TABLE 1–2   Common accessory minerals.

Mineral	Formula	Comments
Apatite	$Ca_5(PO_4)_3(OH,F)$	Contains phosphorus
Pyrite	FeS_2	Main sulfur-bearing mineral in igneous rocks
Calcite	$CaCO_3$	Most abundant carbon-bearing mineral (organic carbon not included)
Magnetite	Fe_3O_4	Common in igneous and metamorphic rocks
Zircon	$ZrSiO_4$	Common accessory in igneous rocks
Sphene	$CaTiSiO_5$	Common accessory in granitic rocks
Gypsum	$CaSO_4 \cdot 2H_2O$	Common sedimentary rock mineral
Clay minerals	Al-silicates with H_2O, Fe, Mg, etc.	Common in sedimentary rocks

ally occurs in nature with charges of -2, 0, or $+6$. In most sulfides the sulfur has a charge of -2 (as in the lead ore galena, PbS). Pyrite (FeS_2) is a notable exception and is the most common accessory sulfide mineral found in common rocks. Native sulfur is common in hot-spring areas near volcanic areas or in other places where hot waters or bacterial action have caused its formation. The ion S^{6+} is usually found in combination with oxygen as SO_4^{2-}, the sulfate ion. The well-known minerals barite ($BaSO_4$), and gypsum ($CaSO_4 \cdot 2H_2O$) are good examples of sulfate minerals; both are of significant economic importance. Another group of anions in nature is the halides (Cl^-, F^-, and Br^-). Chlorides, the first of these, is found in common table salt, the mineral halite (NaCl). Sylvite (KCl), or potassium chloride, is the most important ore for potassium. The most important mineral with the fluoride ion (F^-) is fluorite (CaF_2), or calcium fluoride. Other lesser-known compounds will be mentioned as necessary. Except for iron (which can be Fe^{2+} or Fe^{3+}), all of the ions listed in the preceding formulas are assumed to possess their most common charge.

ROCKS AND THE ROCK CYCLE

Earth is made up of rocks and minerals. The oceans, which cover about two-thirds of the earth's surface, constitute only a very small fraction of the crust and an even smaller fraction of the whole earth. It is easy to introduce the kinds of rocks we find in the earth's crust by use of Figure 1–3, showing the relationships of different rocks as part of a cycle.

Magma, or molten rock formed at depth, crystallizes to form igneous rocks. Magma that reaches the earth's surface is called lava, and it may form either finely crystalline or even noncrystalline, glassy *volcanic* rocks. Magma that crystallizes at depth forms coarse-grained *plutonic* rocks.

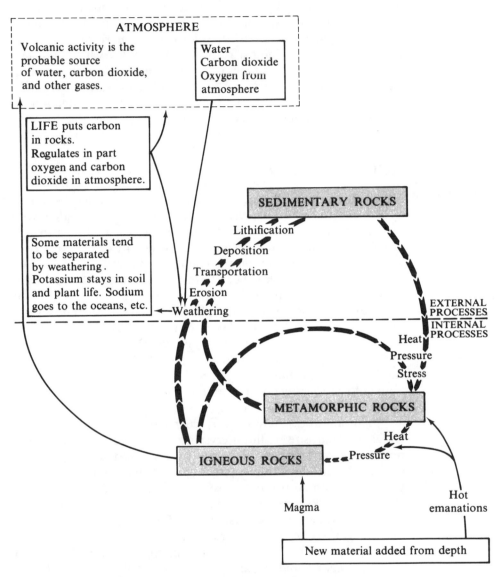

FIGURE 1–3   The middle of this diagram, showing the relationships among igneous, sedimentary, and metamorphic rocks, is what is generally considered the rock cycle. The upper and lower parts of the figure show how material is added to and subtracted from the rock cycle. (Source: From B. Clark Burchfiel, Robert J. Foster, Edward A. Keller, Wilton N. Melhorn, Douglas R. Brookins, Leigh W. Mintz, and Harold Thurman in *Physical Geology*, Merrill, 1982.)

The rocks exposed at the earth's surface to weathering and erosion result in both mechanical and chemical products being added to the oceans and other surface waters. When the mechanically added materials—sediments—settle to the bottom and are compacted, they form *sedimentary* rocks. Chemical precipitates from seawater and inland seas form *evaporites, limestones,* and similar rocks.

When any of these types of rocks are buried and subjected to great pressures and temperatures, though not great enough to cause widespread melting, their mineralogies are changed and *metamorphic* rocks form. Metamorphic rocks can be formed from igneous, sedimentary, or older metamorphic rocks.

IGNEOUS ROCKS

The common types of igneous rocks are given in Table 1–3, a greatly simplified rock chart. Table 1–3 is presented here primarily for readers without a previous geology course. The most abundant rock type on the earth's surface is basalt, which forms most of the oceanic crust (the part of the crust covered by oceans) and is an important rock on the continents as well. If basaltic magma crystallizes at depth, then the rock gabbro forms. Thus Table 1–3 not only lists the igneous rocks in a chemical trend from most silicic (highest SiO_2 content) to basic (low SiO_2) to ultrabasic (lowest SiO_2), but also notes a difference in rocks due to their formation at depth (plutonic) or at or near the earth's surface (volcanic).

Volcanic rocks include those that reach the surface of the earth and those that are emplaced at very shallow depths. Volcanoes spew out a variety of materials, such as ash (finely divided rock), along with water vapor and other gases. Some of the rocks deposited from volcanic eruptions include tuff, volcanic breccia, perlite, pumice, and scoria.

Plutonic rocks may be very coarse grained, such as pegmatites, and vary widely in grain size within a crystallizing magma chamber. A special type of igneous rock is a *porphyry*. Here crystallization started at depth, forming fairly large crystals. Then the setting of the magma chamber changed so that crystallization proceeded with great rapidity, resulting in a sea of fine-grained crystals (or even glass if the rocks were extruded rapidly) surrounding the large crystals. The large crystals are called phenocrysts, and the fine-grained or glassy matrix, the groundmass. Together they form the rock called porphyry. Porphyritic rocks are important for some metal ore deposits such as copper and molybdenum.

The *Bowen reaction series* is important to igneous rocks for, as shown by Norman L. Bowen in 1929, one can derive many types of major igneous rocks from the same assumed parent magma, one with the composition of a basalt. This is shown in Figure 1–4. Even though the assumed starting magma is basaltic, the first minerals to form as the magma is cooled are high-density olivine and low-density, calcium-rich plagioclase. These minerals have one of two fates: They can settle out of the cooling magma, or they can react with it to form new minerals. Olivine that so reacts with the magma, and thus forms the next highest temperature mineral pyroxene, is said to be resorbed by the magma. Similarly, the cal-

TABLE 1–3   Classification of igneous rocks.

Rock (coarse/fine)	Essential Minerals	Important Accessory Minerals
Granite/rhyolite	Quartz, K-feldspar, plagioclase	Biotite, ± hornblende, ± muscovite
Granodiorite/dacite	Quartz, plagioclase, K-feldspar	Biotite, hornblende
Pegmatite	Quartz, K-feldspar, sodic plagioclase	Muscovite, biotite
Diorite/andesite	Plagioclase, hornblende	Biotite
Syenite/trachyte	K-feldspar, plagioclase, feldspathoids	Biotite, amphiboles
Gabbro/basalt	Pyroxene, plagioclase	Olivine
Peridotite	Olivine, pyroxene	± hornblende, ± plagioclase
Anorthosite	Calcic plagioclase	Ilmenite
Dunite	Olivine	Pyroxene
Pyroxenite	Pyroxene	Olivine
Kimberlite	Olivine, pyroxene, ilmenite	Phlogopite, ± diamond, pyrope
Lamproite	Olivine, pyroxene	Pyrope, phlogopite, ± diamond
Carbonatite*	Calcite, other carbonates	Several

*Carbonatitic lavas are known but have no special name.

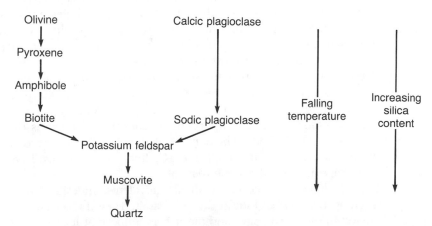

FIGURE 1–4   The Bowen reaction series, showing the general sequence of mineral formation with falling temperature in a basaltic melt. (Source: From Burchfiel, et. al, *Physical Geology*, Merrill, 1982.)

cium-rich plagioclase may rise in the magma or may be resorbed by the magma to form a more sodium-rich plagioclase. In the Bowen reaction series, there are two actual series. The series starting with olivine and continuing to biotite is a discontinuous series, because each new mineral formed by the resorption process is different in chemistry and crystallography from the earlier formed mineral. The plagioclases form a continous series because, even though there is a difference in chemistry, the calcium-rich, high-temperature and sodium-rich, low-temperature minerals are both plagioclases. Below the approximate joining of the discontinuous and continuous series, the minerals potassium feldspar, then muscovite, and finally quartz form. These minerals, though not in a strict sense part of the reaction series, form from the magmas most enriched in silica (SiO_2). Nature shows us that magmatic crystallization is very complex. If olivine does indeed sink in a magma, it forms a rock called dunite. When this happens, the composition of the remaining magma is increased in silica content because the olivine that has settled out has the lowest silica content of the common rock-forming minerals. In this fashion rocks progressively richer in silica can be derived from a parent basaltic magma by a series of chemical and physical processes involving crystal settling, crystal rising, reactions of crystals with magma, squeezing residual liquids out of crystal accumulations, and so on. In this way Bowen showed that, in theory, all major rock types could be explained, at least in part, by his reaction series.

Yet the Bowen reaction series cannot explain the preponderance of high-silica rocks on the continents. It is instead proposed that early in the earth's history some very large-scale differentiation of the planet occurred. During the process, high-silica material was transported up to form a major part of the crust, while the underlying mantle became more silica poor. Thus igneous rocks formed in the continental crust tend to be more silica-rich than those in the oceanic crust, where no high-silica crust was formed.

When igneous rocks are very rich in silica, then quartz appears as an essential mineral. In Table 1–3 the coarse-grained varieties are granite, granodiorite, and pegmatite. Rhyolite and dacite are the fine-grained equivalents of the first two, while pegmatites have no fine-grained equivalents. Pegmatites are important as they usually form from the last residual magma crystallizing, and thus they are commonly enriched in many elements that do not easily enter the structures of the common rock-forming minerals. Elements like niobium, uranium, beryllium, tantalum, and others are enriched in pegmatites.

Granitic rocks are important in economic geology as they are often associated with fluids from which mineral deposits form. These will be discussed as individual elements in Chapter 4. The fine-grained rocks, rhyolite and dacite, are also important for the same reason.

Diorites and their fine-grained equivalents, andesites, are intermediate in composition between granitic rocks and basalt. They do not contain sufficient silica for quartz to form. Andesites and diorites are important for some ore deposits.

Basalts are not commonly hosts for many metal deposits, but special types of their coarse-grained equivalents, gabbros, are. One such special rock is norite, which is commonly enriched in nickel, copper, and cobalt.

Syenites and their fine-grained equivalents, trachytes, are interesting rocks. They contain an abundance of sodium- and potassium-rich minerals, but do not contain quartz as they are somewhat impoverished in silica. They also commonly contain minerals known as feldspathoids, aluminosilicates rich in sodium and potassium that resemble feldspars. Syenites often contain concentrations of elements such as phosphorus, uranium, rare earth elements, and others; and they may be of significant economic importance.

Rocks that are highly impoverished in silica are called ultrabasic. The rocks in Table 1–3 that fall into this category are peridotite, pyroxenite, dunite, anorthosite, kimberlite, and lamproite. The first three of these form in magma bodies as early olivine and pyroxene; because of their density, they settle to the bottom of the chamber. Such rocks may contain economic quantities of chromium, platinum, nickel, and similar metals. Anorthosites are composed mainly of calcic plagioclase, or anorthite. Of low density, this mineral attempts to rise in the magma chamber. Because of the magma's high viscosity, the anorthite is only partially successful, and it is stopped and forms lenses and layers. It is economically important because many anorthosite bodies are mineralized with titanium.

Kimberlite is the most important source of diamond in nature, although some lamproites also contain diamonds. Both rocks form well down in the upper mantle, and they cannot be easily explained by the Bowen reaction series. These rocks are covered in Chapter 6.

SEDIMENTARY ROCKS

Sedimentary rocks include sandstone, shale, limestone, chert, marine evaporites, and nonmarine evaporites. Sedimentary rocks cover two-thirds of the exposed continental crust, and they are of great importance to mineral and energy resources.

Sandstones are composed mainly of fine mineral grains and sometimes rock particles. Quartz is the most common constituent of sandstone due to its great resistance to weathering. The sand grains are cemented by agents such as silica or calcite. Often sandstone has high permeability that allows solutions to move through it with ease. Thus sandstones make good aquifers (water sources), are mineralized as solutions passing through them precipitate minerals, or act as pathway and reservoir rocks for oil and gas accumulation.

While sand grains are deposited fairly near their source localities, finer grained material is carried greater distances. This includes particles of silt to clay size, when they settle out and are compacted, the rocks siltstone to shale form. Shale is the most common sedimentary rock, making up about 70 percent of the continent's sedimentary cover. Shales are important sources for many metals, as aquitards (groundwater barriers), for phosphate ores, and oil-migration barriers, as well as natural barriers to prevent radioactive elements from migrating away from radioactive waste disposal sites.

Limestone forms most commonly by precipitation of calcite from seawater. Mechanical accumulations of calcite-shell secreting organisms, such as reefs, are also important. Limestones are commonly jointed, and thus water and petroleum can migrate through them. Occasionally limestone is replaced by metal-bearing solutions and houses ore deposits, and it is often used for building material.

Chert is a cryptocrystalline rock made up of silica. It commonly occurs interbedded with limestone and occasionally shale.

Evaporites form from the evaporation of alkaline or salt water. Marine evaporites form from evaporation of trapped seawater, and nonmarine evaporites form from evaporation of trapped inland seas or lakes. Both types have major economic importance, and they are covered in detail in Chapter 5.

Some less common rocks are of special interest. Conglomerates are composed of cobbles to pebbles to sand grains deposited close to a source area, and commonly they indicate the start of a new cycle of sedimentation over a newly submerged land surface. They grade upward into sandy rocks. A breccia is like a conglomerate, except that the rock and mineral fragments are angular instead of rounded.

Arkose is a name applied to sandstone that's rich in potassium feldspar as well as quartz, often with abundant clay and iron (reddish) staining. Arkose is an important host rock for some bauxite (aluminum ore) deposits. Graywacke is another impure sandstone rich in clay minerals and volcanic rock fragments that is usually found in deep basins along subduction zones (see section on plate tectonics, below).

It should also be mentioned here that although metal-bearing solutions often are associated with igneous activity, the actual mineralization is due to the favorable porosity and permeability of sedimentary rocks, thus more igneous-derived metallization is housed in sedimentary rocks than in igneous rocks.

METAMORPHIC ROCKS

These rocks form from preexisting igneous, sedimentary, or older metamorphic rocks due to effects of heat and pressure but without major melting. Thus if a rock formed near the earth's surface is buried to great depth, the increase in pressure and temperature causes new minerals to form. These in turn recrystallize to other minerals with further increases in pressure and temperature. *Metamorphic facies* are designated by one or more key minerals that indicate the approximate pressure and temperature regime to which the rocks have been subjected. In contact metamorphism, the heat from an intrusive igneous rock is sufficient to cause recrystallization in the rock that it intrudes, even when the pressure is low. The facies concept tells us that while the mineralogy of metamorphic rocks changes with increasing pressure and temperature, the overall bulk composition of the rock does not, except for water and carbon dioxide. These two (the latter from carbonate minerals such as calcite) are *volatiles* that are driven off by high temperatures. The other chemicals in the rocks are usually unaffected, however.

In Figure 1–5 we see the common metamorphic minerals that form from different parent rocks in response to increased temperature and pressure. For many rocks, the increased pressure imposes a strong directional component upon them, and these are indicated as foliated rocks in Figure 1–5. Thus the sedimentary rock shale, when exposed to low-temperature and low-pressure metamorphism, is transformed into the metamorphic rock slate; with increasing temperature and pressure, it becomes schist. Slate is a very fine grained rock whose individual

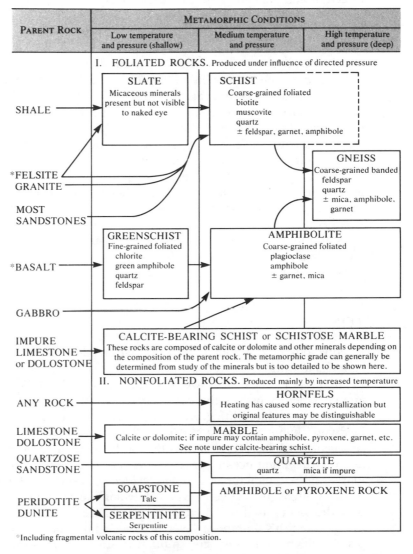

FIGURE 1–5 A generalized chart showing the origin of common metamorphic rocks. (Source: From Burchfiel, et. al, *Physical Geology*, Merrill, 1982.)

minerals are invisible to the naked eye. In schist, however, the minerals are identifiable by eye, and the platy minerals like mica are oriented in a layered fashion
called schistose texture. With still greater temperature and pressure, the schist can
be transformed into an even coarser grained rock known as gneiss. Figure 1–5
also shows that schist can form from felsite (very feldspar- and quartz-rich rock),
granite, and sandstone. Basalt metamorphoses into a special type of schist known
as greenschist, so named because of the presence of green minerals such as chlorite, epidote, and amphibole. The coarse-grained equivalent of basalt, gabbro, goes
directly to amphibolite whereas basalt must go through the greenschist stage before amphibolite. Clay-bearing limestone and dolostone (dolomite or magnesium-
bearing carbonate rock) will form either a calcite-bearing schist or a schistose
marble.

When temperature is more important than pressure, the resulting metamorphic minerals often show little foliation, and are called nonfoliated (Figure 1–5).
When igneous rocks intrude all other rocks, they heat the intruded rock and
cause contact metamorphism, producing the broad group of rocks known as
hornfelses. Limestone and dolostone yield marble, very quartz rich sandstone
yields quartzite, and the ultramafic rocks peridotite and dunite yield either soap-
stone or serpentinite and then amphibole rock or pyroxene rock.

Metamorphic rocks are not as important for mineral and energy resources
as igneous and sedimentary rocks, yet they house metamorphic banded iron formations, high-grade uranium deposits, and many important gold deposits. The
weathering of metamorphic rocks yields valuable accumulations of titanium minerals and industrial minerals such as garnet and magnetite. In other places, contact metamorphism of favorable intruded rocks, such as limestone, yields valuable
ore deposits as well.

PLATE TECTONICS

Anyone looking at a map of the world sees readily that the eastern coast of South
America and the western coast of Africa look like two separated pieces of a puz-
zle, and that if they were moved together they would fit. For many years scientists
argued whether or not the continents were at one time together and became
separated by some drifting process. It was not until the 1960s, after decades of
fieldwork, hypothesizing, and debate, that our modern understanding of this process came together. The key was the earth's hidden seafloor, where the creation
and destruction of oceanic crust allows the continents to move about just as the
map—and a mountain of geologic evidence—says they have. The story of the
theory's discovery is covered in most geology texts and in many other books, so
only the theory itself will be treated briefly here.

The earth's outermost layer is divided into about a dozen giant pieces, or
plates, which are in very slow motion relative to each other. The granitic continents are embedded in these 100-kilometer-thick plates, riding higher than the
denser basaltic ocean floor. When continental and oceanic plates converge, the

lighter continent always overrides the denser seafloor, which moves down into the hot mantle and is destroyed. When two plates move apart, hot basalt rises into the extending zone between them and hardens to form new seafloor.

The zones of *convergence* mostly lie along the sea's great trenches. The zones of *extension* (or divergence) mostly lie along the equally great mid-ocean ridges. A third type of zone, the *transform*, completes the picture: there the two adjoining plates move past each other with neither convergence nor extension.

The discipline of geology that deals with this subject is known as *plate tectonics*. It is shown in somewhat simplified fashion in Figure 1–6. Here we see that the west coast of the United States, as part of the American plate, is trying to override the Pacific plate, with the result that the Pacific plate is being forced downward, or subducted, into the earth's mantle. Such a violent encounter is reflected in earthquakes and volcanism, both of which are common along this subduction zone. In the eastern United States, this coast is about in the middle of the American plate, on the trailing edge, and thus volcanism is absent and earthquakes are rare. How new crust is added at mid-oceanic ridges and how subduction works are shown in Figure 1–7. Plate tectonics is extremely important to modern earth science, and it is equally important for mineral and energy resources. In Chapter 4, plate tectonics is used to account for formation of numerous types of economic mineral deposits.

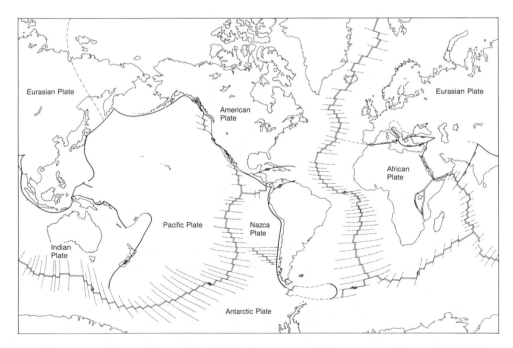

FIGURE 1–6   Plates of the world. Double lines are spreading centers. Lines with barbs are converging boundaries. Single lines are transform boundaries.

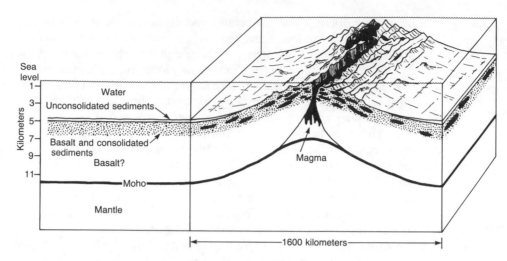

FIGURE 1–7a   Volcanic activity at mid-ocean ridges creates the rocks that underlie the oceans. "Moho" is the seismic boundary between crust and mantle. All of this figure is within the lithosphere.

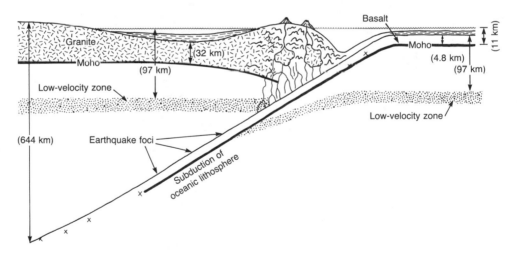

FIGURE 1–7b   Cross section of a subduction zone. Not all features shown here occur at every subduction zone. Drawing is not to scale. (Source: From Burchfiel, et. al, *Physical Geology*, Merrill, 1982.)

THE INTERIOR OF THE EARTH

Geologists know a modest amount about the earth's interior (Figure 1–8). From seismic studies, it is known that the earth has a complex core. The outer part is high-density liquid, probably a nickel-iron alloy, and the inner part is of the same material in solid form. Surrounding the core is dense silicate rock, the mantle, which makes up the largest volume of the earth. Where the crust of the earth meets the mantle is a seismic discontinuity; here, earthquake waves "jump" in velocity because they go from low-density crust to higher density mantle. This jump probably is due to a change in mineralogy. Beneath it, the mantle is further divided into the asthenosphere and the lithosphere. The asthenosphere is somewhat plastic, and the rigid lithosphere over it forms the great plates that control the position of the continents. Between the asthenosphere and lithosphere is another zone where seismic waves are slowed, and the presence of some magma is postulated. This zone is known as the low-velocity zone, and it is presumed that many magmas form here.

MINERAL RESOURCE RECOVERY

Where and how we recover some of our mineral resources is also important. When relatively large amounts of a mineral resource occur close to the surface of the earth, open pit mining is used. Where high-grade ore occurs at some depth, a mine shaft may be sunk and drifts (tunnels) run from the main shaft to the ore zones. In some cases ore is of a high grade, but spotty, and not economically recoverable by either open-pit or underground methods. If, as with uranium or copper, the ore minerals can be dissolved in place by acid or alkaline solutions, then solution mining may be undertaken. These topics are discussed in Chapters 2 and 8.

Many mineral resources occur in rocks of a specific type or age. Commonly, many ore deposits are found within such a setting, which may be referred to as a *metallogenic province*. The porphyry copper belt in the western part of both North and South America is one such well-defined province, as is the sedimentary copper belt in Zambia-Zaire. Other examples include tin provinces in Peru-Bolivia and Korea–Southeast Asia. Most metallogenic provinces have several characteristics in common: alignment with major tectonic elements (such as subduction-zone traces), igneous rocks, rocks of equal age, and rocks with approximately identical chemistry.

Where open or restricted (continental) seas have evaporated, marine or nonmarine evaporite deposits result respectively. The former are valuable for potash, gypsum, and halite; the latter are notable for sodium carbonates, sulfates, and borax. Both are interlayered with limestones and are commonly associated with petroliferous matter and, less frequently, metal sulfide deposits. Other marine sedimentary rocks are rich in phosphorus (for example, in areas such as

FIGURE 1–8   Cross section through the earth. Expanded section shows relationship between the two types of crust and the mantle. The crust ranges in thickness from 5 to 50 kilometers.

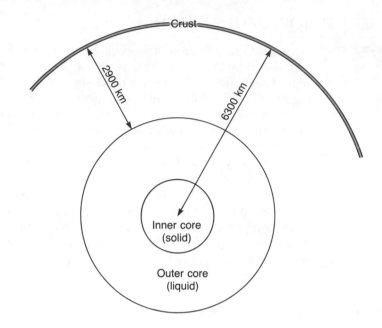

A

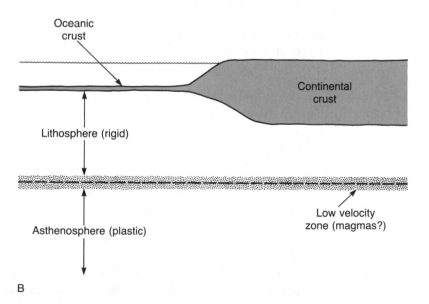

B

Florida, Tennessee, and Utah), while still others may contain economic concentrations of metals. Of the elements crucial to the agricultural industry, potassium (as KCl) is recovered from marine evaporites; phosphate rock is recovered from apatite-rich marine shales; sulfur is most abundant in caprocks of certain marine evaporites; and nitrogen is extracted from the atmosphere.

FUELS AND ALTERNATIVE ENERGY SOURCES

Fuels, discussed extensively in Chapter 7, deserve special mention. While coal reserves are widespread in the United States and are apparently abundant elsewhere in the Northern Hemisphere, they also require the stripping of vast acreage, posing a potentially major threat to the environment. Oil and gas reserves will probably peak early in the twenty-first century on a worldwide basis, then start to decline. Since roughly 30 percent of U.S. energy needs are for transportation—and these needs can only be met by oil and electricity—substitutes for oil must be found for industrial, commercial, and household uses. Coal can be used locally for this purpose, but in the long run may be impractical for many economical and environmental reasons. Alternate sources of power are not as yet fully satisfactory.

We have at our disposal a well-designed nuclear capability, yet some of the public are reluctant to accept nuclear power due to the problem of radioactive waste. It has already been demonstrated that fission-produced radioactive waste can indeed be retained in geologic settings. The Oklo uranium deposit in Gabon contains 2-billion-year-old uranium which sustained a fission reaction for more than 500,000 years. Most of the waste products are still present as the stable daughter products of radioactive decay. Whereas we worry about repositories for some 250,000 years, the Oklo deposit has shown us that radioactive wastes can be isolated for thousands of times this long. (See Chapter 7 for more detail on Oklo.)

Hydroelectric power generation has been with us for a long time, and indeed it will continue to play a significant, but small, role. In short, water is too valuable a commodity to be routinely impounded for power generation. Furthermore, many of the best remaining areas for water-power generation are in wilderness regions, making it neither economic nor wise to even consider using them. Similarly, geothermal energy can be important when natural steam fields are located near metropolitan areas (The Geysers near San Francisco, for example) and when transmission of electric power over great distances becomes more efficient. Thus the potential of geothermal energy in many remote areas in the western United States must await major technological breakthroughs in power transmission. There are several environmental problems with geothermal energy extraction, however. The hot dry rock program is potentially of great importance, especially in the northeastern United States. There, drilling to a depth of 4 to 5 kilometers will return warm water (not steam) from cold water injected into the earth. Along the Atlantic seaboard roughly from New York to Bangor, Maine, space heating accounts for 40 percent of all energy used. The warm water generated by this method could be used for space heating.

Solar energy is at present only of local importance. More and more buildings, including industrial and commercial buildings, are being designed to use energy from the sun. Yet even its most optimistic advocates admit that solar energy will make but a minor dent in the total energy picture by the year 2000. Quite simply, solar energy is still too expensive to compete with natural oil and gas, coal, and nuclear energy. Further, large-scale use of solar energy has its own risk to environmental health associated with it. Fusion energy is another goal for the future; optimistically, it is hoped that experimental plants will be on-line by about the year 2030.

Oil shale is another virtually untapped resource. The amount of petroliferous material stored in these oil-marlstones in nonmarine formations in Utah, Colorado, and Wyoming is impressive; the U.S. Geological Survey includes 2 trillion barrels of oil in its estimate. But there is a drawback to this encouraging picture. Not only are oil shales located in arid to semiarid isolated areas, but some occur in wilderness areas, and recovery of their oil is not now economic. One of the main problems is a shortage of water. Some estimates indicate that mining and processing require three to four barrels of water for every barrel of oil recovered, and these water resources are not available. Transporting the shale elsewhere is less economic than transporting equivalent tonnages of coal, which is itself sub-economic at best. Furthermore, there is potential damage to the environment by mining. We will continue to study promising methods of oil production in which oil can be produced from beneath the ground by dissolving it from the host rock and pumping it to the surface.

WATER

Water resources comprise only that percentage of total precipitation which is withdrawn for use by people. In the United States this amounts to only 9 percent of all the water within our borders. Of this small fraction, most is used for irrigation and for industries; the remainder is used for municipalities. The National Research Council–National Academy of Sciences estimated in 1962 that the United States must find ways to use close to 80 percent of available stream and lake waters by the year 2000, as opposed to about 30 percent now withdrawn.* To do this, recycling of industrial water and other conservation measures must be expanded. However, conservation measures have not been effective enough to keep pace with projected increases in withdrawals. Desalination of ocean water or brines will help, but desalination is expensive, and industrial and agricultural groups have not been cooperative in funding or using experimental systems.

* A. Wolman, *Water Resources Publication 100-B* (Washington, D.C.; National Research Council-National Academy of Sciences, 1962).

It cannot be emphasized strongly enough that water, in terms of both quantity and quality, is the most important resource in the world. Without adequate water, not only agriculture but the entire population will suffer. We must enforce existing environmental controls or implement new ones to ensure an adequate supply of potable water as the growth of this nation and the world continues.

Further, the soils on which we rely for our food and other purposes are not always wisely managed. In Chapter 3 it is pointed out that the United States is losing its soil resources at an alarming rate.

THE ENVIRONMENTAL IMPACT OF MINERAL AND ENERGY RESOURCE EXPLOITATION

As we reap our harvest of minerals and energy, the environment is inevitably affected. Often these effects are adverse, and great care must be taken to minimize or eliminate harm to our delicate environment. While mining, milling, and fuel burning most often attract attention, the most severe environmental problems come from wastes. The media has focused dramatically on the relatively small problem of radioactive wastes, but recently the much larger problem of chemical wastes has been receiving its share of attention. Of our energy sources, coal has damned us with the greenhouse effect, acid rain and toxic gases in the atmosphere, and a staggering number of fatalities per year.

Chapter 8 explores many of the problems of environmental geochemistry as it pertains to public health, with the reminder that there is no "free ride." If we wish to exploit minerals and energy from the earth, then we must be prepared to deal with effects of our actions. Fortunately, we can deal with most of our potentially adverse effects; unfortunately, we often choose not to.

ENVIRONMENTAL GEOCHEMISTRY, HEALTH, AND "ENVIROSCARE"

The impact of chemicals on public health has been of concern for over twenty years, yet, as is emphasized in Chapter 8 and elsewhere in this book, our knowledge of exactly how these chemicals cause adverse health effects under different conditions is meager. While we have some understanding of the effects of certain chemicals on small animals under laboratory conditions, to extrapolate these studies to humans is not always easy. Legislation to deal with potential threats to public health is well meant, but can be ineffective, even self-defeating if it does not reflect all of the technical facts or the uncertainty of our predictions. This book explores some areas of these issues in Chapter 8.

Concern for the environment is admirable and to be encouraged. There are many people with such genuine concern, based on careful assessment of facts. But the legislative aspects of resources and energy are made more complex by a phenomenon that could be called "enviroscare," which is the uncritical exaggeration of a potential environmental issue whether or not there is any real danger. There are many examples more specifically detailed in Chapter 7 under the discussion of nuclear power and related subjects. Another example, that of asbestos, is taken up in Chapter 8.

The bottom line to all environmental problems involving resources and energy is education; only by educating people on a large scale can fact be separated from fiction on these matters and related areas. This book is intended to provide some introductory background material on these topics in order to make the reader more informed concerning our environment, before, during, and after resources are exploited and used.

FURTHER READINGS

Brobst, D. A., and Pratt, W. P., eds. 1973. *United States Mineral Resources*. U. S. Geological Survey Professional Paper 820.

Brookins, D. G. 1981. *Earth Resources, Energy, and the Environment*. Columbus, Ohio: Merrill Publishing Co.

Cameron, E. N., ed. 1973. *The Mineral Position of the United States, 1975–2000*. Madison: University of Wisconsin Press.

Committee on Geological Sciences, National Research Council. 1972. *The Earth and Human Affairs*. San Francisco: Canfield Press.

Committee on Resources and Man, National Research Council. 1969. *Resources and Man*. San Francisco: W. H. Freeman.

Flawn, P. T. 1966. *Environmental Geology*. New York: Harper and Row.

Greenland, D. 1983. *Guidelines for Modern Resource Management*. Columbus, Ohio: Merrill Publishing Co.

Holdgate, M. W., Kassag, M., and White, F., eds. 1984. *The World Environment 1972–1982: A Report by the United Nations Environment Programme*. Vol. 8 in the Natural Resources and Environmental Series. Dublin: Tycooly International.

Kesler, S. E. 1976. *Our Finite Mineral Resources*. New York: Harper and Row.

Landsberg, H. H. 1964. *Natural Resources for U.S. Growth: A Look Ahead to the Year 2000*. Baltimore: Johns Hopkins University Press.

Laporte, L. F. 1975. *Encounter with the Earth: Materials and Processes*. San Francisco: Canfield Press.

Laporte, L. F. 1975. *Encounter with the Earth: Resources*. San Francisco: Canfield Press.

Lovering, T. S. 1943. *Minerals in World Affairs*. Englewood Cliffs, N.J.: Prentice-Hall.

McDivitt, J. F., and Manners, G. 1974. *Minerals and Men*. Baltimore: Johns Hopkins University Press.

National Academy of Sciences. 1975. *Mineral Resources and the Environment*, report prepared by the Committee on Mineral Resources and the Environment.

Skinner, B. J. 1986. *Earth Resources*. Englewood Cliffs, N.J.: Prentice-Hall.

Skinner, B. J., and Turekian, K. K. 1973. *Man and the Ocean*. Englewood Cliffs, N.J.: Prentice-Hall.

U.S. Bureau of Mines. 1985. *Mineral Facts and Problems*. U.S. Bureau of Mines Bulletin 675.

U.S. Bureau of Mines. Annual. *Minerals Yearbook: Metals, Minerals and Fuels*.

2
Ores, Production, and Mining

ELEMENTS AND ORES

Chapter 1 briefly discussed how elements are fixed in the common rock-forming minerals and rocks. How the elements other than the nine most abundant are incorporated into the rock-forming minerals, accessory minerals, and less common minerals is also of prime interest. This is often the key to an element's economic importance. The charge (valence), ionic radius, and affinities of the element must all be considered. A simple example will illustrate this point.

Copper occurs in nature as the native element Cu or as the ions Cu^+ and Cu^{2+}. Although the Cu^+ ion has the same ionic radius as Na^+, copper has a stronger affinity for sulfur than for oxygen or silicates. Hence Cu^+ does not concentrate in the plagioclase feldspar albite ($NaAlSi_3O_8$) but occurs instead in accessory pyrite (FeS_2). This is fortunate because pyrite can be readily separated from its host rock and chemically processed to extract its copper. If the copper were locked up in albite, it would take so much energy to break down the mineral and process it for copper extraction that it would never be economic.

Elements are commonly classified as to whether they have an affinity for sulfur *(chalcophile)* or oxygen as oxides or silicates *(lithophile)*, or whether they occur as gases in the earth's atmosphere *(atmophile)* or as native metals *(siderophile)*. Table 2–1 shows these classifications for 55 significant elements. Some elements are placed in more than one category. Iron, the fourth most abundant element in the earth's crust, is chalcophile, lithophile, and siderophile; its siderophile nature is most evident in the earth's core and in meteorites, where iron occurs in its metallic form. In the earth's crust, iron is concentrated in silicates and sulfides. Iron shows a strong preference for sulfur; thus pyrite is a very common accessory mineral in igneous rocks. But there is much more iron than sulfur present in the earth's crust, so the bulk of the iron is incorporated into silicates such as biotite, pyroxenes, or amphiboles; yet none of these minerals is an iron

TABLE 2–1   Classification of significant elements.

Chalcophile

Copper, lead, zinc, iron, nickel, cobalt, cadmium, mercury, arsenic, antimony, selenium, tellurium, sulfur, molybdenum

Lithophile

Silicon, oxygen, aluminum, iron, sodium, potassium, magnesium, calcium, barium, beryllium, strontium, titanium, zirconium, thorium, uranium, chromium, vanadium, niobium, tantalum, phosphorus, boron, cesium, lithium, hydrogen, fluorine, iodine, chlorine, bromine, carbon

Siderophile

Iron, cobalt, nickel, gold, silver, copper, platinum, palladium, osmium, iridium, rhenium

Atmophile

Nitrogen, oxygen, argon, helium, neon, krypton, xenon, hydrogen

ore. Weathering and other processes must reprocess these original iron-bearing minerals into other materials suitable for iron ore.

Trace-element distribution in rocks is extremely important. If sulfur is present, the chalcophile elements are commonly enriched in sulfide minerals. The lithophile trace elements, and also the chalcophile elements in the absence of sulfur, tend to substitute for those major rock-forming elements to which they are closest in size (ionic radius) or charge. Hence Ni^{2+}, which is close in ionic radius to Mg^{2+} and Fe^{2+}, is enriched in minerals such as olivine substituting for Mg and Fe. Other examples are Ga^{3+} (gallium ion) for Al^{3+}, Cr^{3+} for Fe^{3+}, Co^{2+} for Mg^{2+} or Fe^{2+}, Rb^+ (rubidium ion) for K^+, and so on.

As another example, uranium is present in many rocks as U^{4+}. It is a large ion, close in size to Ca^{2+} and Na^+. Thus it is found in accessory calcium-bearing phases such as apatite and sphene. Because of its high charge ($+4$), it is found in zircon substituting for Zr^{4+} (zirconium ion). However, it is far too big to substitute for Si^{4+} in the silica tetrahedron, hence the amount of uranium found in quartz is vanishingly small. Other examples, as they relate to specific metal occurrences, are given in Chapter 4. The common ionic radii for many elements are to be found in the Appendix.

FORMATION OF ORES

The controlling factors for ore formation are the crustal abundance of an element and the way in which it is incorporated into minerals. In addition to a relatively high concentration of an element in the rock, atoms of the element must be bonded to other atoms in such a way that extraction can be profitable. Aluminum is a good example. Although it is the third most abundant element (by weight) in the earth's crust, only certain minerals such as gibbsite, an aluminum hydroxide abundant in bauxite deposits, can be treated for profitable removal of aluminum. The aluminum in the very common rock-forming minerals such as the feldspars and micas is too tightly bonded to be profitably extracted. Figure 2–1 illustrates how different elements must be present in specific concentrations before they can be properly classified as reserves.

Figure 2–1 also demonstrates that we can arrive at reserve tonnages of mineral elements by multiplying their crustal abundance by a factor ranging between 1 million and 10 billion. Despite obvious uncertainties such as degree of exploration and other factors, this method has worked well for elements like uranium and thorium and has been used successfully in Japan for estimating reserves of several metals.

Sulfides commonly contain economic quantities of metals like copper, nickel, lead, cobalt, and zinc. One of the reasons sulfide minerals are commonly called *ore* is that they can be easily separated from the silicate-oxide barren, or *gangue,* minerals. Consider nickel as an example. Nickel is commonly enriched in very low silica (ultramafic) rocks. If sulfur is present, the nickel will be preferentially enriched in the sulfide minerals. If sulfur is not present, the nickel will be found in rock-forming minerals such as olivines and pyroxenes.

In the first case, the sulfide minerals can be separated from the silicates and oxides by flotation. In this process finely crushed rock is mixed with water to which has been added organic materials, such as kerosene. The organics separate

FIGURE 2–1   Domestic reserves compared to crustal abundance of elements. In general, the reserves (**R**) are equal to the abundance (**A**) times a factor varying from one million (10^6) to one billion (10^9), as illustrated by the 45° slopes. The vertical bars for different elements connect ore that is minable now (dots) to deposits where it is hoped that technological breakthroughs will allow the lower grade sub-ore to be economically mined in the future. (From V. E. McKelvey, U.S. Geological Survey Professional Paper 820, 1973.)

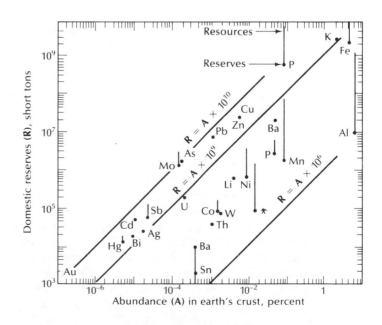

from the water and float to the top and the sulfide minerals, which are fixed onto the organics, also rise to the top. The silicate and oxide waste, or gangue, settles to the bottom. The sulfide-rich organic froth is then skimmed off and roasted to drive off the carbon (as CO_2) and much of the sulfur (as SO_2). The metal-rich residue, or matte, is then further roasted in special furnaces to remove oxygen and the remaining sulfur, and the metal is refined electrolytically.

In the case where nickel is fixed in silicate minerals, the ore grade must be significantly higher than for the nickel-sulfide ores. When nickel is tightly bonded in the silicate structure, a great deal of energy is required to break these bonds and thus allow nickel recovery.

Nature may, however, over long periods of geologic time and under certain conditions, enrich nickel in heavily weathered ultramafic rocks. Such a process is called natural beneficiation.

Thus nickel has to be present in quite different amounts depending on how it occurs in the rocks. Similarly, then, each specific occurrence will have a different concentration factor. This is a common phenomenon for several metals (see Chapter 4 for examples).

METALS AND NONMETALS

For mineral resources we use the term *metal* to describe elements that are used for their metallic properties (conductivity, strength, ductility); the term *nonmetal* refers to those elements and their compounds not used for their metallic properties. Thus sodium (Na), while chemically a metal, is used commercially in sodium compounds based on its nonmetallic properties. Oil, gas, and coal are nonmetals, as are potash (potassium compounds), gypsum ($CaSO_4 \cdot 2H_2O$), sand and gravel, cement, borax, sulfur, nitrogen, and phosphorus. The annual world consumption of nonmetals is substantially greater than metallic elements, as the following quantities show:

Carbon: > 10 billion tons

Sodium and iron: 1–10 billion tons

Nitrogen, oxygen, sulfur, potassium, and calcium: 0.1–1 billion tons

Only below 10 million tons do we encounter the metals which attract so much glamour, including zinc, copper, lead, magnesium, chromium, aluminum, plus the nonmetals hydrogen, fluorine, phosphorus, and barium—all in the category of 1 to 10 million tons per year. Thus iron is the only metal ranked in the top eight elements in order of consumption. Carbon is ranked first because of our great consumption of fossil fuels. Except for carbon and iron, the other top eight elements (N, K, Ca, Na, O, S) are consumed largely in agriculture and related industries. Not listed here are the tremendous amounts of compounds such as sand, gravel, and other types of industrial rocks.

MINIMUM ANTICIPATED CUMULATIVE DEMAND

The acronym MACD had been used to describe resource availability since 1973, when it was first introduced by the U.S. Geological Survey. MACD is the *minimum anticipated cumulative demand* for a particular resource in the United States for the period from 1975 to 2000. The following Roman numerals, or *MACD ratings*, indicate resource availabilities and are assigned to a resource according to its multiple of the MACD:

$$I = 10 \text{ or more times MACD}$$
$$II = 2\text{--}10 \text{ times MACD}$$
$$III = 0.75\text{--}2 \text{ times MACD}$$
$$IV = 0.35\text{--}0.75 \text{ times MACD}$$
$$V = 0.10\text{--}0.35 \text{ times MACD}$$
$$VI = \text{Less than } 0.10 \text{ times MACD}$$

Table 2–2 shows the MACD ratings for some important elements and compounds.

In Table 2–2, an assignment of NE (not estimated) for undiscovered resources may have very different meanings for different commodities. In the case of aluminum, new technology will probably make extraction from clays or low-grade bauxites economic by the year 2000. Hence this NE may be revised to I in the next decade or so. For gold, which is used for purposes of trade payment, resource figures may fluctuate greatly. Gold will continue to be recovered as a by-product of copper mining (20 to 30 percent of U.S. gold production is from this source); but we may import copper to meet our needs rather than mine our own resources. Only if we mine U.S. copper will gold production increase, hence the gold resource MACD rating could be III or IV (or V?), depending on the copper market. Similarly, the extent of uranium mining in the later part of the twentieth century is unknown. If it will be large, then its by-product vanadium may receive an MACD rating of II. If uranium mining should decline, then production of vanadium will also fall and an MACD rating of III might result. For nickel, there is hope that improved technology will allow recovery from weathered, nickeliferous rocks. In that case the undiscovered resources might be ranked III; otherwise the MACD rating for nickel will more likely be IV. The availability of sand and gravel is uncertain because there is relatively little prospecting for this valuable commodity. If the huge sand and gravel accumulations offshore can be economically exploited, then the MACD rating may well be I. If not, the rating will more likely be III—that is, roughly enough to meet demands but not enough to make us complacent about the resource situation.

Industrial materials, especially sand and gravel, cement, and clays, are the largest tonnage commodity in the United States, and next to iron and steel the most valuable commodity (exclusive of fossil fuels). Supplies appear to be ade-

Commodity	Identified Resources	Undiscovered Resources
Aluminum	II	NE*
Asbestos	V	VI
Barite	II	II
Chromium	VI	V
Copper	III	III
Fluorine	V	V
Gold	III	NE
Gypsum	I	I
Iron	II	I
Molybdenum	I	I
Nickel	III	NE
Phosphate	II	I
Sand and Gravel	III	NE
Sulfur	I	I
Titanium	II	II
Uranium	II	III
Vanadium	II	NE
Zinc	II	II

TABLE 2–2   Resource availabilities by MACD.

*NE = not estimated.
SOURCE: U.S. Geological Survey Professional Paper 820, 1973.

quate for the remainder of this century, but careful urban planning will be necessary in the next century. There are too many examples of poor planning that has rendered good quality sand and gravel deposits useless and has created unnecessary transportation problems in some American urban areas (see Chapter 6).

MACD figures can be misleading if we attempt to use them on a worldwide basis. The uncertainty is due in part to the uneven distribution of resources, to uneven population density, to varying degrees of industrialization throughout the world, and thus to widely differing per capita consumption. The United States, for example, with less than 10 percent of the world's population has a per capita consumption four to five times that of any other industrialized nation or continent. Should Asia and Africa rise to the U.S. level of per capita consumption, then these continents would find it difficult to export anything; yet the United States relies heavily on their exports for many commodities. Further, mass starvation and energy shortages may confront us in the twenty-first century. If calculations by the National Research Council–National Academy of Sciences* are correct, then the world's maximum capacity to produce food for its population may be reached by the mid-twenty-first century. If we add to this the probability that oil and gas resources will have peaked early in the twenty-first century and that many mineral resources will have been mined out, then the entire world will face a crisis. This rather pessimistic view is based on the following assumptions: continued high rates of population growth, high rates of food consumption, rapid industrialization of essentially unindustrialized countries, no new major mineral deposit discoveries, and failure to use nuclear or alternative (to fossil fuel) energy sources. Therefore, MACD figures may be rather drastically revised for the twenty-first century. There are simply too many uncertainties at this time to warrant MACD projections beyond the year 2000. Furthermore, world MACD figures must be determined and integrated with domestic figures.

MACD figures are included only as examples of predicting availability of various commodities. The reader is encouraged to examine the examples here in view of newer information (Chapters 4 through 7) for various commodities to see how reliable they are.

ECONOMICS OF ORES

It is not rock chemistry that dictates whether or not ore can be mined. The geologic setting should be such that not too much waste rock (overburden) must be removed; moreover, the ore must be located in areas near enough to markets so that transportation and related costs are not excessive. In short, the economics of extracting the potential ore must be carefully evaluated. Consider the example of a small but rich deposit of some metal located well above the Arctic Circle. If this hypothetical deposit were located more favorably, it would be economic. As it is, the assumed amount of ore would probably be too small and the costs too great to warrant mining and shipping.

Another example could be the numerous small lead-zinc deposits of the Mississippi Valley which also contain high concentrations of mercury, an element in short supply. If rock chemistry alone were considered, these ores would possess enough mercury to warrant their exploitation. Yet the ores are difficult to treat and recovery costs are high. Also, most of these deposits are located in urban areas, agricultural areas, or in areas where uncontrolled mining would affect the water quality of many streams and rivers. Hence the costs to ensure environmental protection would add to the already high costs for recovery. Since the deposits are too small to warrant economic recovery of lead and zinc, reliance on these two metals to cut costs of mercury recovery is unlikely. Thus, this resource must

---

* National Research Council–National Academy of Sciences, *The Earth and Human Affairs* (San Francisco: Canfield Press, 1972).

await technological breakthroughs for profitable extraction without environmental damage.

Rock that contains 0.5 percent copper is not necessarily considered ore. If the copper is present in sulfides, then it may be ore; but if the copper is present in oxides, extraction costs will be too high and thus it is not considered ore. High concentrations in oxidized rocks warrant further exploration by deeper drilling; that is, if the deeper rocks are less oxidized, the possibility of finding copper in sulfides is greater. The nature of the rocks drilled should be very carefully examined. If pyrite (a sulfide) with a low copper content is encountered, then the higher concentrations of copper in the oxidized rocks above it must be due to copper coming there from some other original source or due to impoverishment of sulfur in the deep rocks. The present-day economic geologist uses many tools for exploration and very sophisticated techniques to evaluate rocks for their economic potential.

BY-PRODUCTS

Nature's substances are not pure. When a pure major substance is desired from a rock, its separation involves removing impurities, many of which become important by-products.

Many metals are typically found in complex ores, although the ore is classified on the basis of one or few major metals (for instance, copper ore). During the milling and refining processes, these other metals are recovered. By-products are important sources for many metals produced in the United States.

By-products of major significance and their major ores include: gold, silver, and molybdenum from copper ore; gallium from aluminum ore; silver from lead ore; cadmium and thallium from zinc ore; and cobalt from nickel ore. In the United States virtually 100 percent of domestic production of certain metals comes from by-products (see Chapter 4). Important by-products come from other ores. Uranium is removed for environmental safety from phosphate rocks in Florida, for example; and vanadium and molybdenum are important by-products from uranium ores. Similarly, sulfur is obtained as a by-product from petroleum as the sour crudes are purified (see Chapter 7). Should oil shales be exploited for their petroliferous content in the future, the mineral dawsonite, a potential aluminum ore, will be recovered as a by-product. Finally, when seawater is evaporated to recover salt (sodium chloride), the residual brine can yield the by-products magnesium, potassium, and bromine.

RECOVERY METHODS

The key to recovery of any mineral commodity from the earth is extraction at a profit. Some commodities, such as bromine and magnesium, can be extracted as elements directly from seawater (Chapters 4 and 5), others as compounds (halite

View of the Oklo uranium mine, Republic of Gabon. This view not only shows a typical small open pit mine, but several of the 1975 attendees of the Conference on the Oklo Phenomenon (Natural Fission Reactor) sponsored by the International Atomic Energy Agency, the French Atomic Energy Commission, and the Republic of Gabon. Nature sustained a fission reaction at this site some two billion years before the twentieth century (see Chapter 7 for details).

or common table salt) by evaporation of seawater, and others from the atmosphere (nitrogen and argon). Yet most of our mineral resources require extraction from the earth.

Major mining methods include conventional underground mining, open pit mining, and solution mining. Of lesser importance are placer mining and seafloor mining. Ore bodies result from a complex interplay of geochemistry and rock

View of part of the open pit at the Highland uranium mine, Powder River Basin, Wyoming. Here the ore is located very near the surface and it is economic to recover the ore by open pit methods.

deformation. Rocks, and their ores, commonly become increasingly complex with increasing depth, and ways to economically recover ore are in turn often complex.

Underground Mining

First it must be determined that the element or compound is concentrated sufficiently to be extracted during milling. Next, a specific mining method is chosen. The diagram in Figure 2–2 shows a case in which open pit mining could not be used because of the very high waste-to-ore ratio that would result. Thus underground mining is usually undertaken for ore not close to the surface. Examples include copper vein deposits, sporadic uranium deposits, and lode gold deposits. Advantages of this method are that (1) once the main shaft is complete, it is relatively inexpensive to develop the underground workings and (2) detailed drilling from the shaft or tunnels is inexpensive compared with surface drilling. A major disadvantage, of course, is that the ore must be of high enough grade and in enough quantity to justify the cost of the underground mining.

In general, if pockets of ore are close to the same depth, then a vertical or inclined shaft is sunk as close as possible to the center of the ore pods (Figure 2–2). *Drifts* (tunnels) off this main shaft are then excavated, usually just under the ore pods so that they can be mined from two or more levels. On mountainsides, horizontal tunnels called *adits* are driven into the ore body. Occasionally the ore is oriented in such a way that tunnels are driven in at odd angles to extract the ore.

When the ore body is so voluminous that a large volume of rock (far in excess of drift dimension) is extracted, a large "room" called a *stope* is created. If the rock is hard and resistant, stopes may need little accessory support. However,

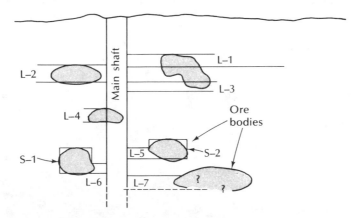

FIGURE 2–2   Underground mining. In this method a main shaft is sunk to a predetermined depth, and tunnels (called *drifts* or *adits*) from this main shaft are used to reach the ore bodies at various levels (L–1, L–2, etc.). When the ore is thicker than the tunnels, stopes (S–1 and S–2) are excavated, provided the rock can withstand excavation or can be supported by other methods. (Source: Brookins, *Earth Resources, Energy, and the Environment*, Merrill, 1981.)

Headframe of the mine shaft at the Section 30 Mine in the Grants Mineral Belt, New Mexico. High-grade uranium ore is stockpiled in front of the frame awaiting shipment to the mill. The actual mine workings are at a depth of 300 meters.

the rocks will commonly give way under the pressure of the overlying rock cover; cave-ins can be prevented by leaving rock pillars or by erecting supports of timber or other materials. Limiting factors here include the maximum volume of rock that can be extracted without causing a cave-in, regardless of accessory support. Rock pillars are often not feasible because they themselves represent ore-grade rock. In such cases these pillars may be left in place until an area is effectively mined out. Then, one by one, the most distant pillars are removed systematically, progressing toward the main shaft, and the mined-out areas allowed to cave in. This practice is normally only done when such caving will not cause excessive subsidence at the surface and when it has been determined, to the best of the mining engineers' knowledge, that no more ore remains in the outlying limits.

Underground mining is very complex. Not only must maximum efficiency for ore recovery be sought, but safety factors must also be vigorously enforced at all steps. In mines where explosive gases like methane or radioactive gas like radon can accumulate, sophisticated ventilation systems are installed. Even where only nonflammable, nonradioactive rock dust is involved, good ventilation is always essential to assure worker health.

View of high-grade uranium ore associated with clay (dark areas) and organic matter at an underground mine in the Grants Mineral Belt, New Mexico. The scale is indicated by the miner's hard hat at bottom right. It is concentrations of ore pods such as this that the explorationist tries to find by drilling from the surface some 200 to 1500 meters above.

Mine disasters are still too common, especially in coal mines, even though the safety record of underground mining has improved significantly in the last 25 years. Coal mining still presents the greatest immediate hazards from explosions and caving as well as long-term hazards from inhaling carcinogens (including radioactive materials enriched at low levels in coal). Modern uranium mines have one of the best underground mining safety records because the required workings are less extensive and because radiation and safety regulations are rigidly enforced by the federal government. The potash and salt underground mining industry and the limestone mines can claim similar safety records.

Another interesting facet of underground mining is that much of the waste rock is not brought to the surface and dumped. Instead it is used as backfill for support of underground workings where the ore has been mined out.

Open Pit Mining

Open pit mining is used primarily where the ore deposits are relatively close to the surface, the ore is disseminated through the rocks, and the host-rock characteristics are such that the walls will be self-supporting. The pit walls must be designed so that they will remain stable during the mining operations.

Deciding whether to mine ore by open pit or underground methods is often difficult. Figure 2–3 shows a hypothetical massive ore body close to the surface. As illustrated, the limits of the ore must be carefully determined prior to excavation. Should too small a pit be cut initially, it is difficult, time-consuming, and costly to attempt to widen the pit. Conversely, too large a pit is wasteful. Too much rock is excavated and unlike most underground mining, the waste rock is deposited elsewhere on the surface, causing more disturbance of the original surroundings.

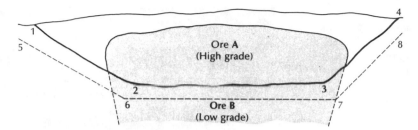

FIGURE 2–3   Open pit mining. One type of open pit mining is illustrated here. The ore body **A** has been outlined by drilling and found to be relatively close to the surface. The solid line 1–2–3–4 indicates the most favorable width-to-depth conditions for the open pit. Although ore continues below the line 1–2–3–4, developing the pit along the dotted line 5–6–7–8 would move too much barren rock and increase mining costs, making ore recovery unprofitable. The more or less uniform body shown is typical of some porphyry copper deposits. (Source: Brookins, *Earth Resources, Energy, and the Environment*, Merrill, 1981.)

Figure 2–4 shows more hypothetical ore, some of which occurs in smaller bodies, which is not interconnected and has an overall depth-to-width ratio greater than unity. Underground methods must be used to mine this ore body **B** because too large an open pit would have to be made to remove the ore. Ore body **A,** because of its massive nature and shallow depth, can be mined by open pit methods.

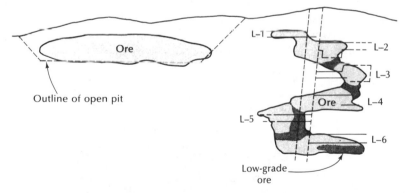

FIGURE 2–4   Economic mining methods. Because ore body **A** is tabular and near the surface, it can be mined by open pit methods, as indicated by the dashed line. At a much greater depth, it would be more profitable to sink a vertical shaft to the ore level and recover the ore by a series of tunnels and larger mined-out areas (stopes) where the ore thickens. Ore body **B,** a more or less discontinuous ore encountered from near the surface to fairly great depths, should be mined by underground methods. Although the initial expense to sink the main shaft is considerable, once the shaft is down it is inexpensive to develop the ore body by drilling underground. (Source: Brookins, *Earth Resources, Energy, and the Environment*, Merrill, 1981.)

Boxes of drill core of high-grade (dark gray to black) uranium ore and low-grade (light colored) ore from the Oklo mine, Gabon.

Solution Mining

Occasionally ore that is soluble in acids or bases occurs in rocks (usually sedimentary rocks), as shown in Figure 2–5. If the ore is widely distributed in one or more layers or is too sporadic to warrant open pit mining, then solution mining may be undertaken.

Solution mining may also be recommended where the ore is of low grade or the ground too weak to support underground workings. This method's advantages are low capital requirements, negligible rock excavation, and no vast amounts of tailings as waste. Solution mining works best for deposits that cannot be economically mined by either open pit or underground methods, and care must be taken that all solutions stay within the limits of the ore horizon. In this

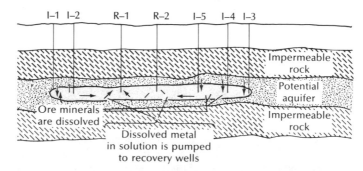

FIGURE 2–5   Solution mining. For some ore, such as copper and uranium ores, the ore minerals can be dissolved by acid- or alkaline-leach injected under pressure through a series of injection wells (I–1 to I–5) and recovered by pumping from recovery wells R–1 and R–2. As shown here, the injection wells surround the recovery wells. (Source: Brookins, *Earth Resources, Energy, and the Environment*, Merrill, 1981.)

method several injection wells are drilled into the ore-bearing rock, which must be relatively permeable for the method to work. The ore metal is dissolved and pumped to the surface from several centrally located recovery wells. This method works well for ores that are readily soluble and where the solutions, either acidic or basic depending on rock chemistry, will not remove much of the gangue (waste rock). If too much gangue is dissolved, the ore horizon becomes more permeable and dissolved material may be lost into aquifers that provide water for irrigation or consumption. Thus environmental impact must be carefully assessed even though no mechanical extraction of rock is planned except surface scraping for milling and supporting facilities.

Solution mining has been used on a small scale for many years, sometimes quite by accident. In the 1930s slightly acidic rainwater percolating over waste from copper mining at Ray, Arizona, dissolved some of the remaining copper and redeposited it electrolytically on cans in a garbage dump at the end of the waste rock dump. The method was so efficient that for a while a major cost for subsequent copper mining of dump materials was the cost of scrap. Even now solution mining is used in the copper industry, often after fracturing the rocks by explosives to allow water to infiltrate the rocks more easily and dissolve the copper. The solutions containing copper are channeled down a gradient where they can be recovered by displacing another metal from scrap. The copper-coated scrap is then milled with ease.

Some low-grade uranium deposits are excellent candidates for solution mining. Acidic solutions dissolve only the uranium minerals, which may constitute only 0.1 percent or less of the rocks, and the uranium is thus removed with little change of the leached rocks. This method does not expose miners to radiation or

Very large (1 1/2 meter diameter) drill core from the Callahan Mine, Harborside, Maine. The large core was drilled for two purposes: (1) as the main shaft excavation for underground operations; (2) in order to closely examine the complex rocks in the area which, at the surface, are largely covered by forest, water, or glacial debris.

dust, and no dump materials result. The method works best for small, low-grade deposits where neither underground nor open pit methods are feasible. At present, testing is under way to see if solution mining can be used for deep uranium ore. Again, effects on the environment will be carefully assessed prior to any mining by this method.

Special types of mining and milling are discussed in Chapter 4 for several metals, in Chapter 5 for nonmetals, and in Chapter 7 for energy resources.

FURTHER READINGS

Bateman, A. M. 1950. *Economic Mineral Deposits,* 2d ed. New York: John Wiley & Sons.

Buddington, A. F. 1933. Correlation of kinds of igneous rocks with kinds of mineralization. In *Ore Deposits of the Western States* (Lindgren Volume), 350–85. New York: American Institute of Mining and Metallurgical Engineers.

Kesler, S. A. 1983. *Our Finite Mineral Resources.* New York: McGraw-Hill.

Lamey, C. A. 1966. *Metallic and Industrial Mineral Deposits.* New York: McGraw-Hill.

Lindgren, W. 1933. *Mineral Deposits,* 4th ed. New York: McGraw-Hill.

Mauss, E. A., and Ullmann, J. E., eds. 1979. *Conservation of Energy Resources.* New York: The New York Academy of Sciences.

Nobel, J. A. 1955. The classification of ore deposits. *Economic Geology* (special issue): 155–69.

Park, C. F., and MacDiarmid, R. A. 1970. *Ore Deposits,* 2d ed. San Francisco: W. H. Freedman.

Ridge, J. D., ed. 1968. *Ore Deposits of the United States 1933–67,* vols. 1 and II. New York: American Institute of Mining, Metallurgical and Petroleum Engineers.

Thomas, L. J. 1973. *An Introduction to Mining.* Sydney, Australia: Hicks, Smith and Sons.

3
Water, Soils, and the Greenhouse Effect

INTRODUCTION

Water is the world's most valuable resource. The agricultural industries, population growth and distribution, industrial expansion, and exploitation of most mineral commodities are controlled by the availability of water. It was no accident that when the first samples were returned from the moon, there was an intensive search of these rocks for structurally bonded water, the hope being that its presence would allow the development of technology to successfully extract this water for possible lunar bases. Unfortunately, the rocks are totally anhydrous, and thus water will have to be transported to the moon should bases be built there.

This chapter looks at the hydrologic cycle, at the world, and then the water resources picture for the United States. Environmental aspects of water are treated here in some detail, and the pollution, in particular, of existing water resources is given a hard look. It will be found that many of our waterways and other water reservoirs, including groundwater, are highly polluted, with the human contribution regrettably high in this picture.

This chapter also discusses soils in the United States since, coupled with water availability, soil types dictate crop growth and thus food supply. Again, there are many unfortunately negative sides to soil exploitation as well. Soil erosion is accelerated by unwise farming methods and other anthropogenic causes, and many soils are highly contaminated with toxic materials, in part due to heavy uses of pesticides.

Finally, the greenhouse effect is covered in this chapter because of its pronounced effect on the future water budget for the world. The United States will, apparently, be very severely affected by the greenhouse effect, especially since there is no effort whatsoever to cut back on accelerated coal combustion or on its emission controls, and the CO_2 content (as well as for other gases) of the atmosphere continues to steadily increase at about 1 percent per year. The effects of the greenhouse effect are not certain, but it appears that the United States could lose much of its crop yield from the Great Plains states due to greatly reduced precipitation, and overall water resources in the United States will diminish greatly.

THE HYDROLOGIC CYCLE

The hydrologic cycle is illustrated in Figure 3–1. In this figure we see that most of the water budget of the earth is controlled by the oceans. The oceans contain about 1,320,000 petaliters (PL) (1 PL = 10^{15} liters) of water, far more than any other reservoir (see Table 3–1). Some 360 PL is lost by evaporation to the atmosphere each year, and, of this, some 323 PL rains or snows directly back into the oceans (about 90 percent). The remaining 37 PL is carried over land masses as clouds; and an additional 62 PL is accumulated in the atmosphere by evapotranspiration. This total of 99 PL is returned as precipitation (rain and snow). To close the cycle, some 37 PL is returned to the oceans, mainly by surface runoff from streams and rivers (33 PL) and the remainder by subsurface flow (4 PL).

The situation for the United States is shown in Figure 3–2. Of the 150,000 billion liters per day (BLD) held as water vapor in the air that crosses the conterminous United States, some 16,000 BLD reaches the surface by precipitation. Most of this, 10,500 BLD, is evaporated from wet surfaces or transpired from

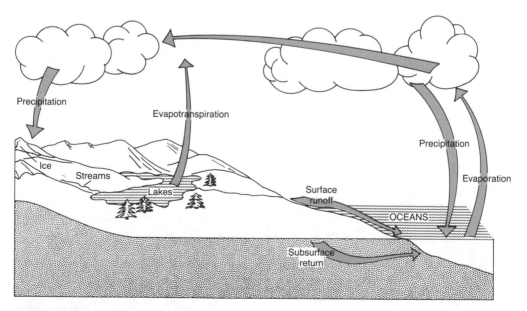

FIGURE 3–1   The hydrologic cycle.

TABLE 3–1   The earth's water.

Type	Location	Volume (liters)	Percent of Total
Surface Waters	Freshwater lakes	1.25×10^{17}	0.009
	Saline lakes, inland seas	1.04×10^{17}	0.008
	Stream channels	1.00×10^{15}	0.0001
Subsurface Waters	Vadose water	6.7×10^{16}	0.005
	Groundwater (to 750 meters)	4.17×10^{18}	0.31
	Groundwater (below 750 meters)	4.17×10^{18}	0.31
Other Reservoirs	Icecaps, glaciers	2.9×10^{19}	2.15
	Atmosphere	1.3×10^{16}	0.001
	World's oceans	1.32×10^{21}	97.2

SOURCE: U.S. Geological Survey Water-Supply Papers, 1972.

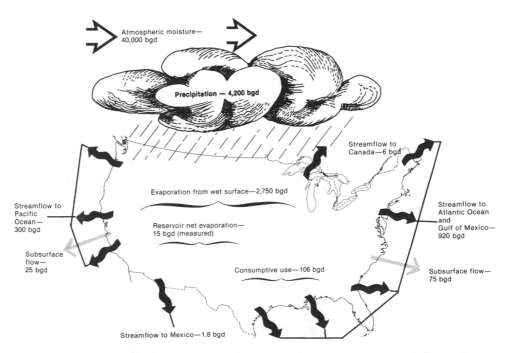

FIGURE 3–2   Water budget of the contiguous United States; numbers are billion gallons per day. (Source: U.S. Water Resources Council)

leaves. The remaining 5500 BLD is the amount that reaches groundwater, streams, and lakes; is evaporated from reservoirs; or reaches the oceans by runoff. Only about 2500 BLD is available for human use.

Surface Water

Surface water accounts for 65 percent of water use in the United States, groundwater for 20 percent, and saline water for 15 percent. The saline water comes from bays, estuaries, and the ocean. It is used for cooling purposes in electrical generating plants, for manufacturing and, in small amounts, as supply to desalination plants. Surface water occurs in lakes, swamps, rivers, marshes, and artificial reservoirs. Reservoirs vary tremendously in size. The largest reservoirs as of 1984, each with over 2 million acre-feet capacity, total 31. (One acre-foot equals 1.03 million liters.) There are about 1600 reservoirs with capacities from 5000 to 2 million acre-feet and about 47,500 reservoirs with 50 to 5000 acre-feet. The most abundant, of course, are the 1.8 million small farm ponds and small reservoirs with less than 50 acre-feet capacity. This last group provides about 2 percent of the water stores in the United States, whereas the 31 giant reservoirs provide 41 percent. Sixteen hundred in the next category store 37 percent, and 47,500 in the third category store 20 percent. This information is summarized in Table 3–2.

Groundwater

Groundwater represents the third largest reservoir of water for the world (see Table 3 1). Groundwater is water which occurs in water-saturated media, that is, at and below the water table. Vadose water is the water held in unsaturated media, that is, above the water table. Groundwater is divided into shallow and deep, with the cutoff rather arbitrarily set at 750 meters (2500 feet). The U.S. Geological Survey estimates that there are equal amounts of groundwater in both

TABLE 3–2 Summary of manmade reservoirs in the United States, "1975."

Reservoir Size (Thousand Acre-Feet)	Number of Reservoirs	Normal Storage Capacity (Thousand Acre-Feet)	Percentage of Total Normal Storage Capacity
More than 10,000	5 } 31	117,000 } 191	25 } 41
2,000 to 10,000	26	74,000	16
5 to 2,000	1,600*	168,000*	37*
0.05 to 5	47,500*	91,000	20*
Less than 0.05	1,843,000*	10,000*	2*
Total (approx.)	1,892,131	460,000	100

*Approximate values.
SOURCE: U.S. Water Resources Council.

sub-reservoirs, each of perhaps nearly 4200 PL capacity. Yet most of this ground-water is not available as a resource. The deep groundwater is difficult to extract, unpredictable in terms of flow, recharge, and other factors, and in general is unexploited. The shallow groundwater is a major resource for the United States and the world.

Most recoverable groundwater is from *aquifers*—rocks, soil, or gravel that are saturated with water and sufficiently permeable to transport and yield water in economic quantities. Aquifers may be thin or thick, and of a great many phys-ical shapes, but most are bedded units. They receive their water through down-ward percolation from the surface environment. Groundwater also provides about one-third of surface stream flows through natural springs and other seepage.

Within the conterminous United States, the groundwater resource to 750 meters totals about 33,000 to 59,000 trillion gallons (125 to 223 PL), far and away much greater than all surface rivers, lakes, and streams combined. But much of this is either not recoverable or polluted. Recharge is also a major prob-lem. Overdrafts have been common from aquifers, especially in the western and midwestern United States. Groundwater provides a significant amount of the fresh water used in the United States, and nearly 70 percent of this (220 billion liters per day) is used for irrigation.

There are many areas in the United States where groundwater is of concern. In the western and midwestern states, one of the most significant aquifers is the Ogallala Formation (Figure 3–3), which is an important water resource for many states. Yet there are severe problems with the Ogallala and other aquifers. The amount of water withdrawn has greatly exceeded the rate of recharge, so that many wells have dried up and others produce far below capacity. This has caused great problems to farmers and ranchers, although they are in part to blame be-cause the situation is due to excessive withdrawals. Since the rate of water re-newal is very slow, it may be many decades before the Ogallala again produces the quantity of water it did some 30 years ago.

Other groundwater problems exist. In coastal areas, overpumping of fresh water has allowed encroachment of salt water. Even in the nation's interior, some aquifers communicate with salty groundwater, and if the fresh water is pumped out too rapidly, the saline water enters and mixes with the aquifer water. Further, pumping water from deep aquifers often means that inferior quality or polluted surface and near-surface water may more rapidly percolate into the deep aquifer. Finally, surface streams depend on groundwater during dry periods, and if groundwater is drastically overdrawn less surface water is available. Much of the crop-growing areas of Nebraska, Kansas, Colorado, New Mexico, Oklahoma, and Texas are being affected by the depletion of the Ogallala aquifer.

Groundwater protection should be an integral part of every state's planning, but it is all too infrequent. If an area is researched before heavy pumping begins, then poor water management may be avoided. For example, in the sand plains surrounding the Little Plover River drainage near Adams, Wisconsin, the U.S. Department of the Interior and others conducted an in-depth study of the subsur-face hydrology of the area, as well as pumping tests to determine how the surface and subsurface flows should be affected by an increased number of wells. This

FIGURE 3–3   Map showing limits of Ogallala aquifer in the western United States. Overdrawing of water has lowered water levels by more than fifty feet over much of the aquifer.

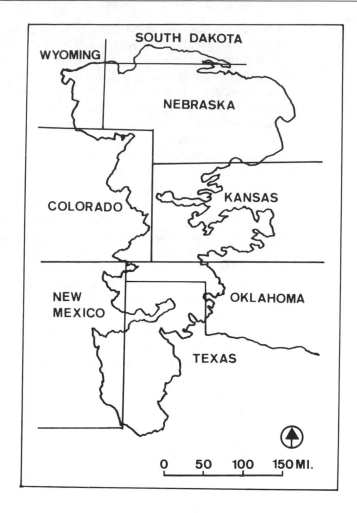

allowed them to properly space and monitor wells to ensure a reliable water supply. Unfortunately, this is the exception and not the rule. Once a source of water is identified, under most conditions, all interested parties seem to do their best to exploit it to the maximum as soon as possible. This leads to a predictable shortage of water as the supply is overmined, and the next step is seeking expensive alternatives to provide water or, in many cases, ceasing the activities for which the water was planned.

WATER RESOURCES REGIONS IN THE UNITED STATES

The U.S. Water Resources Council, for reasons of water budget and planning, has divided the United States into 21 regions (see Figure 3–4). Eighteen of the regions make up the conterminous United States, while Alaska, Hawaii, and Puerto Rico are each a separate region. Of the eighteen, only four have an assured supply greater than the demand: New England (1), Mid-Atlantic (2), South Atlantic–

Gulf (3), and Ohio (5). However, even within these regions the water is not uniformly distributed and there are local shortages. In the summer of 1986, for example, severe drought in the Southeast caused tremendous problems for those in agriculture and related industries. In the Great Lakes (region 4), Tennessee (6), Upper Mississippi (7), Lower Mississippi (8), Souris-Red-Rainy (9), Missouri (10), and Arkansas-White-Red (11), the ratio of supply to demand is about even, so that in years of drought the supply is inadequate and in wet years there may be a surplus. The remaining regions, all in the West—Texas-Gulf (12), Rio Grande (13), Upper Colorado (14), Lower Colorado (15), Great Basin (16), Pacific Northwest (17), and California (18)—have supply-to-demand ratios less than one, and in some cases well below one. Outside the conterminous United States, Alaska (19) and Hawaii (20) have an adequate to good water supply, while the Caribbean (21) is adequate to marginal.

Table 3–3 shows a summary of these 21 regions with their approximate size, estimates of their water supply, source of waters for withdrawal, main uses of water, and existing problems. No region is without problems.

Lake Erie (region 4) is a classic example of excessive pollution, although all the Great Lakes are polluted to some degree. Pollutants have been added to the lake from tributaries and coastal areas; from industrial and domestic discharges, shipping operations, recreational use, and atmospheric fallout. Heavy metals, for example, have been accumulating in both Lake Erie waters and bottom sediments for a very long time. The Detroit River is responsible for large amounts of copper,

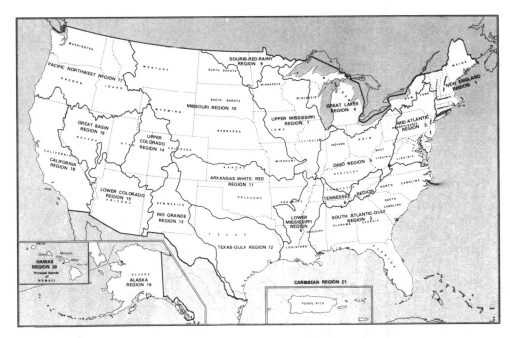

FIGURE 3–4   Water resources regions of the United States. (Source: U.S. Water Resources Council, 1978b)

TABLE 3–3 Summary data of U.S. water resource regions.

Region	Area (Thousands of km^2)	Supply	Sources*	Largest Consumer	Problems
1. New England	180	Adequate	S,L,CL	Industry	Local supply, pollution flooding, water conflict
2. Mid-Atlantic	270	Adequate	S,L,CL	Manufacturing, SEGP*	Local supply, flooding salt water intrusion, erosion, sediment, pollution, dredging, waste disposal
3. South Atlantic–Gulf	700	Good	S,L,CL,GW	SEGP, paper-pulp	Salt water intrusion, pollution, local supply
4. Great Lakes	345	Adequate	L,S	SEGP, manufacturing	Quality, erosion, sediment, extreme pollution
5. Ohio	410	Good	S,GW,L	SEGP, manufacturing	Quality, acid mine drainage, flooding, erosion, sediment, pollution
6. Tennessee	110	Good	S,L,GW	SEGP	Quality, floods, pollution, erosion, sediment
7. Upper Mississipi	465	Good	S,L,GW	Irrigation, SEGP	Supply, quality, flooding, erosion, sediment
8. Lower Mississippi	270	Adequate	GW,S,L,CL	Irrigation	Supply, flooding
9. Souris-Red-Rainy	140	Adequate	L,S,GW	Irrigation	Pollution, flooding

10. Missouri	1325	Marginal to inadequate	R,S,GW	Irrigation	Supply, quality, flooding, erosion, sediment
11. Arkansas-White-Red	630	Marginal	S,L,R,GW	Irrigation	Supply, quality, erosion, sediment
12. Texas-Gulf	460	Marginal to inadequate	S,L,GW,CL,R	Irrigation	Supply, quality, erosion, sediment
13. Rio Grande	355	Inadequate	S,R,GW	Irrigation	Salinity, supply, pollution, erosion, sediment, flooding
14. Upper Colorado	265	Inadequate	S,R	Irrigation, exports	Supply, salinity, quality, sediment
15. Lower Colorado	400	Inadequate	S,GW, imports	Irrigation	Supply, salinity, erosion, flooding. sediment
16. Great Basin	360	Inadequate	S,GW	Irrigation	Supply, quality, flooding, pollution, salinity, sediment
17. Pacific Northwest	700	Adequate	S,CL	Irrigation	Supply, quality, sediment, estuary pollution
18. California	430	Inadequate	S,L,GW,CL, imports	Irrigation	Supply, flooding, quality, erosion, sediment
19. Alaska	1520	Adequate	S,L,CL,SI	Manufacturing	Local supply, flooding, management
20. Hawaii	16	Good	S	Irrigation	Quality, flooding, pollution
21. Caribbean	8	Marginal	S	Irrigation	Management, supply

SOURCE: U.S. Water Resources Council.

*Abbreviations: S, streams and rivers; L, lakes; CL, coastline; GW, groundwater; R, reservoirs; SI, snow and ice; SEGP, steam electric generating plants.

lead, and zinc entering the lake, while these same three elements can also be found in sewage, atmospheric fallout, shoreline erosion, dredging residues, and the smaller tributaries of Canada and the United States. Cadmium is added by atmospheric fallout from surrounding industries. Organics are added by industrial operations, from sewage, and from several other sources. Nutrients, in some excess, of carbonaceous matter, phosphorus, and nitrogen are added in enough quantities to cause partial eutrophication—excessive growth of aquatic vegetation and stress upon the ecosystem. Mercury is added mainly from polluted Lake Huron via the Detroit River; its source is presumably the chlor-alkali plants around the shore. Of the Great Lakes, Lake Erie and perhaps Lake Ontario, near its southern shore, are the most polluted. Lake Huron has heavier pollution in its eastern half, whereas Lake Superior and Lake Michigan are only moderately polluted. The Great Lakes water resources region is one of major pollution concern.

Similarly, the Lower Mississippi region (8) is very much affected by anthropogenic factors. The entire deltaic environment of the Mississippi River has been disturbed by overimpoundments to divert waters, by dredging operations to increase ship transport, by drilling for offshore petroleum, by unevenly distributed sediment, by uneven erosion, by salt-water intrusion, and other factors. Chemical pollutants also present growing problems.

Water supply is a critical issue in all regions, even if locally, but the problems of the Lower Colorado region (15) are especially acute. Roughly half the population of the western United States depends on the Colorado River for its water. Because the Colorado River has insufficient water to meet agricultural demands, groundwater is excessively pumped from Nevada and Arizona to make up the difference, resulting in groundwater depletion. This is decreasing the river flow in some areas and dropping the groundwater level some 8 to 12 feet per year. Part of the problem lies with the Mexican Water Treaty of 1944, by which 1.5 million acre-feet per year is committed to Mexico from the Colorado River as well as other rivers in the United States. Due to excessive depletion it may not be possible to meet this commitment in the not-too-distant future. A combination of anticipated increased use for Colorado River water, coupled with excessive overdrafts, lack of replenishment, some salinity and other pollution problems, and possible upstream damming, paint a bleak picture for the Colorado region.

Farther west, the California region (18) is beset with problems. This region also relies on the Colorado River water for a substantial part of its supply, and future supplies are not necessarily guaranteed. Pollution by pesticides, herbicides, and natural salts, especially selenium-bearing salts, poses a major problem in the San Joaquin Valley; erosion and sedimentation, coupled with dredging and dredging spoils, indicate a major problem for the Sacramento Valley.

WATER USES

Table 3–4 shows the total water withdrawals and consumption for 1975 and 1985 and projections for the year 2000. Interestingly, total fresh-water use will

TABLE 3–4 Total withdrawals and consumption, by functional use, for the 21 water resources regions in "1975," 1985, and 2000.

Functional Use	Total Withdrawals (million gallons per day)			Total Consumption (million gallons per day)		
	"1975"	1985	2000	"1975"	1985	2000
Fresh Water:						
Domestic:						
Central (municipal)	21,164	23,983	27,918	4,976	5,665	6,638
Noncentral (rural)	2,092	2,320	2,400	1,292	1,408	1,436
Commercial	5,530	6,048	6,732	1,109	1,216	1,369
Manufacturing	51,222	23,687	19,669	6,059	8,903	14,699
Agriculture:						
Irrigation	158,743	166,252	153,846	86,391	92,820	92,506
Livestock	1,912	2,233	2,551	1,912	2,233	2,551
Steam Electric Generation	88,916	94,858	79,492	1,419	4,062	10,541
Minerals Industry	7,055	8,832	11,328	2,196	2,777	3,609
Public Lands and Others[1]	1,866	2,162	2,461	1,236	1,461	1,731
Total Fresh Water	338,500	330,375	306,397	106,590	120,545	135,080
Saline Water,[2] Total	59,737	91,236	118,815			
Total Withdrawals	398,237	421,611	425,212			

[1]Includes water for fish hatcheries and miscellaneous uses.

[2]Saline water is used mainly in manufacturing and steam electric generation.

SOURCE: U.S. Water Resources Council.

decrease somewhat between 1985 and 2000 due to recycling, conservation, and more efficient water-use technology. When discussing water uses in general, it is important to distinguish between water used and then returned to its source, and that which is consumed and not returned. Transpiration, for example, is a consumptive process. Figures 3–5 and 3–6 show the data of Table 3–4 plotted on pie

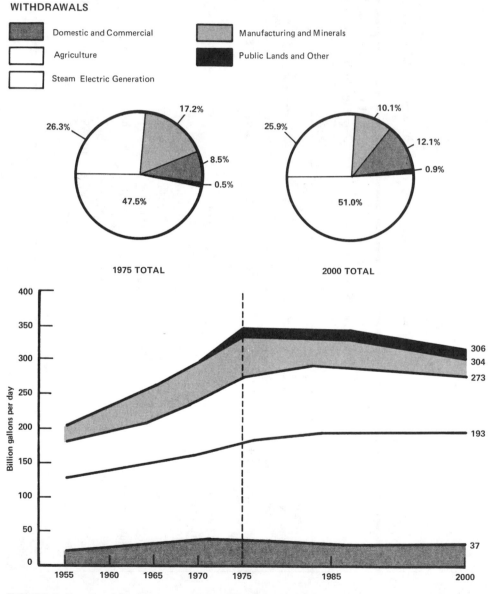

FIGURE 3–5   Total fresh-water withdrawals in the United States, 1975 and 2000 (projected), by functional use. (Source: U.S. Water Resources Council, 1978b)

diagrams and graphs. Figure 3–6 shows that the most significant gains in consumption are for the categories of manufacturing and minerals processing and for
steam electric generating plants.

Fresh-water uses are divided among domestic, commercial, manufacturing,
agricultural, steam electric plants, minerals industry, and public lands and others

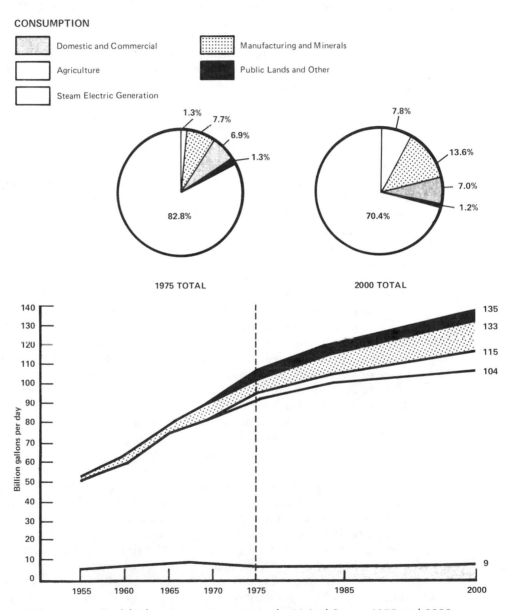

FIGURE 3–6   Total fresh-water consumption in the United States, 1975 and 2000
(projected), by functional use. (Source: U.S. Water Resources Council, 1978b)

(Table 3–4); most saline water is used by industries. About 75 percent of the domestic water is used for drinking, food preparation, washing and sanitary needs. Depending on the area, outside activities such as lawn watering and swimming pools are considered in this category. The remaining 25 percent is for fire protection and other municipal activities. Water conservation in the domestic category is simple. Such methods as more efficient water taps and toilet designs are easy to implement and they work, but overall have relatively little impact on total water use and are not of great national significance.

Commercial and manufacturing uses of water, typically concentrated in areas of high population density, are currently being handled more efficiently. Recent pollution controls now require plants to recycle water from one to several times, and over 90 percent is returned to surface water sources. Much of this returned water meets environmental standards, and thus does not pollute the surface sources.

In the United States, a tremendous amount of water—about 80 percent of total consumption—is used in agriculture-based industries. About 1 percent is used for watering livestock, and the rest goes toward irrigation. Water withdrawals for irrigation alone will increase to over 50 percent of the U.S. total by the year 2000, up from 47 percent in 1975. The complex farm picture in the United States assumes, in most scenarios, a nearly constant source of water, yet water from available aquifers has decreased. Some current farm practices, aimed at quick crop yield, fail to consider and correct for erosional factors. The old contour type of farming, while not often efficient, keeps the land surface fairly stable. The modern circular-planted crops use more water and do not effectively prevent erosion. Any time erosional processes are allowed to encroach on farmlands, there is a net loss of available water due to increased evaporation, local runoff, and seepage. A very real question is whether the use of copious amounts of water for irrigation of crops already in oversupply is justified. Further, inefficiency and waste of water have been known to be encouraged by federal projects.

As is pointed out later in this chapter, the greenhouse effect will further disrupt agricultural use of water.

Steam plants producing electricity consume fairly large amounts of water (Table 3–4), but due to more efficient recycling, the amount withdrawn should decrease by 2000 even though the total amount consumed will increase slightly due to more on-line power plants (see Figures 3–5 and 3–6). The mining and treatment of coal, oil shale, and so forth account for about 5 percent of water used for energy generation. In some areas, water simply is not readily available for large-scale energy production. As is discussed in some detail in Chapter 7, one of the problems preventing easy exploitation of oil shales in the western United States is the lack of water for processing. The nearest available water for such use is either from already overcommitted supplies or from wilderness areas.

Minerals industries, that is, mining of metals, nonmetals, and fuels, consume a small but significant part of the U.S. total (Table 3–4). Fuels in particular, especially due to this nation's reliance on coal, will account for over 60 percent of water consumption in this category by 2000 due to coal slurry lines and coal cleaning. The amounts consumed for nonmetals and metals will stay at their present levels.

Instream use of water is also important. Hydroelectric power, while expected to increase only slightly by 2000, is an excellent form of energy. The water is not consumed, it is not polluted, and the lifetimes of the power stations, with low maintenance costs, are long. However, as is pointed out in Chapter 7, one limiting factor for hydroelectric power in the United States is the lack of available water. Other instream uses are for fish and wildlife, recreation, and some navigation.

DRINKING WATER QUALITY

The Safe Drinking Water Act of 1974 (Public Law 93-525) was passed to prevent drinking-water-related health problems and to ensure a good supply of safe drinking water for the future. While the public is perhaps almost complacent about drinking water now that so many good steps have been taken to adhere to PL 93-525, the problem of polluted drinking water from the various sources discussed earlier in this chapter remains. The passage of the Toxic Substances Control Act of 1976 (PL 94-469) was intended to keep our drinking water safe.

At the same time we are faced with problems concerning the safe disposal of toxic materials. The United States produces over 30,000 different chemicals, and not very much is known about long-range—or in many cases even short-range—health effects of many of these substances. The National Academy of Sciences (see references at the end of this chapter) has studied this problem in some detail and in 1980 listed 22 known carcinogenic chemicals which might pollute waters. Yet there are many chemicals that may be toxic at lower levels than those detected by monitoring laboratories. Certainly, many carcinogens may not be initially hazardous at very low concentrations. Such substances can gradually accumulate in plants and animals and thus enter the food chain. Further, some chemicals once introduced into water react with other substances to form new carcinogenic or toxic compounds. Elements like cadmium, lead, mercury, and even copper, iron, and zinc may be toxic to lethal at low levels in fish, for example. The environmental aspects of these and other metals are further discussed in Chapter 8. Chemicals such as cyanide, PCBs, PBBs (polychlorinated biphenyls), phenols, toluene, and others are especially problematic as they do not degrade, and they are known to concentrate with time. Because of this knowledge, fishing is banned in parts of the United States (some of the Great Lakes and the Hudson River, for example). Chloroform may be a byproduct of chlorine from chlorinated water which reacts with humic matter on stream bottoms. Cannon and Hopps (1968) have pioneered efforts to relate health and disease trends to water geochemistry (see references at the end of this chapter). While this is discussed more fully in Chapter 8, suffice it to say here that not only are such links established, but the problem is very large. Figure 3–7 identifies areas with known problems of drinking water quality in the United States. Cannon, Hopps, and others argue

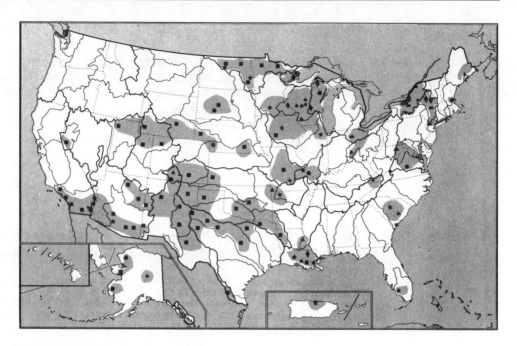

Explanation

Area problem

▨ Area in which existing or potential pollution of domestic water supply was reported

☐ Unshaded area may not be problem-free, but the problem was not considered major

Specific sources of pollution

● Industrial chemicals other than chlorinated hydrocarbons

◆ Chlorinated hydrocarbons from treatment processes and energy development

▲ Heavy metals (e.g., mercury, zinc, copper, cadmium, lead)

★ Coliform and other bacteria

■ Saline water

⬠ General municipal and industrial wastes

Boundaries

── Water resources region

── Subregion

FIGURE 3–7   Areas of reported drinking-water pollution problems, as identified by governmental study teams. (Source: U.S. Water Resources Council, 1978b)

that monitoring of water for inorganic constituents is often satisfactory, whereas monitoring for bacteria and organics is not. Yet the combined effect of various organics in water systems may be of much greater concern than the inorganic constituents. Remedies include better planning, more efficient wastewater management and treatment of solids, and better monitoring equipment. Unfortunately some spills are always likely to take place, therefore much care must be taken to prevent them.

POLLUTION

Surface waters are especially vulnerable to pollution. It has almost been tradition to dump unwanted materials into lakes, rivers, and oceans. Yet these same waters are considered valuable as sources for drinking, domestic, and commercial use. It is common to refer to pollution as point source or dispersed. *Point-source* pollution means the source is restricted to a specific area or a restricted area of known contamination. Waste from a metropolitan area, if untreated, is a good example of a point source. So would a specific chemical from a specific plant, factory, or military operation. The U.S. Congress passed the Federal Water Pollution Control Act Amendment in 1972 (Public Law 92-500) which requires more stringent controls on suspected point-source contamination. Figure 3–8 shows point-source problem areas. Point-source discharges are of many types. Municipal wastes are often not treated to the same degree, the limiting factor essentially being economic. Typically, waste is treated to just meet the standards set, and often the levels of a particular pollutant will be slightly above the standard. While legal, this raises the possibility of damage to the surface water during the above-average pollutant discharge. Fecal coliform bacteria, many toxic substances, excess nutrients, and oxygen-consuming materials are typical of or indicative of pollutants. The fecal coliform bacteria can accompany disease-causing bacteria and organisms linked to salmonellosis, dysentery, cholera, and typhoid fever, among others. Many of the harmful organisms are eliminated by use of chlorine, but this very treatment may introduce by various reactions a series of chlorinated hydrocarbons, including the carcinogen chloroform. Discharge of metals from specific operations is common. For example, in the late 1960s, arsenic was detected in the water of the Kansas River at Lawrence, Kansas. It was finally traced to the Fort Riley Military Reservation 120 miles to the west, where the source was identified as cheaply made detergents. An inferior grade sulfuric acid, made from metal-mining sulfide byproduct (see Chapter 5), was used for making the detergents, and this sulfuric acid contained slightly elevated arsenic content. Waste detergent was then discharged into the Kansas River, raising its arsenic content well above background levels (background is due normally to mineral sources and animal population). The remedy was to switch to detergents with acceptable levels of arsenic. The point here is that no intentional pollution of the Kansas River was implied, yet too often apparently innocuous materials are discharged that may contain unsuspected harmful substances.

Excess nutrients and oxygen-consuming materials greatly affect the aquatic life of surface waters. *Cultural eutrophication* is the term used for the buildup of nutrients when caused by human activities. It causes nutrient-consuming plants to flourish, but it also chokes off the oxygen supply, thus killing off other plants and most animal life. The source of these nutrients are both point source and dispersed.

Another point source of concern is the discharge of cooling water from electric power plants. This raises the temperature of the local surface water, and can lower the dissolved oxygen content. Again, both plant and animal life are threat-

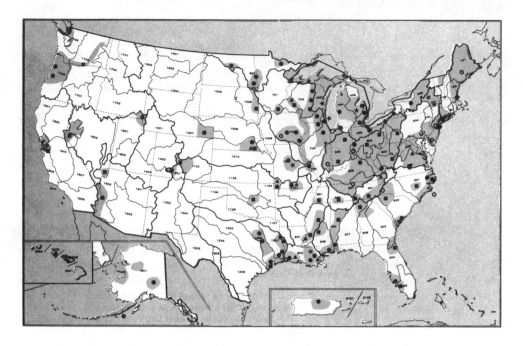

Explanation

Area problem

Area in which significant surface-water pollution from point sources is occurring

Unshaded area may not be problem-free, but the problem was not considered major

Specific types of point-source pollutants

● Coliform bacteria from municipal waste or feedlot drainage

★ PCB (polychlorinated biphenyls), PBB (polybromated biphenyls), PVC (polyvinyl chloride), and related industrial chemicals

▲ Heavy metals (e.g., mercury, zinc, copper, cadmium, lead)

■ Nutrients from municipal and industrial discharges

○ Heat from manufacturing and power generation

Boundaries

—— Water resources region

—— Subregion

FIGURE 3–8   Areas of surface-water pollution problems from point sources (municipal and industrial waste), as identified by governmental study teams. (Source: U.S. Water Resources Council, 1978b)

ened. More efficient cooling towers have lessened but not eliminated this problem, especially in areas of intensive use of surface water.

Just identifying the point source and taking measures to prevent future contamination is not always enough. Chemicals like PCBs (polychlorinated biphenyls) and heavy metals like nickel and chromium may remain in the contaminated waters long after the source of pollution has been brought to a halt.

There are many *dispersed* sources, both human and natural (Figure 3–9). In the recent past not as much attention was given to these as was given to point sources, primarily because point sources are easier to zero in on. Yet there are many severely polluted dispersed-source areas. The runoff from areas affected by urban activities, logging operations, mining, agriculture, and so forth are, for the most part, somewhat easy to identify. The selenium concern in the San Joaquin

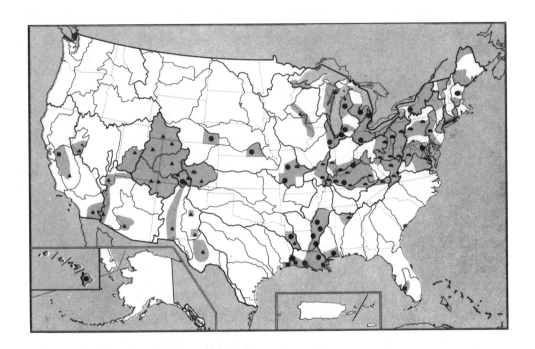

Explanation

Area problem

Area in which significant surface-water pollution from nonpoint sources is occurring

Unshaded area may not be problem-free, but the problem was not considered major

Specific types of nonpoint-source pollutants

●   Herbicides, pesticides, and other agricultural chemicals

▲   Irrigation return flows with high concentration of dissolved solids

■   Sea-water intrusion

◆   Mine drainage

Boundaries

——— Water resources region

——— Subregion

FIGURE 3–9   Areas of surface-water pollution problems from nonpoint (dispersed) sources, as identified by governmental study teams. (Source: U.S. Water Resources Council, 1978b)

Valley of California is a good example: Excess selenium has been added to waters from irrigating selenium-bearing soils, with an apparent adverse effect on animal and plant life. A natural source is the contamination of surface waters by dissolution of surface salts during flooding, as is common in areas of high surface salt content.

Agricultural activities are the main source of dispersed contaminants, such as excess nutrients and fertilizers, herbicides, pesticides, and other undesirable materials. In the United States large amounts of oxygen-consuming materials, 66 percent of phosphorus, and 75 percent of nitrogen discharged to streams comes from agricultural activities. This situation can be remedied by better soil conservation measures and would include less soil loss through improved streambank stabilization, better application of herbicides and pesticides, and the increased use of settling ponds to remove materials prior to discharge. Forestry operations such as logging promote rapid erosion by the removal of the natural vegetative cover, and increase runoff of excess nutrients to nearby surface waters. Careful logging can prevent erosion from happening.

Drainage from abandoned mine sites can be a major source of contamination in some areas of the United States and elsewhere. This problem is made more difficult to correct because determining ownership and thus responsibility is often virtually impossible. By far the most severe cases come from areas affected by the acid mine drainage of coal mining operations, covered in more detail in Chapter 7.

Increased salinity of surface water can result from several processes. When fresh water is overdrawn in coastal areas, saline waters often encroach on the aquifer. In areas of extensive use of fertilizer and irrigation, soluble salts leach from the crop land and are discharged to the surrounding surface waters, increasing their salinity. These saline waters must be treated before reuse to reduce the salt level, thus increasing the expense involved.

Pollution of groundwater is a major problem throughout the world. Since by definition groundwater is out of sight, and effects of pollution are not easily recognized, pollution is harder to detect in groundwater than in surface water. Moreover, treatment to remedy the situation is more costly and time-consuming. Pollution has both natural and human sources. Naturally high salt content may be due to dissolved elements and ions (magnesium, iron, sulfate, nitrate, arsenic, fluoride, phosphorus, chloride, and others) being brought into the aquifer. In some cases, natural radioactivity levels may be high. Natural pollution may be easier to identify than to remedy since often no single source can be pinpointed for treatment.

Human pollution includes such things as drilling of oil and gas wells through aquifers and allowing petroleum, gas, and brine to migrate into fresh groundwaters. In areas where toxic chemicals and other hazardous wastes were disposed in deep wells, some of these substances migrated into overlying groundwater. Landfills made from industrial and municipal wastes are a likely source of many contaminants which reach the groundwater by seepage. No controls were in effect while many of these landfills were accumulating, and the very locations of a good number are not known. Even the overall magnitude of this problem is unknown. Pollution of groundwater from cesspools and septic tanks is also a problem of presumably large concern, especially where urbanization has encroached on previously rural areas. Further, leached salts from irrigation of croplands can seep into groundwater carrying with them dissolved fertilizers, herbicides, and pesticides.

Figure 3–10 shows the areas of the conterminous United States where groundwater pollution has been identified. Since the United States derives 50 percent of

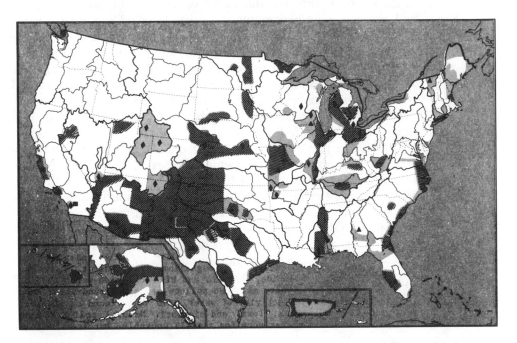

Explanation

   Area problems

   Significant ground-water pollution is occurring

   Salt-water intrusion or ground water is naturally salty

   High level of minerals or other dissolved solids in ground water

   Unshaded area may not be problem-free, but problem was not considered major

   Specific sources of pollution

   ● Municipal and industrial wastes including wastes from oil and gas fields

   ◆ Toxic industrial wastes

   ▲ Landfill leachate

   ■ Irrigation return waters

   ◤ Wastes from well drilling, harbor dredging, and excavation for drainage systems

   ★ Well injection of industrial waste liquids

   Boundaries

—— Water resources region
   Subregion

FIGURE 3–10   Areas of groundwater pollution problems, as identified by governmental study teams. (Source: U.S. Water Resources Council, 1978b)

its drinking water from these sources, it is not a small problem. The U.S. Water Resources Council reported some 800 cases of waterborne diseases due to groundwater sources in 1975. Many factors that result in groundwater pollution may be traced to salt-water intrusion, lowered water tables, seepage of surface and near-surface waters to depths, and other factors—all related to groundwater overdrafts.

Before remedial treatment of most groundwaters is possible, the mechanisms for transport, the sources for contamination, and the regional hydrology and geology must be carefully assessed. Detailed sampling, careful analyses of waters collected, and sophisticated computer modeling are all necessary parts of such projects. There should be careful siting of landfills and septic tanks. The Environmental Protection Agency, for example, estimates that one-third of all septic tanks are so poorly sited that they represent a potential hazard for groundwater contamination. Basically, if potential sources of contamination can be identified, some preventative work may be much more beneficial than the expensive, time-consuming remedial work.

Solution mining involves the dissolution underground of soluble ore minerals, such as certain ores of uranium or copper. Chemicals, sometimes acids and sometimes bases, are added to dissolve the ore minerals, and the metal-rich liquid is then pumped to the surface. By law, the rocks affected must be treated to return them as much as possible to their original condition. This can be done by adding appropriate salts to neutralize the acid or base. The picture is made complex by many factors that include losing some materials not pumped to the surface and mineral reactions that release non-ore metals to solution. Pollution problems are discussed further in Chapter 8. In most cases, solution mining is not undertaken if there is a nearby area using the aquifer for water supply.

FLOODS

Floods are a natural phenomenon that can occur anytime during the year and in just about any area. In the United States roughly 85 people die each year from floods; still, most people are very blasé about the probability of a flood affecting them. Property and related damage amounts to several billion dollars a year; damage is possibly more noticeable in urban areas, but actually more widespread in lands used for farming and ranching. Floods may be a consequence of tidal storms (in coastal areas), rapid snowmelt, or isolated torrential downpours which commonly result in flash flooding, as in the western United States. Nevertheless it is a common fact that areas prone to flooding are especially ripe for development. Floodplains possess a moderate topography, are usually fertile farmlands, or are easy to develop as urban centers with easy transportation access. Human impact on floods is indirect. Thus if forestry operations alter the watershed drainage, increased runoff may promote flooding. If water is diverted from channels, the effect may be to promote flooding wherever the water is taken.

If the sediment load that a river carries to the ocean is interrupted by dams or other impoundments, the riverbed downstream is severely affected, causing lower water tables, sometimes pollution, and impoverishment of some cropland. In north-central New Mexico, for example, the building of Cochiti Dam and Reservoir has been successful in controlling much of the flooding downstream along the Rio Grande, and the reservoir serves as a water supply and recreation area. Yet downstream, the effect of varying the sediment load has been to cause a near-surface zone of groundwater pollution in the Albuquerque area that has begun entering and contaminating privately owned wells. Albuquerque is the urban center for one-half million people, about one-third of New Mexico's population. Nationwide, major areas of flooding are noted in Figure 3–11, on page 62.

It is not easy to prevent flooding. Nature may release large amounts of water over very short intervals of time, and most efforts to thwart the effects, say, of a hundred-year flood are doomed to failure. Further, a so-called hundred-year flood is a statistical gimmick, which over a prolonged period of time is realistic. Yet, near the Mississippi River, two hundred-year floods occurred in just one month during 1973. Flash flooding is even more unpredictable. In many communities construction knows no bounds, nor is geology-based city planning much of a factor. Some areas are scraped level to fill ditches and arroyos (in the Southwest) so that construction may proceed unhampered. When even a moderate storm occurs, some of the water enters the old channel, and because it has just been filled and not yet compacted, subsidence of buildings can and does occur. At greater risk are those homes built along the old channels and down a steep gradient near areas of high relief.

EROSION AND SEDIMENTATION AND THEIR EFFECTS

Erosion and sedimentation have severe effects on lands worldwide. The effects include gullying and channeling, erosion of streambanks, and the shifting of sand along coastlines. Human aspects of both erosion and sedimentation are pronounced.

Soil loss from irrigated cropland is on the order of 20 tonnes per hectare, and sometime three times as much. As mentioned earlier, the modern circularly plowed field, while more efficient for irrigation, promotes erosion and runoff. Forest removal also promotes erosion and soil loss. Overgrazing of pasturelands removes the vegetative cover and promotes erosion. Road construction, the largest earth-moving operation in the United States, has a pronounced effect on erosion. As these lands are eroded, the sediment load to surrounding surface water is greatly increased, as is the scouring action of the water. Sediment-laden water can more readily destroy beaches, channel banks, and even shorelines. The deposition of sediment in rivers causes slowing and spreading of the river flow. The sediment lowers the water capacity of lakes and reservoirs; the sediment behind the Aswan Dam of Egypt is a classic example. Further, the sediment carries pollutants such as phosphates and pesticides, which can affect the aquatic plant and animal life.

The most pressing problem is the loss of good land by erosion; because the effect of this is most often felt by a secondary party, there are not good economic

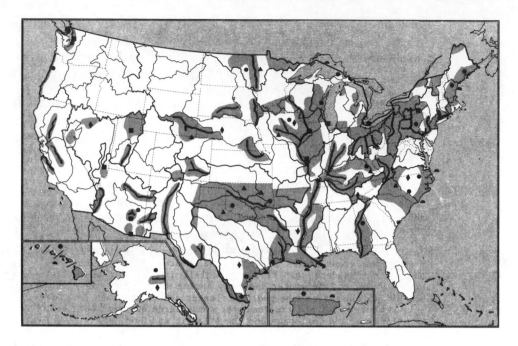

Explanation

Area problem

Area in which flooding causes major damage to agricultural, urban, and other developments

Unshaded area may not be problem-free, but problem was not considered major

Major streams and tributaries with periodic overbank flooding

Reasons for intensified flood damage

● Urban, suburban, and industrial development on flood plains

■ Accelerated runoff from urban areas

▲ Inadequate upstream watershed management

♦ Loss of flood control from inadequate structural systems

◖ Tidal effects along coasts during storms and hurricanes

Boundaries

Water resources region

Subregion

FIGURE 3–11   Areas where flooding problems have been identified by governmental study teams. (Source: U.S. Water Resources Council, 1978b)

incentives to keep soil loss to a minimum. Successful ways to prevent some loss to erosion include contour plowing, leaving some crop residue on the ground, alternating crops with strips of pasture, and carefully planning irrigation.

Surface-mined areas must now include a plan for land reclamation, but mined areas affected before reclamation law are problematic, especially in coal-mining areas.

Dredging operations are necessary in waterways such as the Mississippi River near New Orleans to keep channels open. This classic delta area has been overdredged for many reasons; because of this, coupled with excess levees built for flood control, some river diversion, and other factors, the possibility exists that a good deal of the present shoreline may drown in the foreseeable future. Dredging also can have an adverse effect on aquatic and some plant life. Marshes and other wetlands are areas of natural erosion, and further dredging would only promote subsequent encroachment of the seas.

Bays, estuaries, and nearby coastal waters are valued waterways. Not only are they often major routes for transportation, but they contain abundant wildlife resources and habitats for fish, and provide for recreation. Of utmost concern is the buildup of toxic wastes from industrial operations. The finely suspended sediment load is a prime site for things such as pesticides, heavy-metal oxyhydroxides, and other potentially harmful substances. These are consumed by plants and some bottom animals and thus enter the food chain. The problem is especially acute in the Mid-Atlantic states, in New England, and in the Great Lakes. It is multiplied and accelerated by the common practice of dumping in nearby waterways because on-land sites are too expensive. Further, spills of oil and other pollutants (see Chapters 7 and 8) have seriously damaged valuable coastal areas. The impact of building, roads and other construction, and related activities is often to speed up this degradation of our coastlines.

To remedy this situation, a long-range plan for the ocean waters out to the 200-mile (320 kilometer) limit of the Exclusive Economic Zone is now being undertaken. This vigorous program will attempt a reasonable and realistic environmental assessment of the current situation, and develop a plan for the careful exploitation of offshore resources.

DESALINATION

Desalination is the process by which salts are removed from water. There are about 360 desalination plants operating in the United States, but their total output is only 0.15 percent of the supply. These facilities, found in California, Florida, and Texas with some others in the southwest and northeast, are common along coastal areas where they provide water for industrial operations. Roughly 25 percent of the plants provide some water to municipalities as well.

Worldwide desalination has been proved successful in areas with very warm water, such as the Persian Gulf and nearby waters, and in areas where the water is very cold. In the case of very warm water, desalination methods such as flash distillation (which induces boiling by reducing air pressure) are used. Israel has demonstrated that desalination of salt-bearing waters is economic, and their agricultural industry in large part depends on desalination. Kuwait and Saudi Arabia are now developing desalination programs, and it is presumed that other petrodollar-rich countries in the area will follow. In northern England, at the port city of Ipswich, the water is extremely cold. The seawater is trapped and an organic compound, n-butane, added to cause freezing. Ice consists of pure water, hence the ice is separated out and the solvent remains with the newly concentrated brine. Unfortunately, in areas where this method works best, there is commonly a good water supply. Standard distillation methods become economic in areas of low water resources and heavy tourism, such as the Caribbean.

In the United States, as in many other parts of the world, the groundwater is highly saline, in many instances saltier than seawater. Waters in communication with these brines may continue to build up their dissolved salt content to over 100,000 ppm (10 weight percent). Ocean waters contain about 35,000 ppm (3.5 weight percent) dissolved solids. The temperature of these groundwaters is not very high, and desalination must deal with not only a very high dissolved content but also a lower starting temperature than desirable. Special desalination methods are not economically practical when, in order to comply with the U.S. Public Health Service Standards set in 1962, drinking water must contain less than 500 ppm dissolved salts, and the Environmental Protection Agency has recommended that drinking water contain no more than 250 ppm dissolved salts.

In future years these same economics may make desalination practical, however. In many regions of the United States water supply is inadequate to marginal at best, and demands on available water are increasing. Further, the Colorado River Compact, which guarantees Mexico a fixed amount of water each year, may not have enough to meet domestic demands alone. In this area, and perhaps others, desalination may be the only alternative. Combined solar water evaporative systems may work in some areas, but scale may be a problem. Nuclear or conventional power plants may be designed to make desalination methods useful and practical. The Soviet Union is currently building a breeder reactor (see Chapter 7) for desalination purposes. In the absence of other readily available water resources, and with the assumption that new wilderness areas will not have their waters exploited, then desalination may be a viable route to choose.

CLOUD SEEDING

As mentioned at the beginning of this chapter, the United States only receives as precipitation about a tenth of the clouds that pass over the land. For many years attempts have been made to seed clouds with silver iodide, frozen carbon dioxide, and other materials that may help water vapor nucleate (form drops) and cause rain. The efforts involving winter clouds have been moderately successful, achieving increased precipitation from a few percent to perhaps 20 percent. The seeding of winter clouds in the western United States has helped increase snow in the mountainous areas, thus helping assure adequate snowmelt to replenish reservoirs in the spring.

Seeding of summer clouds is much more problematic. The U.S. Water Resources Council reports that in some areas cloud seeding may actually have decreased the precipitation. Yet in North Dakota and in Florida projects have, for the most part, been somewhat successful. Our understanding of how to seed clouds to produce a given amount of precipitation for a more or less specific area is meager. The situation may be made more complex in the future. Twenty-six states now have legislation addressing weather modification by cloud seeding or in anticipation of other methods. However, interstate agreements on this issue are not well defined. What, for example, is the legal responsibility of a state that induces precipitation by cloud seeding that causes flooding in a neighboring state? Clearly both the legal and physical aspects of cloud seeding have a long way to go.

WATER CONSERVATION

Conservation of any resource is desirable; in the case of water, it is necessary. The United States is not in a position to be complacent about its current water resources, and with the very real greenhouse effect zeroing in on the world, the United States may find itself in a very undesirable position. Water conservation can work.

The largest consumer of water in the United States is agriculture. Irrigation alone accounts for 83 percent of fresh-water consumption. The U.S. Water Resources Council reports that a reduction in water withdrawals could save some 20 to 30 percent. Methods to do that include more efficient distribution, more care taken with farming methods, better impoundments, ditch coverings to prevent evaporation, better ditch and pipe linings, more efficient, possibly computerized, gates and release systems, and the implementation of land treatment. At present, however, the profit margin of many farms is very narrow and many operate presently at a loss, hence the economic incentive to make changes is small.

Water savings in the commercial and domestic categories are comparatively small, but should be undertaken nevertheless. Water used for activities such as lawn watering, car washing, and swimming pools should be examined. Efficient shower fixtures, toilet flushing mechanisms, and so on can also reduce the demands on water.

Conservation of water in industrial and manufacturing operations has been successfully undertaken. Water is efficiently recycled in some operations several times, and is adequately treated before being released. More conservation by industry can be realized, however, through the design of more efficient waste treatment devices.

Steam electric generating plants, vital to the United States and world economy, use large amounts of water, most of which is recycled. Conservation measures may be able to further reduce the amount needed per plant, but may be offset by an increasing number of plants. Whether fired by gas, coal, or nuclear reactors, these plants require large amounts of water withdrawal to dispose of the waste heat that results from conversion of heat energy to electricity. Fortunately, only a few percent of the available fresh water is required for these operations. Nuclear plants require more water than the others (see discussion of this problem in Chapter 7), but again, water loss is under 2 percent. Most of this country's steam electricity-producing plants are located in the eastern half of the country. The combined water regions of Ohio, South Atlantic-Gulf, Mid-Atlantic, and the Great Lakes account for over 50 percent of total energy consumption for electricity in the United States, and fortunately these regions have adequate to good supplies of water. Conservation can be made effective and efficient by ensuring less evaporation of water during the cooling and exchange facets of the operations, coupled with more efficient cooling towers prior to environmental release.

Conservation in the mining and related fields, while not of the same significance as other categories such as irrigation, can nevertheless be more efficient. For discussion, the mining category is subdivided into metals, nonmetals, and fuels. The metals sector uses roughly 11 to 12 percent withdrawals and consumption of water, primarily for ore washing and processing. Covering these operations prevents loss by evaporation, and more efficient recycling of waste-pile slurry water to the front end of the ore processing is possible. Nonmetals are heavily weighted by the crushed stone and sand and gravel industries; the water they use is not as well conserved as may be possible. Withdrawals and consumption for nonmetals are about 54 and 25 percent, respectively. While the withdrawal factor cannot be realistically lowered, the consumption figures can. Fuels account for 35 percent of mineral withdrawals, but over 60 percent of the consumption is due to secondary petroleum recovery and other uses within the oil and gas industries (see Chapter 7). Coal processing demands an increasing amount of water for washing and transporting coal slurries. Conservation of water consumed for the fuel industries possibly will not be attempted. The price of petroleum is quite volatile, and when high, crude oil is produced in the most expeditious manner without regard for the environment. Water used for coal washing is normally so highly acidic and contaminated that it cannot realistically be used again. Coal slurries, once the coal has been separated, leave highly toxic and carcinogenic, very finely divided, colloidal and dissolved constituents. Presently, the wastewater is impounded and allowed to evaporate as there is no incentive to recover the water.

Oil shale may not be processed in the United States in the foreseeable future (see Chapter 7). If, however, it is attempted, then additional water problems will arise: Not only does it take three barrels of water to process one barrel of oil from shale, but oil shales are located mainly in semiarid to arid parts of the western United States. Overdamming of already highly exploited rivers is not reasonable or realistic, and opening of wilderness areas to provide the water is unwise. The presence of pollutants (toxins and carcinogens) in wastewater from pilot plant operations prevents reuse or disposal into streams. Thus conservation measures are not likely.

In summary, water conservation should be attempted at all possible times and in all categories. The potential conservation is greatest in agriculture, but is also substantial in industrial, manufacturing, and steam electric power generating plants. How to implement conservation measures in these sectors is, of course, another matter. Finally, it does not appear that conservation measures will be attempted for the fuels industries.

SOILS IN THE UNITED STATES

Soils are very definitely major resources of any nation. Coupled with an adequate water supply, soils are the medium in which crops are grown, and it is essential that a major commitment to soil management be made.

Soil is that unconsolidated material at the earth's surface which is capable of supporting the growth of plants. Soils form ultimately by weathering of rock; they may be residual if formed in place over their source, or transported if the source is far removed from the soil. Most soil scientists prefer a simple description of soils by a soil profile (Figure 3–12). The A horizon is the topsoil, or zone of leaching, which is underlain by the B horizon, a subsoil or zone of accumulation. These are underlain by the C horizon, the zone of partially altered parent material.

Factors of importance in soil formation are climate, topography, parent material, plant and animal activity, and time. Of these, the first two, climate and topography, are the most important. The references at the end of the chapter list background readings on soils and their formation and classification.

Soils in the conterminous United States are divided into types I to VIII, according to their ability to yield crops or pasture plants without themselves being degraded (Table 3–5). Types I, II, and III are, in general, favorable for cultivating crops, type IV is only suitable for certain crops, and types V through VIII are not at all suited for crops. Total croplands in the United States, in terms of the eight soil classes, are given in Table 3–6 (see page 69).

Cropland in the United States has decreased from 449 million acres in 1958 to 421 million in 1982 with only minor fluctuations (see Table 3–7 on page 70).

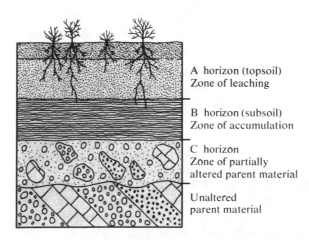

FIGURE 3–12   The major horizons found in a typical soil profile. (Source: Brookins, 1982)

TABLE 3–5   Classification of U.S. soils.

Class	Description/Characteristics
I	Very well suited for cultivated crops, pasture, forest, range, and wildlife. Land is nearly level, and the erosion hazard from wind and rain is low. Soils hold water well, and either contain the necessary nutrients for plant growth or are highly responsive to fertilizer. They are deep, well drained, and easily worked.
II	Less favorable for crop growth than class I but still good. Require careful soil conservation practice, and good soil management. Favorable in most cases for cultivated crops, pasture, forest, range, and wildlife. Generally found in more sloping terrain than class I types, and more susceptible to erosion.
III	Can be used for cultivated crops, pasture, forest, range, and wildlife, but are not as receptive to cultivated crops as types I and II. Subsoil is not very permeable, and slopes are steeper than I or II. Erosion is a greater problem as well. Commonly more alkaline than types I or II.
IV	Can be used for cultivated crops with difficulty. They support some crops, but not many. Steep sloped, not very permeable, alkaline, very sensitive to local climatic fluctuations, and highly susceptible to erosion. Suited for forest and pasture use.
V	Normally used for pasture, forest, range, and wildlife and rarely used for crops. Soils are gently sloped, but possess undesirable qualities such as occurring in frequently wetted bottom lands, or as rocky mountain valley flatlands, areas of short growing season, and ponded areas.
VI	More steeply sloped than type V, their physical setting makes them impossible to cultivate by contour plowing. Not receptive to fertilizers. Main use is for pasture, forest, range, and wildlife.
VII	Only use is for forest and wildlife, with very limited grazing. Soils are extremely sensitive to local climate conditions.
VIII	Not used at all for crops. Found in badlands, rock outcrops, sandy beaches, river wash, mine tailings, and other nearly barren lands. Only use is for recreation, wildlife, or possibly aesthetic purposes.

SOURCE: "National Agricultural Lands Study: Interim Report No. 4 —Soil Degradation: Effects on Agricultural Productivity." (U.S. National Association of Conservation Districts, Washington, D.C.)

crease is due in part to urbanization (which accounted for some 61 million acres in 1967 and 74 million acres in 1982). Urbanization is greater in the Northeast than in the rest of the nation as over 80 percent of cropland is in rural areas. Much of the prime cropland is located in the northern plains, southern plains, corn belt, lake states, and delta states. As discussed later in this chapter, almost all of these states will be adversely affected by the greenhouse effect. Further, the Ogallala aquifer, which provides a good deal of water to the plains states (Figure 3–3), is becoming increasingly depleted. The greenhouse effect will delay replenishment of the aquifer.

SOIL EROSION

Most soil erosion is due to the action of wind or water. Water-induced erosion is classified as rill, sheet, gully, and streambank erosion. Rill erosion removes soil in narrow channels formed from concentrated flow, while sheet erosion removes soil uniformly in a thin layer. While gullies and streambank erosion are more noticeable, sheet and rill erosion remove much more topsoil per year. Wind erosion is a major factor in areas of high winds and low rainfall, and occurs mainly in the southwest and midwestern plains states.

In 1982 the above erosional types accounted for some 6.5 billion tons of topsoil erosion. Sheet and rill erosion amounted to 3.4 billion tons, and gully erosion, streambank erosion, and erosion from roadbanks and construction sites accounted for another 1.1 billion tons. Wind erosion (2.0 billion tons) accounted for the rest. The areas most prone to erosion are shown in Figure 3–13 on page 71.

TABLE 3–6   Distribution of soil types in areas of the conterminous USA (percent).

Area (States)	Types I–III	Type IV	Types V–VIII
Pacific: California, Oregon, Washington	25	13	62
Mountain: Nevada, Idaho, Utah, Montana, Wyoming, New Mexico, Colorado, Arizona	18	12	70
Southern Plains: Texas, Oklahoma	44	13	43
Northern Plains: North Dakota, South Dakota, Kansas, Nebraska	59	11	30
Corn Belt: Ohio, Indiana, Illinois, Iowa, Missouri	76	10	14
Lake States: Wisconsin, Minnesota, Michigan	58	23	19
Delta States: Louisiana, Mississippi, Arkansas	55	11	34
Southeast: Alabama, Georgia, Florida, South Carolina	46	23	31
Appalachian: Tennessee, North Carolina, Virginia, Kentucky, West Virginia	43	13	44
Northeast: Maryland, Delaware, New Jersey, Pennsylvania, New York, Massachusetts, Vermont, New Hampshire, Maine, Rhode Island, Connecticut	36	10	54

SOURCE: Batie, 1983.

TABLE 3–7   Total land and water in the conterminous USA and U.S. Caribbean lands.

	Millions of Acres	
	1977	1982
Land		
Nonfederal		
Forest	370	394
Rangeland	408	406
Cropland	413	421
Pastureland	134	133
Other	166	133
Total Nonfederal Land	1491	1487
Federal	402	404
Total Land Area	1893	1891
Water		
Small Water	9	10
Coastal and Boundary Water	48	48
Census Water	38	39
Total Water Area	95	96
Total Land and Water	1988	1987

SOURCE: U.S. Council on Environmental Quality, 1984.

So-called fragile land presents a special problem. This land is named because it is not well suited for long-term use as cropland. In some states farmers have plowed this land for quick-yield crop growth, with the result that pronounced erosional effects render the land unfit for easy crop cultivation for perhaps decades. The fragile lands are all in soil classes IV to VII, and their use increased over 9 percent between 1977 and 1982 (compared to an increase of only 1 percent for types I–III). This unwise use of these fragile lands also occurred during a time when federal programs were trying to reduce surplus crop yields. Many farmers, aware of falling cattle prices, decided to attempt a quick profit of short-term crops such as wheat on the fragile lands. The result was not only more surplus (the exact opposite of the intent of the federal programs), but also an increase in the erosion rates of 15–20 tons per acre (33–44 tonnes per hectare), or five to seven times the erosion rate of the first three classifications of soil.

Erosion can be controlled by a series of conservation systems, including contour plowing, streambank protection, field windbreaks, conservation tillage, waterways, diversion channels, and other means. Conservation tillage is especially good for erosion control. This method uses a maintenance cover of plant growth during the noncrop seasons to prevent erosion by maintaining moisture content, providing cover against wind and water.

Yet soil erosion continues, resulting in two major environmental effects: First, soil resources are depleted. Second, water resources such as lakes, rivers, and streams are degraded by the inflowing soil particles. Sediment pollution is considered a major problem in the United States and elsewhere where there is a high degree of urbanization and construction. Other environmental aspects of soils, including erosion and chemical additives, are treated in Chapter 8.

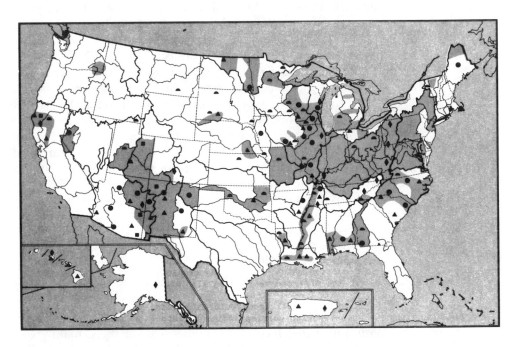

Explanation

Area problem

▨ Area in which erosion or sedimentation is a significant problem

☐ Unshaded area may not be problem-free, but the problem was not considered major

Nature of erosion or sedimentation problems

● Cropland or rangeland erosion or sedimentation

◆ Urbanization, mining, or industrial and highway construction

■ Natural erosion of stream channels

⬥ Shoreline, streambank, or gully erosion

▲ Sedimentation of farm ponds, lakes, water supply, and flood-control channels

Boundaries

── Water resources region

── Subregion

FIGURE 3–13   Areas where erosion and sedimentation problems have been identified by governmental study teams. (Source: U.S. Water Resources Council, 1978b)

THE GREENHOUSE EFFECT

Many islands in the Caribbean, and elsewhere, have an idyllic, soft temperature range, never too hot or too cold. Other places such as outposts above the Arctic Circle have very harsh weather conditions. The responsible factor for these extremes, and everything in between, is climate. We have always been complacent about climate, assuming it is essentially a constant. Yet the geologic record tells us that past climates have been both warmer and colder, but we have passed this off as something beyond our control.

Today, however, we have initiated actions that will have potentially drastic effects on our climate. We have been burning fossil fuels as a source of energy at an ever increasing rate, and the product of carbon dioxide (CO_2) has been accumulating in the atmosphere at a pronounced rate. A 1979 report injected the first hard numbers into the early debate, showing that the CO_2 content of the earth's atmosphere had increased 9 percent from 315 to 345 ppm, between 1958 and 1979.* This trend has been verified from stations around the world where CO_2 is monitored, and is proportional to the amount of fossil fuels burned. The prestigious National Research Council of the National Academy of Science undertook a detailed study of the CO_2 problem and concluded that carbon dioxide levels may reach 600 ppm, almost a 100 percent increase, within the next 50 to 100 years. What are the consequences of this increase? The following statements, modified from the National Research Council's 1983 report *Changing Climate*, are an overview.

1. Carbon dioxide (CO_2) has been shown to be accumulating in the earth's atmosphere at a rapid rate.

2. Much if not most of the CO_2 increase is due to the burning of fossil fuels (coal, oil, and gas). Deforestation of lands may also contribute to the increase.

3. With anticipated increased burning of fossil fuels, mainly coal, as a major source of energy, the CO_2 level of the earth's atmosphere should reach over 600 ppm within the period 2050–2075. There is a 1 in 20 chance that this may happen by 2035.

4. The effect of increased CO_2 in the atmosphere means that there will be an accompanying increase in mean surface temperature for the earth.

5. A doubling of the atmosphere's CO_2 content will result in a net temperature increase of somewhere between 1.5° and 4.5°C.

6. Other gases, such as methane, nitrous oxide, freon, and ozone, being added to the atmosphere may have a significant reinforcing effect on the warming trends expected just from CO_2 accumulation alone. Ralph Cicerone of the National Center for Atmospheric Research (NCAR) has calculated that these

---

* See "Changing Climate," U.S. National Academy of Science Press, 1983.

trace gases have already reached 60 percent of the effect of the increased CO_2 levels.

7. For a 3° to 4°C increase in temperature, the sea level would rise about 70 cm (just over two feet), with severe implications for coastal areas.

8. Climatic conditions will change worldwide. Many areas will be considerably more arid than at present, and a few areas will receive more precipitation (Figure 3–14).

9. Rainfall overall will be expected to decrease worldwide. Due to warming, evaporation of rainfall will increase and thus decrease the amount of water runoff.

10. Some benefits will occur with increased CO_2 content of the atmosphere, especially for some plants because of increased water use efficiency and photosynthesis. Some countries, such as the USSR, will benefit from the increased warming trends, making much of their barren, arid land more suitable for farming.

11. There are large areas of uncertainty involving all these tentative conclusions, but only the magnitude, not the validity, of the above statements is in question.

The CO_2 effect is a fairly clearcut scientific conclusion. The effects of the trace amounts of other gases are much more difficult to assess. Freon, a group name for a series of chlorofluorocarbons, is used as propellants in aerosol spray

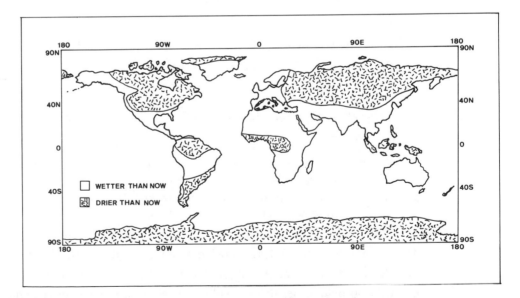

FIGURE 3–14 Possible consequences of global warming, based on paleoclimatic reconstruction when Earth surface temperatures were approximately four degrees (celsius) warmer. Modified from several sources (DOE, EPA).

cans, as solvents, and as refrigerants. Use of freon in aerosol spray cans has been banned in the the United States but is still common in many other countries. Although freon is a greenhouse gas, the freon ban is due to recognition of the fact that atmospheric accumulation of freon (and other gases) is attacking the earth's protective ozone layer. Only human activities put these gases into the atmosphere.

In the case of methane, the picture is less clear. Methane has a short residence time in the earth's atmosphere, but the flux from land to atmosphere is increasing. This may be due to tremendous increases in termite population, possibly to cattle ranching, or mining, but this is not certain (again, the magnitude of these effects is uncertain).

The increase in nitrous oxide is also uncertain, but presumably it is due in part to the increased use of nitrogenous fertilizers. Interestingly, Cicerone of NCAR predicted that these trace gases combined will have a greater effect on the warming of the earth's surface than CO_2 buildup in just 50 years, meaning simply that the period of time which it will take to result in a 3° to 4°C increase could be moderately early in the next century.

Ozone (O_3) is a normal constituent of both the stratosphere and the lower troposphere. In the troposphere, the ozone content has been increasing due to emissions from burning fossil fuels. Ozone, like CO_2 and other trace gases, is a greenhouse gas, and its accumulation in the troposphere facilitates the warming trend.

There are several important feedbacks to the predicted global warming. We can divide these feedbacks into those that promote cooling and those that promote warming. The cooling feedbacks are:

1. Plants may grow faster in a CO_2-rich atmosphere, and thus absorb more CO_2.
2. Since worldwide humidity will increase, there will be increased cloud cover.
3. The tropics will be, to some degree, more reflective due to deforestation.

The warming feedbacks are these:

1. There will be a faster temperature rise in the higher latitudes that will reduce the temperature differences between the poles and the equator, reducing the transfer of CO_2 to ocean depths (Figure 3–15).
2. As the oceans warm up, their ability to keep CO_2 dissolved is lessened, thus resulting in loss of CO_2 to the atmosphere.
3. There will be increased humidity as more water evaporates.
4. Deforestation will release considerable CO_2 to the atmosphere.
5. Glaciers will retreat, and the reflectivity of the poles will increase.
6. Carbon held in permafrost will be released due to warming trends.

While it is difficult to attempt to quantify these feedbacks, it is apparent that the warming feedbacks will far outweigh the cooling feedbacks (Shepherd, 1986).

FIGURE 3–15 Schematic diagram showing the buildup of carbon dioxide in the earth's atmosphere concomitant with a temperature increase. The axes are intentionally unlabelled.

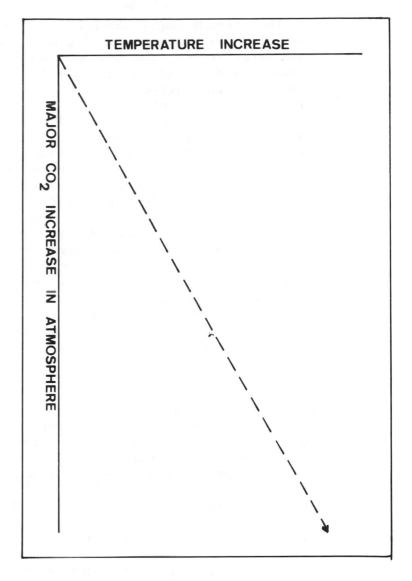

The effects of the global warming will be varied, although the most serious effects may be felt in the tropical countries in Africa and Asia. The picture is also not good for the United States. Scientists predict that only in the extreme south-eastern and northwestern corners of the conterminous United States will precipitation remain at current levels, or possibly increase somewhat. Must of the rest of the country will experience much warmer temperatures. There will be less rainfall, and more of the rain will be lost by evaporation than previously due to the warmer ground surface. This, in turn, will decrease the water available to streams, in turn decreasing the amount available for agriculture and other purposes, and decreasing the rate of renewal to groundwater.

The National Research Council estimated in *Changing Climate* that only a 1°C temperature increase, accompanied by a 10 percent decrease in precipitation, could lead to a 10 percent decrease in spring wheat crops, a 4–5 percent decrease in winter wheat, a 5 percent decrease in soybeans, and a 4 percent decrease in corn. For more pronounced temperature increases, the precipitation would be less and the yields still further reduced.

The United States relies heavily on tapped runoff for domestic water. It is known that runoff decreases rapidly with increases in temperature, and runoff further diminishes with decreased precipitation. The NRC further estimates that for the seven water resources regions of the western states (regions 10–15 and 18; see Figure 3–4), a 2°C increase in temperature, with 10 percent decrease in precipitation, would cause between 40 to 76 percent shortages in water supply! These alarming figures are due in part to the arid nature of much of this land.

The Colorado River, used extensively by many western states (about 15 percent of the water needed by California comes from the Colorado), would suffer a 40 percent decrease in flow by a 2°C warming and a 10 percent decrease in precipitation. Similar effects would be felt for the Rio Grande and other major rivers in these seven regions.

But what about the breadbasket regions in the midwestern United States? Many of these areas have water budgets near balance at the present time, meaning that years of drought create deficits in the water supply system that are balanced by surpluses in wet years. The effect of warming, coupled with decreased precipitation and therefore decreased available water supply, will tip the balance to the deficit side in these regions. Temperature increases will also result in crop damage and other deleterious effects.

During summer months, often temperatures reach and remain at high levels for short to moderate periods, commonly resulting in heat-related negative health effects, excessive energy demands, and so on. The predicted greenhouse effect will bring a dramatic increase in the frequency of such heat waves, and the United States must be ready for such an eventuality.

Sea level is expected to rise significantly due to the greenhouse effect, with estimates of 70 centimeters (over two feet) obtained for an increase of surface temperatures of 3–4°C. It is assumed that an increase in temperature will result in melting of Arctic ice, but very little of the antarctic ice. Should the predictions be off, and the ice of the Antarctic start to melt, more rapid and drastic sea level changes would result, perhaps as much as several meters. The times predicted for this melting to occur and the seas to rise vary from a few decades to over a hundred years; but the calculations all point to an increase in mean sea level due to the greenhouse effect. How will we deal with an increase in sea level? We can build barriers along metropolitan coastal areas, dependent on supplies of natural materials, rock stability, economics, and other factors. But more serious here are factors such as intrusion of salt water into coastal community water supplies, flooding of estuaries and marshes, and the effects inland of storms at sea. Even now, when El Nino (a Pacific climate perturbation that occurs every several years) hits coastal California, severe damage is done; a modest increase in sea level will amplify that effect. If defenses in the form of barriers or dikes aren't practical, then coastal cities must retreat. Given the time over which sea level is expected to increase, it is possible that such a retreat could be orderly and without too much undue hardship.

What can be done to either alter or live with the greenhouse effect? Although this is a truly international problem, international cooperation has always been difficult to obtain. Only by reducing the need for large-scale production of fossil fuels can the greenhouse effect be decreased, and even then it can only be decreased, not stopped. It is already too late to reverse it totally, at least for perhaps hundreds of years.

FURTHER READINGS

Batie, S. S. 1983. *Soil Erosion: Crisis in America's Croplands?* Washington, D.C.: The Conservation Foundation.

Berner, R. A., and Berner, E. K. 1987. *The Global Water Cycle: Geochemistry and Environment.* Englewood Cliffs, N.J.: Prentice-Hall.

Biswas, A. K. 1970. *History of Hydrology.* Amsterdam: North-Holland Publishing Co.

Bridges, E. M., and Davidson, D. A. 1982, *Principles and Applications of Soil Geography.* New York: Longman Press.

Davis, S. N., and Dewiest, R. J. M. 1965. *Hydrology.* New York: Wiley.

Dewiest, R. J. M. 1966. *Geohydrology.* New York: John Wiley & Sons.

Environmental Quality. 1984. 15th Annual Report of the Council on Environmental Quality.

Fox, C. S. 1949. *The Geology of Water Supply.* London: London's Technical Press.

Gross, M. G. 1980. *Oceanography,* 4th ed. Columbus, Ohio: Charles E. Merrill.

Hunt, C. A., and Garrels, R. M. 1972. *Water, the Web of Life.* New York: W. W. Norton.

Leopold, L. B. 1980 *Water: A Primer.* San Francisco: W. H. Freeman.

Liss, P. S., and Crane, A. J. 1983. *Man-made Carbon Dioxide and Climatic Change.* Norwich, England: Geo Books.

Loveland, D. G., and Greer, D. 1986. *Groundwater: A Citizen's Guide.* Washington, D.C.: League of Women Voters Education Fund.

McGuinness, C. L. 1963. *The Role of Ground Water in the National Water Situation.* U.S. Geological Survey Water-Supply Paper 1800.

National Academy of Science—National Research Council. 1982. *Carbon Dioxide and Climate: Second Assessment.* Washington, D.C.

National Academy of Science—National Research Council. 1983. *Changing Climate.* Washington, D.C.

National Association of Conservation Districts. 1981. *Soil Degradation: Effects on Agricultural Productivity.*

National Association of Conservation Districts. 1981. *Final Report* (National Agricultural Lands Study).

Shepherd, M. 1986. The greenhouse effect and global warming: *Electric Power Research Institute Journal,* June: 4–15.

Turekian, K. K. 1968. *Oceans.* Englewood Cliffs, N.J.: Prentice-Hall.

United Nations Publication. 1976. *Water Desalination in Developing Countries.*

U.S. Department of Agriculture. 1955. Water. In *The Yearbook of Agriculture.*

U.S. Department of Agriculture. 1980. *Agriculture and the Environment: First Annual Report:* Washington, D.C.

U.S. Department of Energy. 1984. *U.S. Crude Oil, Natural Gas, and Natural Gas Liquid Reserves: Annual Report.* Report DOE/EIA-0216(84).

U.S. Water Resources Council. 1978a. *The Nation's Water Resources, 1975–2000: Vol. 1, Summary.*

U.S. Water Resources Council. 1978b. *The Nation's Water Resources, 1975–2000: Vol. 2, Quality and Related Land Considerations.*

Walton, W. C. 1970. *Groundwater Resource Evaluation.* New York: McGraw-Hill.

Wolman, A. 1962. Water Resources Publication 1000-B. Washington, D.C.: Research Council—National Academy of Sciences.

Wolman, A. 1965. Metabolism of cities. *Scientific American* 213:30.

4
Metals

INTRODUCTION

Elements with the metallic properties of ductility, malleability, high thermal conductivity, high electrical conductivity, and strength are classified as metals. Metals, especially iron and steel, are key to all highly industrialized nations and also to the developing nations. Metals are often referred to as *abundant*, if they constitute more than 0.1 weight percent of the earth's crust, and *scarce* if they are less than 0.1 percent. The abundant metals are iron, aluminum, magnesium, titanium; all others are scarce. This chapter will discuss the metals in order of their importance, as reflected in demand, to the United States. Therefore, the order of discussion will be iron, aluminum, copper, zinc, lead, manganese, magnesium, chromium, titanium, nickel, cobalt, molybdenum, tin and tungsten, gold, silver, platinum-group elements, mercury, and other minor metals. Some of the elements discussed in this chapter have more use as nonmetals (as chemicals or minerals) than as metals, but are nevertheless covered here for convenience. This group includes magnesium and titanium, in particular, although there are significant nonmetal uses for many other metallic elements.

Before these elements are discussed individually, however, it is useful to discuss some aspects of metal ore deposits in terms of plate tectonics and time. This introductory material is fundamental to understanding why metal ore deposits are located where they are, and it can provide empirical guides for speculating on future exploration for metal deposits. In addition, a brief overview of metals will follow.

PLATE TECTONICS AND METAL DEPOSITS

The theory of plate tectonics, discussed briefly in Chapter 1, is extremely useful to qualitatively explain the occurrences of many types of metal ore deposits, and a few examples will be given in this section. Metal deposits discussed later in this chapter will refer back to this section.

Figure 4–1 shows an idealized model of convergent and divergent plate boundaries. Processes affecting the genesis of rock and mineral deposits commonly reflect the type of plate tectonic setting involved. The abundant literature on this subject has been nicely summarized by Sawkins (1984), and the descriptive material here follows his approach.

Not covered here are the deposits due to fumarolic activity on the seafloor. These are discussed under "Seafloor Metal Deposits" at the end of this chapter. Other metal deposits, which are not directly a consequence of plate tectonic activity, will be discussed separately.

While the following discussion is brief, the reader will nevertheless note that a great many metals are mentioned in conjunction with different plate tectonic settings and processes. While recognition of the type of plate tectonic setting by the geologist is no guarantee that a particular ore deposit may be found there, the types of possible deposits, and guides to exploring for them, are very much narrowed down. An understanding of such settings provides the exploration geologist with a powerful tool for searching for new metal deposits.

Convergent Boundaries

Metal deposits related to the principal arc part of convergent plate boundaries (Figure 4–1) include porphyry coppers, some copper breccia pipes, many skarn and vein deposits, hot-spring deposits, and massive magnetite deposits. In general, the principal arc is characterized by linear belts of plutonic rocks and overlying volcanic rocks, with the latter commonly stripped by erosion. These occur above actively subducting slabs that dip steeply into the mantle.

Of this group of deposits, the porphyry coppers are the most striking. These derive their name from the porphyritic, fractured, granitic rocks mineralized with

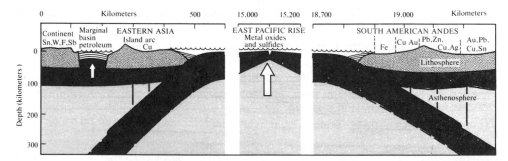

FIGURE 4–1   An idealized diagram showing the relationship between the East Pacific Rise (divergent plate boundary), Pacific margins (convergent plate boundaries), and metallic ore deposits. (Source: Burchfiel et. al., *Physical Geology*, Merrill Publishing Co., 1982)

copper and other elements that occur parallel to active subduction zones around the world (Figure 4–2). Most fall in the age range of 150 Ma to 15 Ma. In addition to copper, they contain abundant iron, usually molybdenum, and gold and silver. This group also contains numerous copper deposits in brecciated rocks found between the plutonic and volcanic rocks, often with other metals. Skarn deposits are characterized by silicate and other minerals formed in carbonate host rocks due to action of fluids from igneous rocks. These deposits are commonly rich in iron, tungsten, tin, copper, and some gold. Complex vein deposits of copper-gold-silver are also associated with principal arc systems, a direct consequence of hydrothermal fluids generated in response to the igneous activity in these zones. Closer and at the surface are hot-spring deposits, commonly rich in gold, mercury, antimony, and arsenic. Massive magnetite deposits are also found parallel to some subduction zones.

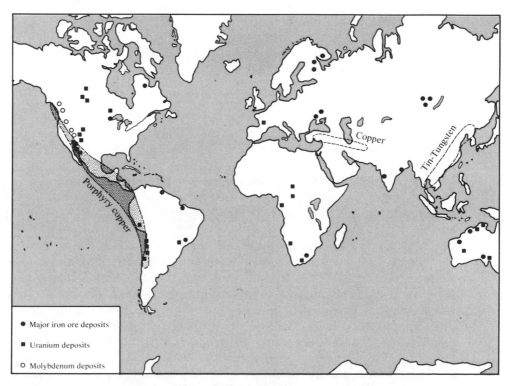

FIGURE 4–2   Worldwide metal ore distribution. The porphyry copper, molybdenum, and tin-tungsten belts are spatially associated with igneous rocks and major subduction zones. Iron and uranium do not show any regularity in their spatial relationship to orogenic belts or igneous rocks. Most of the iron deposits are Precambrian bedded sedimentary types, whereas the uranium deposits are found in sedimentary, metamorphic, or igneous rocks. (Source: Brookins, 1981)

The inner side of the principal arc commonly contains contact-metasomatic deposits of zinc, lead, silver, and sometimes copper, as well as vein deposits. Some of the plutonic rocks are rich in tin and tungsten.

Rift zones near convergent plate boundaries are characterized by high-silica rocks, such as rhyolite porphyries, and numerous caldera settings with diversified rocks. Major molybdenum deposits (like the Climax deposit of Colorado) are spatially related to the Rio Grande rift, and many deposits contain tungsten, tin, uranium, tantalum, and other lithophile elements. This type also includes the Kuroko-type massive sulfide deposits rich in zinc, lead, and copper. The Colorado Mineral Belt contains not only the lithophile elements, but numerous vein deposits of the base and precious metals as well. Also included here are the massive sulfides found in greenstone belts, such as that at Kidd Creek, Canada, as well as some gold deposits, most notably the ore body at Homestake, South Dakota.

Most if not all of the deposits of the principal arc systems are epigenetic, that is, the ore minerals were formed after the host rocks to the ore had formed.

Collisional processes are well known in plate tectonics. The Himalayas are perhaps the most striking example of a collisional event. Magmatic rocks accompanying such collisions are commonly anatectic (formed by melting of existing rock), and granitic varieties may contain tin, tungsten, and copper, while other occurrences contain uranium.

Divergent Boundaries

Divergent plate boundaries also result in a complex series of metal deposits. Ophiolites formed in such settings are occasionally rich in copper (Cyprus) and chromite (Greece).

Intracontinental processes are characterized by hot spots, localized zones of igneous activity rising from beneath the overlying plate. Some of the intracontinental rocks are rich in tin, and titanium is enriched in anorthositic rocks in many areas.

This group also contains the major ultramafic complexes such as the Bushveld complex (South Africa), the Stillwater complex (Montana), and the Sudbury irruptive (Canada). The Bushveld is well known as the world's largest reserve of chromium, platinum-group elements, and vanadium-bearing magnetite. The Stillwater complex contains abundant low-grade chromite and economic amounts of platinum-group elements. The Sudbury irruptive is a major source of nickel, copper, and some cobalt. Uncertainty surrounds the origin of the Bushveld and Sudbury rocks: It has been argued that these occurrences are the consequence of meteoritic impact, the metals contained there perhaps being derived from the meteoritic source. Others contend that the metals are terrestrially derived, and that if a meteorite did indeed strike these sites, its role was to induce the terrestrial magmatism.

Also included in this group are the carbonatites, found in close association with alkalic igneous rocks. These rocks, characterized by igneous calcite, are commonly rich in lithophile elements such as niobium, titanium, the rare-earth elements, phosphorus (in apatite), fluorine (in fluorite), and some copper, uranium, thorium, and others.

Continental rifting environments are also divergent, and these contain some important metal deposits. Numerous copper deposits are known, which include the very important rift-related stratiform deposits of copper, zinc, and cobalt found in central Europe (Kupferschiefer-type), the Zambian copper belt, and elsewhere. The rifting processes are presumed to produce abundant basaltic rocks, whose metals are released by weathering and other processes, then transported to and fixed in sedimentary rocks. While copper is often the dominant element in such deposits, cobalt is very rich in the Zambian belt, and nickel and zinc are often enriched in some of these rocks.

The very important Mississippi Valley-type lead-zinc deposits are also included here. These consist of carbonate host rocks with mineralization rich in lead and zinc.

Advanced stages of rifting often result in sedimentary-rock-hosted massive sulfide deposits, such as those at Sullivan, British Columbia, and Mount Isa, Australia. The host rocks are commonly siltstone and shale, although a wide variety of rocks may be impregnated with sulfides. These deposits are rich in zinc, lead, copper, and usually silver. Some uranium deposits are related to these advanced stages of rifting as well.

METAL DEPOSITS THROUGH GEOLOGIC TIME

Meyer, in his elegant 1981 essay "Ore-Forming Processes in Geologic History," pointed out that different periods of earth history are marked by different types of metal deposits. He recognized four broad divisions of geologic time: the Archean Eon, here taken from the earliest known rocks at 3.8 Ga to the Archean–Early Proterozoic boundary at 2.5 Ga; the Early Proterozoic Era, 2.5–1.8 Ga; the Middle and Late Proterozoic Eras, up to about 570 Ma; and the Phanerozoic Eon, from 570 Ma to the present. In addition, three transition periods are recognized: That marking the boundary between the Archean and Proterozoic at 2.5 Ga and that between the Late Proterozoic and Phanerozoic at about 570 Ma are well-recognized changes in tectonic styles, with the boundaries defined by these changes. The transition at 1.8 Ga is different in that this represents a period of increased oxygen content of the atmosphere such that chemical sediments of iron could no longer form. The following brief discussion is modified from Meyer's essay.

In the period from the earliest known crust (approximately 3.8 Ga) to 2.5 Ga several important types of metal deposits occur. Included here are the chromium deposits of layered intrusives of Greenland, Zimbabwe, India, and the Stillwater Complex of Montana. The nickel-rich, copper-poor ores in layered intrusives from Australia and elsewhere also formed during this time interval. Massive sulfides in volcanic rocks, mainly andesite-rhyolite, formed then as well, including the large Kidd Creek, Canada deposit, and other places in Canada and Australia. Banded iron formations became increasingly abundant during this period, with major occurrences in Greenland, South Africa, Venezuela, India, Australia, Canada, and the United States. Major gold occurrences were formed in Zimbabwe, South Africa (Witwatersrand and others), India, Brazil, Australia, Canada, and the Homestake mine in the United States. Most of the world's gold reserves are found in these deposits.

The period between 2.5 and 1.8 Ga saw a continued increase of banded iron formations, with major deposition in South Africa, Australia, Canada, Brazil, the USSR, and the United States. The Great Dyke of Zimbabwe with its large chromium deposits formed in this period, as did the famous Bushveld Complex with its chromium and platinum-group elements. The large nickel-copper sulfide deposits at the Sudbury irruptive formed at this time, as did the smaller occurrences in Canada. Massive sulfides in volcanic rocks formed in Canada, Sweden, and at Jerome, Arizona, and elsewhere in the United States. Minor tin-tungsten mineralization occurred in granites associated with the Bushveld complex. This period also marks the first sandstone-type uranium deposits (Gabon).

The period 1.8 Ga to about 570 Ma saw the end of ore deposits in ultramafic to mafic layered intrusives, although this period does include the Muskox chromite deposits of Canada as well as nickel sulfide ores in parts of Canada and Scandinavia. This time period is also marked by the formation of most of the world's titanium occurrences in anorthosites in layered intrusives, such as those in the United States, Canada, the USSR, Norway, and Angola. This is the only period in earth history when these titanium ores were generated. Massive sulfides in volcanic rocks during this period were rare, but massive sulfides in sedimentary rocks became very important. Copper-rich ores formed in the United States (White Pine), Africa (Zambian-Zaire belt), Canada, and Australia. Lead-zinc ores formed at several places in Australia (Mount Isa, Broken Hill, others), Canada (Sullivan, B.C.), and the United States (Franklin, N.J.). Most of the world's large unconformity-type uranium deposits formed during this period, including the deposits in Canada (Saskatchewan, others) and Australia. Gold ores were absent during this period, as were porphyry coppers except for the deposit at Haib, Namibia. Tin-tungsten mineralization was minor, with occurrences in Namibia and Brazil.

The Phanerozoic (since 570 Ma) is marked by the greatest diversity of ore deposits. In large part this is due to the change in tectonic style, plus a large amount of recycling of material due to subductive and other plate tectonic processes. While layered intrusive chromites are absent in the Phanerozoic, Alpine-type chromites are common. This group encompasses occurrences in the Philippines, Turkey, Cuba, Norway, the USSR, Australia, Cyprus, and the small deposits in California. While important, these deposits are dwarfed by those in layered intrusives.

Nickel ($\pm$ copper) sulfides in layered intrusives are also not abundant, and include deposits in parts of Canada, South Africa, and the USSR.

Massive sulfides are extremely important. Large deposits in volcanic rocks are found in several rock types: ophiolites (Japan, Turkey, Cyprus, Greece, USA, Norway); andesite to rhyolite (Spain, Australia, USA, USSR, Canada); Kuroko-type (New Guinea, Japan, Fiji, Solomon Islands, others). Massive sulfides in sedimentary rocks include the copper deposits of the Kupferschiefer of central Europe, smaller deposits in Bolivia and in the United States as well as lead-zinc deposits in clastic sediments in England, Ireland, Poland, the USSR, and Germany. Mississippi-type lead-zinc deposits formed in the United States, the USSR, Italy, Australia, and elsewhere at this time.

The major red-bed copper deposits formed in France, Germany, the United States, Canada, north Africa, and elsewhere. Although important, these deposits are dwarfed by the banded iron formations. Major uranium mineralization in this period includes the "porphyry uranium" Rossing deposit in Namibia and the vein-type deposits in Zaire, Czechoslovakia, France, Portugal, and Germany as well as the sandstone-type deposits of the western United States (New Mexico, Wyoming, Texas), Australia, Niger, and Somalia. Gold deposits are smaller than Precambrian ores, but nevertheless important. Noteworthy are sites including the Mother Lode of the United States and deposits in the USSR and Australia.

The Phanerozoic saw the formation of the world's most important porphyry copper deposits. These include many deposits in Chile, the United States (Bingham Canyon, Butte, Ely, Silver City, Tyrone, Morenci, others), Canada, New Guinea, Malaysia, Peru, Argentina, the USSR, and elsewhere. Also included here are the large molybdenum deposits in the United States (Climax, Colorado).

Finally, much of the world's tin and tungsten deposits occur in Phanerozoic rocks, with noteworthy occurrences in South China–Korea, Malaysia-Burma-Thailand, England, Bolivia, Nigeria-Sudan, Untied States skarn deposits, Tasmania, and others.

In summary, it is probable that many shallow deposits of the Phanerozoic had counterparts in the Precambrian, but erosional, metamorphic, and plate tectonic processes have obliterated them. Especially vulnerable would have been hot-spring, shallow vein, and near-surface massive sulfides, and much of the porphyry copper and tin-tungsten granites. Nevertheless, well-recognized differences over time in types of metal deposits are noted. These are presented in abridged form in Table 4–1. It is apparent that geochronologic study (dating of rocks and minerals by radiometric means) can assist in the exploration of certain metal ore deposits.

METALS: DEMAND AND SUPPLY

It is instructive to study the U.S. demand (consumption) figures for 15 metals and for iron ore between 1975 and the year 2000 (Table 4–2). The data are from the United States Bureau of Mines (USBM). There are several points of interest in Table 4–2. The recession that affected the United States and much of the rest of the world in 1982 is clearly reflected in the data shown. Steel and iron demand dropped 28 percent between 1981 and 1982, and as of 1987 had still not recovered to pre-1982 demand. There are many reasons for this including world glut, antiquated mills, high labor costs, environmental control costs, poor economy,

TABLE 4–1 Summary of some metal occurrences with time

Time Period and Metal Occurrence	Location
Archean: 3.8 Ga to 2.5 ± 0.1 Ga	
Chromium in layered intrusive rocks	Zimbabwe, Greenland, India, Stillwater (Montana)
Nickel sulfide ore in layered intrusive rocks	Western Australia, Canada
Nickel-copper sulfide ores in layered intrusives	Zimbabwe
Massive sulfides in volcanic rocks	Canada (Kidd Creek, etc.) Australia
Banded iron formation	Greenland, Africa, Venezuela, India, USA, Canada, Australia, USSR
Gold ores (various)	India, South Africa (Witwatersrand), Brazil, Canada, USA (Homestake), Australia
Porphyry copper	Finland, Canada
Tin-tungsten	Transvaal, Zimbabwe
Early Proterozoic: 2.5 ± 0.1 Ga to 1.8 ± 0.1 Ga	
Chromium in layered intrusive rocks	India, Zimbabwe, South Africa (Bushveld)
Nickel-copper sulfide ores in layered intrusives	Canada (Sudbury)
Massive sulfides in volcanic rocks	Canada, USA (Jerome, Ariz., etc.), Sweden
Banded iron formation	South Africa, Australia, USSR, USA, Brazil
Gold ores (various)	Canada, South Africa, USSR, Ghana
Minor tin-tungsten	Africa (Bushveld-related granites)
Middle and Late Proterozoic: 1.8 ± 0.1 Ga to 570 ± 50 Ma	
Minor layered chromites	Canada
Nickel sulfides in layered intrusive rocks	Canada, Scandinavia
Titanium ores in anorthosites	USA, Canada, USSR, Norway, Angola
Massive copper sulfides in sediments	USA, Canada, Africa (Zambian-Zaire belt), Australia
Massive lead-zinc sulfides in sediments	Canada (Sullivan), Australia (Mt. Isa, Broken Hill), USA
Vein and unconformity uranium deposits	Canada (Saskatchewan), Australia
Porphyry copper	Namibia
Tin-tungsten	Namibia, Brazil

TABLE 4–1   Continued

Time Period and Metal Occurrence	Location
Phanerozoic: 570 ± 50 Ma to Present	
Alpine-type chromites	Norway, USSR, Cuba, Turkey, Philippines, Australia, USA (minor)
Nickel sulfides in layered rocks	Canada, USSR, South Africa
Massive sulfides in ophiolites	Cyprus, Turkey, Japan, Philippines, Greece, Yugoslavia, USA, Norway
Massive sulfides in volcanic rocks	USA, USSR, Spain, Australia, Canada, etc.
Massive sulfides: Kuroko-type	Japan, New Guinea, Fiji, Solomon Islands
Massive copper sulfides in sediments	USA, central Europe (Kupferschiefer), Bolivia
Massive lead-zinc sulfides in sediments	England, Poland, Ireland, USSR, Germany
Mississippi Valley-type lead-zinc	USA, Italy, USSR, Australia, north Africa
Red-bed iron ores	France, Nigeria, USA, Germany, Canada, north Africa
Vein and related uranium deposits	Namibia (Rossing), Zaire, Czechoslovakia, Germany, France, Portugal
Sandstone and related uranium deposits	USA (New Mexico, Wyoming, etc.), Australia, Niger
Gold-quartz veins	USA (Mother Lode, Calif.), USSR, Australia
Porphyry copper	USA (Bingham Canyon, Butte, etc.), Chile (Chuquicamata, etc.), Mexico, Canada, Peru, Argentina, USSR, New Guinea, Malaysia, etc.
Rift-type molybdenum deposits	USA (Climax, Colo.)
Tin-tungsten	South China–Korea, Bolivia-Peru, England, Malaysia-Burma-Thailand, Spain, USA

SOURCE: Modified from Meyer (1981).

and other factors (discussed later in this chapter). As iron and steel go, so go many of the other metals, and the data in Table 4–2 show that almost every industrial metal had a sharp decline between 1981 and 1982. The two exceptions are titanium, used primarily for nonmetallic purposes, and gold, which is much more volatile in price and much less subject to the iron and steel industry.

TABLE 4–2   Demand pattern for major USA metals, 1975–2000

Metal	Units	1975	1980	1981	1982	1983	1984	1985	2000
Iron-steel	Million st	138	109.5	119.8	86.1	94.2	110.2	101.3	160
Iron Ore	Million st (Fe)	77.2	64.3	66.4	38.6	44.8	45.8	44.2	60
Aluminum	Thousand mt (Al)	4079	5223	5209	4811	5442	5279	5127	9260
Copper	Thousand mt	1467	2175	2278	1761	2020	2155	2139	2800
Zinc	Thousand mt	1331	951	1146	869	1005	1036	961	1400
Lead	Thousand mt	1167	1038	1142	1112	1141	1102	1050	1600
Nickel	Thousand t	219.3	205.8	206.0	180.5	204.4	218	230	350
Cobalt	Thousand t	6.4	7.8	5.7	5.2	7.1	8.1	8.2	15.5
Chromium	Thousand st	410	587	510	319	329	437	378	815
Manganese	Thousand st	1133	1029	1027	672	668	627	655	920
Titanium	Thousand st	452	478	517	515	523			750
Molybdenum	Thousand st	27.7	30.5	30.8	18.6	21.5			35.5
Magnesium	Thousand t	1008	916	916	693	792			960
Tin	Thousand mt	53.7	46.7	52.5	30.3	38.2			46.4
Tungsten	Thousand mt	6.3	9.9	10.3	6.1	6.5			19.0
Gold	Thousand tr oz	2870	1031	1666	1979	1680			3700

NOTE: st, short ton (2000 pounds); t, long ton (2200 pounds); mt, metric ton; lb, pounds; tr oz, troy ounce. Data include nonmetal use for aluminum, zinc, magnesium, and titanium.
SOURCE: U.S. Bureau of Mines, *Minerals and Materials* (monthly); *Mineral Facts and Problems*, 1985 ed.

The projected figures in Table 4–2 are for the most probable consumption in 2000 as estimated by the USBM. Despite the obvious uncertainties that go into this kind of prediction, the trends are positive for all metals, and strongly so for some. At present in the United States, production of most metals is somewhat precarious due to competition from foreign sources (many from industries subsidized by their respective governments), lack of quotas on imports, illegal dumping (selling below cost) in some instances, high labor costs, inefficient old mills (in the case of iron-steel), and cost of environmental controls. What this means simply is that for many metals, while our consumption will steadily increase, we will become increasingly dependent on imports from foreign countries to meet our needs. This could have a pronounced effect on the United States in the future should a widespread war break out, or sea lanes be closed to shipping.

Table 4–3 shows data for reserves and demands for several metals from 1984 through 2000. These data clearly show those elements for which the United States will be strongly dependent on imports; included are chromium, cobalt, tungsten, manganese, mercury, nickel, niobium, platinum-group elements, silver,

TABLE 4–3   Metals reserves and demand, 1984–2000

Metal	U.S. Reserve / U.S. Demand	World Reserve / World Demand	Metal	U.S. Reserve / U.S. Demand	World Reserve / World Demand
Aluminum	8 MT / 5 kt	21,000 MT / 15 kt	Magnesite*	10 Mt / 14 Mt	2800 Mt / 110 Mt
Antimony	90 kt / 440 kt	4000 kt / 1000 kt	Manganese	0 / 14 Mt	1000 Mt / 110 Mt
Arsenic	50 kt / 310 kt	1000 kt / 590 kt	Mercury (flasks)	140,000 / 700,000	4 million / 3.7 million
Beryllium	28 kt / 6.9 kt	420 kt / 12 kt	Molybdenum	3000 kt / 550 kt	6000 kt / 1800 kt
Cadmium	90 kt / 75 kt	560 kt / 350 kt	Nickel	300 kt / 3800 kt	58 Mt / 18 Mt
Chromium	0 / 8000 kt	360 Mt / 74 Mt	Niobium	0 / 100 kt	4500 kt / 435 kt
Cobalt	0 / 195 kt	4000 kt / 600 kt	Platinum group (troy oz)	1 million / 34 million	1 billion / 130 million
Copper	57 Mt / 37 Mt	340 Mt / 170 Mt	Silver (troy ox)	930 million / 1.9 billion	7.9 billion / 5.4 billion
			Tin	20 kt / 700 kt	3.1 Mt / 3.9 Mt
Gold (troy oz)	80 million / 52 million	1.3 billion / 670 million	Titanium	8.1 Mt / 11 Mt	190 Mt / 42 Mt
Iron	3700 Mt / 900 Mt	71,000 Mt / 9900 Mt	Tungsten	150 kt / 230 kt	2800 kt / 970 kt
Lead	21 Mt / 12 Mt	95 Mt / 61 Mt	Zinc	22 Mt / 19 Mt	170 Mt / 130 Mt

*Magnesium recovered from seawater not included.

NOTE: Reserves and demands reported in thousand short tons (kt) or million short tons (Mt) except where noted.

SOURCE: Data from J. D. Morgan, U.S. Bureau of Mines; also published in *Mining Engineering,* April 1986, quoted with permission.

tin, and titanium. Even when the reserve-to-demand ratio is favorable, the United States is expected to import major amounts of these and many other metals. In many cases this is necessary to continue favorable trade relationships with various countries. Importation of a critically needed element coupled with other commodities guarantees the United States of supply. In our dealings with Zaire and Zambia, for example, we import not only badly needed cobalt, but also considerable copper. In other cases, environmental aspects have caused curtailment of production of metals, and we thus rely on imports. This is the case for cadmium, arsenic, and others. Interestingly, this increases the overall background levels of these materials in the United States. It is apparent that the United States has become more and more dependent on worldwide economic factors and that this trend is accelerating.

The data shown in Table 4–3 indicate that at least to the year 2000, reserves are very adequate to meet demand. For some metals, such as mercury and lead, a shortage will likely occur early in the next century, while for tin, world reserves may be adequate to meet demand. These reserve figures are considered fairly conservative, and do not include subeconomic or other resources. In addition, there is always the possibility that new metal deposits will be found.

Table 4–4 shows a different perspective on the metals, listing the proportion of our supplies that is imported, as well as the countries providing most of the material. Many of these countries are outside the western hemisphere and, again, there could be problems of closing sea lanes, for example, that could adversely affect the U.S. supply.

TABLE 4–4   Import reliance for selected metals

Metal	Percent	Exporting Countries
Manganese	100	South Africa, Brazil, France, Gabon
Niobium	100	Brazil, Canada, Thailand
Aluminum*	97	Australia, Jamaica, Guinea, Suriname
Cobalt	95	Zaire, Zambia, Canada, Norway
Platinum-Group Elements	92	South Africa, UK, USSR
Chromium	73	South Africa, Zimbabwe, Yugoslavia, Turkey
Tin	72	Thailand, Malaysia, Bolivia, Indonesia
Zinc	69	Canada, Peru, Mexico, Australia
Nickel	68	Canada, Australia, Botswana, Norway
Tungsten	68	Canada, China, Bolivia, Portugal
Silver	64	Canada, Mexico, Peru, UK
Mercury	57	Spain, Algeria, Japan, Turkey
Cadmium	55	Canada, Australia, Mexico, Peru
Copper	27	Chile, Peru, Mexico, Canada
Gold	31	Canada, Uruguay, Switzerland
Iron Ore	22	Canada, Venezuela, Liberia, Brazil
Iron and Steel	22	European Economic Community, Japan, Canada

*Includes primarily bauxite and alumina.
SOURCE: Data from U.S. Bureau of Mines.

IRON ORE

Iron is the fourth most abundant element in the crust of the earth. It is often concentrated into deposits containing up to 70 percent iron oxide. The major types of iron deposits contain one or more of several important iron minerals: hematite, magnetite, goethite, and siderite (see Table 4–5 for details).

By far the most important type of ore deposit for iron is the banded iron formations (BIF) of Precambrian age. These deposits are chemically formed sedimentary rocks that were laid down from roughly 3.5 Ga to 1.8 Ga. Their formation deserves some special mention. In the Archean and Early Proterozoic, the atmosphere contained less oxygen and more carbon dioxide than later in the Proterozoic and to the present. At about 1.8 Ga there was a transition in the earth's atmosphere from oxygen poor and carbon dioxide rich to oxygen rich and less carbon dioxide rich. Before 1.8 Ga the conditions were favorable for iron as Fe^{2+}-bearing minerals (mainly clay minerals and some amphiboles). In large sedimentary basins, very extensive deposits of iron silicates, usually interlayered with silica and limestone, resulted. The silica was transformed into chert (cryptocrystalline quartz) during diagenesis. Because the chert and limestone layers are light and the iron minerals dark, the rock has a distinctly banded appearance. Magnetite also formed in the iron bands, and with burial or metamorphism, hematite formed as well.

When the BIF have been strongly metamorphosed, the iron content of the rocks increases due to the driving off of water and carbon dioxide from the iron silicate and limestone layers. In some places the non-iron minerals have been leached, thus enriching the iron content, and in other places the rocks have been subjected to iron addition concomitant with silica removal. This is a natural beneficiation which works to our advantage. In many instances, however, the rocks have only undergone burial and folding but without any strong metamorphism. In these cases the iron content (as iron oxide) may only be 10 to 25 percent or so, well under ore grade of about 30 to 35 percent. These ores, called taconites, represent the bulk of BIF in the United States. For many decades these were subeconomic, until it was found that when crushed and mixed with certain clay minerals, the iron minerals were electrostatically attracted to the clay and formed small pellets. These pellets could be separated by magnetic means, yielding an enriched product containing 60 percent or so of iron oxide. Further, the pellets are fairly uniform in size, thus making them ideal for iron furnace feed. This method has proved so successful that taconites now represent the most abundantly mined iron ore in the United States, accounting for some 70–75 percent of iron ore consumed in the United States each year.

BIFs are commonly quite thick as well as laterally extensive, ranging from 30 to 600 meters thick. Due to their great age, many of the BIFs have been complexly folded, often with the result that their original bedding planes are nearly vertical. Because of their favorable dimensions, the BIFs are mined by open pit methods. All continents contain them, and they account for most of the major

TABLE 4–5 Main ore minerals for metals

Element	Mineral	Formula	Percent Primary Metal
Iron			
	Magnetite	Fe_3O_4	72
	Hematite	Fe_2O_3	70
	Goethite	$FeO(OH)$	63
	Siderite	$FeCO_3$	48
Aluminum			
	Gibbsite	$Al(OH)_3$	35
	Boehmite	$AlO(OH)$	45
	Diaspore	$AlO(OH)$	45
	Kaolinite	$Al_2Si_2O_5(OH)_4$	21
	Nepheline	$NaAlSiO_4$	19
	Anorthite	$CaAl_2Si_2O_8$	19.4
	Alunite	$KAl_3(SO_4)_2(OH)_6$	19.6
	Dawsonite	$NaAl(CO_3)(OH)_2$	17.5
Copper			
	Chalcopyrite	$CuFeS_2$	36.4
	Chalcocite	Cu_2S	80
	Covellite	CuS	66
	Azurite	$Cu_3(CO_3)_2(OH)_2$	55
	Malachite	$Cu_2(CO_3)(OH)_2$	57
Zinc			
	Sphalerite	ZnS	67
Lead			
	Galena	PbS	87
Chromium			
	Chromite	$FeCr_2O_4$	46.5
Nickel			
	Pentlandite	$(Fe,Ni)_9S_8$	Variable, high
	Garnierite	Hydrous Ni silicate	Variable, high
Cobalt			
	Linnaeite	$(Co,Ni)_3S_4$	Variable, high
Titanium			
	Rutile	TiO_2	60
	Ilmenite	$FeTiO_3$	31.6

TABLE 4–5   Continued

Element	Mineral	Formula	Percent Primary Metal
Magnesium			
	Magnesite	$MgCO_3$	29
	Dolomite	$CaMg(CO_3)_2$	13
	Brucite	$Mg(OH)_2$	42
Manganese			
	Pyrolusite	MnO_2	63
	Rhodochrosite	$MnCO_3$	48
Tin			
	Cassiterite	SnO_2	79
Tungsten			
	Scheelite	$CaWO_4$	64
	Wolframite	$FeWO_4$	60.5
Molybdenum			
	Molybdenite	MoS_2	75
Gold			
	Native Gold	Au	100
Silver			
	Argentite	Ag_2S	90.5
	Native Silver	Ag	100
Mercury			
	Cinnabar	HgS	86
Arsenic			
	Arsenopyrite	$FeAsS$	46
	Orpiment	As_2S_3	61
	Realgar	Sb_2S_3	70
Beryllium			
	Bertrandite	$Be_4Si_2O_7(OH)_2$	15
	Beryl	$Be_3Al_2Si_6O_{18}$	5
Rare-Earth Elements			
	Baestnasite	REE-carbonate	Several percent
	Monazite	REE-phosphate	Several percent
	others	Silicates, oxides	Several percent
Niobium			
	Columbite	Complex Nb oxy-salt	Several percent

occurrences noted in Australia, the USSR, South Africa, Canada, South America, and the United States (Table 4-6). The reserves of iron ore represented by the BIF are extremely large, constituting some 4 billion tons in the United States alone. The USBM estimates worldwide reserves of 72 billion short tons of contained iron to the year 2000, relative to a projected demand of 9.9 billion short tons (Table 4–3). This is a very fortunate situation for future industrialization of the world community.

Banded iron formations are not found in rocks younger than 1.8 Ga or so. This is presumably due to the fact that Fe^{2+} was no longer as stable, and was more readily oxidized to the insoluble ferric ion, Fe^{3+}. So the processes that allowed Fe^{2+} to be transported freely in the surface environment prior to 1.8 Ga ceased at this time.

Sedimentary rocks also account for another important group of iron deposits, the oolitic, red-bed deposits. These deposits take their name from the presence of abundant hematite, and often the presence of oolites (smooth, layered, rounded grains). These deposits formed in restricted sedimentary basins, presumably under local reducing conditions which allowed some iron as Fe^{2+} to be

TABLE 4–6 World iron ore reserves (millions of short tons contained iron)

Continent/Country	Reserves	Resources	Total
North America			
Canada	12,000	17,000	29,000
United States	4,000	15,600	19,600
South America			
Brazil	18,000	11,000	29,000
Venezuela	1,400	2,500	3,900
Europe			
France	1,800	1,800	3,600
Sweden	2,200	800	3,000
USSR	31,000	26,000	57,000
Africa			
Liberia	700	100	800
South Africa	1,200	1,800	4,000
Asia			
China	3,000	4,100	7,100
India	6,200	2,500	8,700
Oceania			
Australia	11,800	8,200	20,000
Other (all continents)	9,300	22,600	31,900

SOURCE: U.S. Bureau of Mines.

transported into the basin but which was immediately oxidized and precipitated as Fe^{3+}-bearing minerals (goethite or hematite). These basins are small, typically 1 to about 100 kilometers in diameter, and the minable layers often less than 10 meters thick. They are less iron-rich than the BIF, and they contain a high amount of phosphorus, making them difficult to beneficiate. Nevertheless, this type of iron deposit, found in the Late Proterozoic and Phanerozoic sedimentary rocks, is very important in many parts of the world. This type of ore accounts for much of the iron ore mined in France (Alsace-Lorraine), Germany (Rhine Valley), Egypt and elsewhere in northern Africa, Pakistan, Newfoundland, and in the United States (Clinton, N.J.; Birmingham, Ala.).

Iron, as mentioned above, is easily oxidized to Fe^{3+} at the earth's surface. Hence when Fe^{2+}-bearing rocks with a high iron content are weathered, Fe^{3+}-minerals form and are fixed in the rocks as a residual deposit. If the conditions favor laterites, then iron laterites can form. While such residual deposits are impure and commonly small, they are nevertheless an important source of iron in some Third World Countries.

Historically, some of the most important iron deposits were those in igneous rocks and in contact-metasomatic replacement deposits. Magnetite is often fixed by magmatic segregation into layers, such as magnetite-rich layers in anorthositic rocks. In granitic rocks, magnetite is often concentrated into layers or injected into surrounding rocks; the magnetite-bearing granites of Sweden and the well-known Kiruna iron ores are of this type. While important, they are in many places mined out and the remaining ore dwarfed by the BIFs. When limestone is intruded by igneous rocks, the result is often a dissolution of the limestone and replacement by iron minerals such as magnetite. Such deposits are small, very rich, and still a resource where found. In the United States, the Cornwall, Pennsylvania deposit, which provided iron ore for the American colonies, is of this type.

IRON AND STEEL

Together iron and steel production and consumption dwarf that of all other metals combined, accounting for some 95 percent of all metals in the world. Iron is most commonly produced from iron ore by use of the blast furnace (Figure 4–3). In this furnace, iron ore is mixed with limestone, coking coal, and small amounts of manganese. The limestone acts as flux and removes many impurities, while the coke serves both as a fuel and as the reductant (to reduce $Fe^{3+,2+}$ minerals to native iron). The manganese further removes impurities. Preheated air is injected at the bottom of the furnace to ignite the coke. The output from the blast furnace is pig iron, named from the shape of the large ingots. A small amount of iron is produced by a process known as direct reduction.

While some pig iron is used directly, the vast majority of it is used for the production of steel and cast iron. Roughly 100 million short tons of steel and cast iron (Table 4–2) is consumed in the United States each year, with steel account-

FIGURE 4–3   Blast furnace for cast iron. Low-grade iron ore is crushed, mixed with clay, and rolled to get iron-ore enriched pellets that can be separated by magnetic means. The other essential ingredients—limestone flux and coking coal—are combined with the iron pellets before hot air is added. Slag is driven off, and molten cast iron remains (Source: Brookins, 1981)

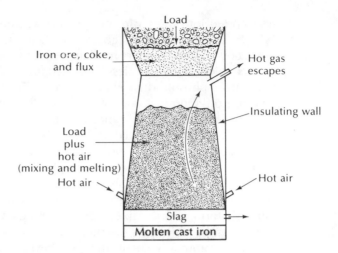

ing for 85 percent and cast iron 15 percent. Figure 4–4 shows a flow chart for iron and steel. Pig iron has a modest carbon and silica content, cast iron a lower carbon content, and steel the lowest carbon content. The cast irons (called gray, white, ductile, and malleable), while less important than steel, can be machined and cast into many complex shapes more easily than steel, hence their importance.

Important types of steel include the normal carbon steels (with 0.05 to 1.25 percent carbon plus small amounts of silicon and manganese), the alloy steels to which other metals have been added, and stainless steel, which contains 12 to 60 percent chromium. The foundries for steel and iron differ in that higher temperatures are required for the steel foundries, and the product contains less carbon than cast iron.

Scrap plays an important part in the iron and steel picture. As indicated in Figure 4–4, scrap is a prime feed for steels and irons, and the produced scrap is used as new feed. Obsolete scrap (automobiles, old equipment, and so on) constitutes the remainder. Some 62 million tons of scrap was used for steel and iron production in 1983.

The iron and steel produced in the United States is primarily used for construction (31%), transportation (27%), machinery (18%), appliances and equipment (7%), oil and gas industries (6%), containers (6%), and other purposes (5%). While other metals could substitute for iron based on performance criteria, they are not as inexpensive as iron and steel. The outlook for domestic iron and steel is not favorable, but it is not bleak either. Several developing countries (Korea, Taiwan, and Brazil, for example) have initiated modern steelmaking capacity, and the production facilities in Japan and several European countries are superior to those in the United States. Most large steel plants in the United States use somewhat outdated facilities and are less energy efficient than their foreign counterparts. Labor costs in most foreign countries are substantially lower than those in the United States. Many foreign countries subsidize their steel industries to allow them to sell below production costs when the price decreases. Finally, new environmental concerns have increased domestic costs: The U.S. Bureau of Mines reported that 16 percent of the iron and steel industries' capital went for environmental concerns between 1973 and 1981. When these and other factors are considered, it is safe to say that while the overall trend for production in the United States will increase in the next 15 years or so, the United States will increase its imports from about 15 percent (1981) to 25 percent (2000), according to the USBM. Scrap will be an increasingly important part of the steel industry. Small steel mills operating with an electric furnace (which takes scrap but not iron ore) that turn out a much less diversified product than the big plants are becoming increasingly important and will continue to do so. These mills, for example, can turn out steel from scrap for $300 to $400 a ton as opposed to $1000 a ton for a traditional large plant. As long as cheaper steel is available from foreign sources, the domestic iron and steel industries will suffer.

Environmental concerns affecting the iron and steel industries are of several types. Clean-air and other regulations have a more pronounced effect on pig-iron production and on steel and iron foundries than they do on finishing operations. It is likely that future finishing mills in the United States will use raw iron or steel

FIGURE 4–4   Iron material flow diagram. (Source: U.S. Bureau of Mines)

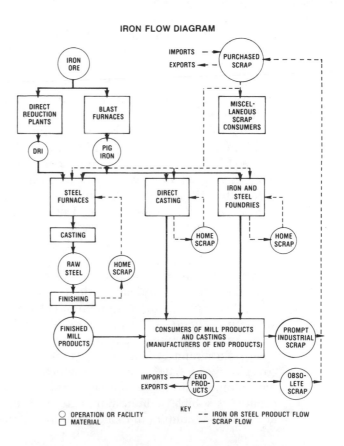

produced abroad. Blast furnace slag and other wastes are voluminous in some cases, and if not properly impounded, can pose environmental problems. One area where iron ore producers have been unnecessarily affected is in the treatment of iron ores near the Great Lakes. Following the asbestos scare of the 1970s (described in Chapter 8), just about every fibrous mineral in the world was classified as an asbestoslike material, and thus of "great environmental concern." The concern for the iron ore producers was that waste rock from northern Minnesota was being dumped into Lake Superior, and some of this waste material was presumably being detected in the water supply of Duluth and other cities. Not only was the lake dumping stopped, but the mining companies had to build huge impoundments for iron ore wastes with elaborate covers, at a cost of hundreds of millions of dollars. As is discussed in Chapter 8, the amphibole-dominated waste rock is not a health hazard at all, and simple land dumping would be quite sufficient for proper disposal.

ALUMINUM

Aluminum is the third most abundant element by weight in the earth's crust, yet most of it is bonded to oxygen in silicates and is extremely difficult to extract. Consequently, little of the world's aluminum is produced from the most common rock-forming minerals, the aluminosilicate feldspars.

Only iron surpasses aluminum in use as a metal, and uses for aluminum continue to increase. As a lightweight yet strong metal, aluminum is ideal for construction materials and transportation vehicles. Other well-known uses include beverage cans, electrical equipment, and military equipment.

About 8.5 million tons of aluminum metal was consumed in the United States in 1980—about one-third of the world consumption according to the U.S. Bureau of Mines. Most aluminum consumed in the United States is produced domestically, although we rely heavily on imports of aluminum ore (Table 4–4). Before the mid-nineteenth century, rare minerals such as cryolite (Na_3AlF_6) were used because the aluminum could be extracted with existing technology. At the time aluminum was not considered to be a useful iron substitute. Although bauxite was discovered in the 1800s, aluminum production remained at a low level until technological advances made aluminum extraction profitable. The importance of aluminum was fully recognized with the coming of World War II, and since then the industry has continued to thrive.

The principal ore of aluminum is bauxite. Bauxite, more properly called an aluminum laterite, is formed under tropical weathering conditions. Unlike silicates (where aluminum substitutes for silicon and both are tightly bonded to oxygen), aluminum in bauxite ore is less strongly bonded to oxygen. It can be recovered by the Bayer process, in which the ore is leached under pressure and at high temperature in the presence of a basic, sodium-rich solution. The resulting product is soluble sodium aluminate, and most impurities remain behind as solids. Hydrated aluminum oxide (alumina) is then precipitated.

Aluminum is produced by reduction of the alumina in a molten bath of synthetic (or natural) cryolite by electrolysis. This is a very expensive process, and roughly 3 percent of U.S. electrical energy consumption is due to production of alumina and aluminum.

Bauxites are aluminum-rich laterites. The combination of torrential downpours, rapid runoff and oxidation, high temperature, slightly acidic conditions, coupled with favorable rock chemistry and permeability, will leach out most of the common rock-forming elements such as potassium, sodium, calcium, magnesium, and also silica if the parent rock is silica-impoverished. Iron, if present, will oxidize and remain in the residual laterite. The aluminum released by destruction of parent aluminosilicates also remains in the laterite, and usually forms gibbsite, although boehmite and diaspore also form in many deposits (Table 4–5).

Nepheline syenites (or their chemical equivalents) are favorable source rocks for bauxite due to their high aluminum content and low silica content, and often low iron content. For rocks such as granite, the silica content is so high that the rocks weather easily to kaolinite plus quartz but not to gibbsite. Kaolinite contains only 21 percent alumina whereas gibbsite contains 40 percent alumina (Table 4–5). The technology to recover aluminum from kaolinite has been developed by the U.S. Bureau of Mines, but it is not economic at this time.

Bauxite can also form in limestone environments, especially if the limestone is argillaceous (clayey) and the clay minerals are kaolinitic. This is illustrated in Figure 4–5. The tropical weathering promotes the dissolution of the limestone, creating karstic conditions (prone to caverns and sinkholes), and the same weathering alters the clay minerals to gibbsite.

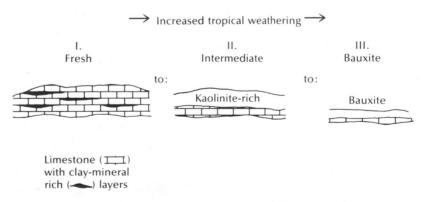

FIGURE 4–5 Formation of bauxite from limestone. Some bauxites form from rocks which contain relatively aluminum-rich layers. Most limestones are impure, and the clay minerals they contain are typically aluminum-rich. When exposed to tropical weathering, the limestone itself is readily removed and kaolinite is formed or concentrated from preexisting clay minerals. Further weathering results in the formation of bauxite deposits as shown. This is typical of the deposits in the Caribbean islands. (Source: Brookins, 1981)

A great variety of rock types can, under conditions for laterite formation, form bauxites. In the United States, the marginally economic deposits of Arkansas formed from nepheline syenite, while the even more marginally economic deposits in Georgia and Alabama formed from Tertiary clays and sands.

Bauxite derived from argillaceous limestone accounts for the ores in Jamaica, Haiti, and the Dominican Republic. Ores in Suriname and Guyana occur on kaolinite developed at the contact between Precambrian basement rocks of the Guyana Shield and Tertiary sediments. In Brazil bauxite has formed on unconsolidated sediments.

The very large bauxite deposits of Wiepa, Queensland, Australia developed from a lithologically discontinuous arkose with kaolinite-rich parts altering to gibbsite. The Darling Range bauxite in Western Australia is developed on Precambrian crystalline rocks. Together, these two deposits account for roughly 20 percent of the world's total reserves (Table 4–7).

The largest bauxite deposits in the world (Table 4–7) occur in Guinea in western Africa, where gibbsite has formed due to weathering of Devonian schist and arkose. In Europe, most of the deposits are developed in karstic limestone.

Iron and other insoluble elements, if present in the source rocks, will be fixed in the laterite that forms, thus laterites have the potential to house a great many different minerals. These are discussed where appropriate later in this chapter.

Bauxites form at or near the surface, hence they are mined by conventional open pit methods. In places where the ore is high grade only at the surface, the material is scraped off.

Besides kaolinite, alunite and even anorthite (in anorthosites) have proved to be successful source materials for aluminum if necessary. At present these materials represent a subeconomic resource, as does the mineral dawsonite, which might be recovered as a product of oil shale processing. Dawsonite is an attractive mineral because it is water soluble, hence recovery costs would be low.

Environmental problems from alumina and aluminum production involve the slurry ("red mud") produced by the Bayer process, and fluorine from the electrolysis process.

About a ton of red mud slurry is produced for each ton of alumina produced. The red mud slurry is pumped to settling areas. It contains large amounts of iron, titanium, phosphorus, and other elements that may have been present in the laterite. If no toxic concentrations of any element are found, this material, when dried, is suitable for fill and other industrial purposes. The slurry contains caustic soda which must be prevented from leaking from the impoundment areas, and as dewatering takes place the surface dust may be windblown, hence covers must be installed.

The fluorine emissions from electrolysis are regulated by the EPA, and set at no more than 2.1 to 2.2 pounds (about 1 kilogram) of fluoride particulate per ton of aluminum produced. A series of hoods and scrubbers recover most of the fluoride, which is then recycled back to the electrolysis cells. Other gases, such as sulfur dioxide, are effectively removed by these scrubbers as well.

TABLE 4–7   World reserves
of bauxite

Continent/Country	Million Metric Tons
North America	
United States	38
Dominican Republic	30
Haiti	10
Jamaica	2000
South America	
Brazil	2500
Guyana	700
Suriname	575
Venezuela	235
Europe	
Greece	600
Hungary	300
USSR	300
Yugoslavia	350
Others	92
Africa	
Guinea	5600
Cameroon	680
Ghana	450
Sierra Leone	140
Others	10
Asia	
India	1000
Indonesia	750
China	150
Others	60
Australia	4440
Total	21,010

SOURCE: U.S. Bureau of Mines.

In addition to the red mud and fluorides, secondary plants may release oils, chemicals, and graphite, which can pollute streams if not carefully monitored.

The aluminum outlook for the United States is problematic. The world market is very competitive, and many foreign producers are subsidized by their governments, thus giving them a decided advantage over their U.S. counterparts.

Metal prices are somewhat unstable as well, and in recent years there has been a worldwide overproduction of aluminum. Environmental pressures have also increased the cost of domestic production. The most significant cost increase, however, has been in the cost of energy to produce alumina and aluminum. Nevertheless, it is anticipated that the production will increase by about 30 to 40 percent by the year 2000, although almost all of the production will come from imported bauxite and alumina.

Recycling of aluminum, especially for aluminum beverage cans, is well known. Old scrap recycling is less efficient; the U.S. Bureau of Mines estimates that 15 to 20 percent of old scrap is recycled whereas close to 50 percent of aluminum beverage cans are recycled, and even close to 90 percent in states with beverage deposit laws.

The only important by-product of aluminum is gallium. Gallium is an important metal for the growing solar industries where it is used in photovoltaic cells, and its future growth looks very promising.

COPPER

Copper is a relatively rare element, with a crustal abundance of only 50 ppm. Despite this low crustal abundance, copper is the third most abundantly consumed metal behind iron and aluminum. Copper is one of the oldest metals, with evidence for its use predating 6000 BCE.

In today's world, copper is used predominantly by the electrical industry, with abundant use also for roofing, plumbing, instruments, utensils, coinage, and other purposes. Consumption in the United States was 2.1 million metric tons in 1985 (Table 4–2). Of this, 55 percent was from domestic production, 26 percent was from recycled scrap, and 19 percent was imported, most from Chile, Peru, Canada, and Mexico. The United States could easily be self-sufficient in copper as reserves are abundant.

World reserves, as shown in Table 4–8, are very large. The largest reserves are found in Chile, with the United States second, and the combined ores of Zambia and Zaire third. Not included in the reserve figures of Table 4–8 are the reserves in the recently discovered Olympic Dam deposit in Australia or the potential copper resources from ocean-floor nodules (see "Seafloor Deposits," this chapter). Some 58 countries produce copper in commercial amounts.

Copper minerals are numerous, but the most important ore minerals are chalcopyrite, chalcocite, covellite, bornite, and enargite (Table 4–5). The copper carbonates malachite and azurite are also important. These minerals often form as alteration products of copper sulfides, and are used by explorationists to zero in on potential copper deposits.

There are relatively few types of major copper deposits. The most dominant type is porphyry copper, described briefly earlier in this chapter. These deposits consist of mineralized porphyritic granitic rocks, along with accompanying vein, hydrothermal vein, and skarn deposits in well-defined belts in the principal arc

TABLE 4–8 World copper reserves

Continent/Country	Million Metric Tons
North America	
United States	57
Canada	17
Mexico	17
South America	
Chile	79
Peru	12
Europe	Several
Africa	
Zambia	30
Zaire	26
Asia	
Philippines	12
Other	14
Oceania	
Australia	8
New Guinea	6

SOURCE: U.S. Bureau of Mines.

systems of subduction zones (see Figure 4–2). Thus the porphyry coppers are found in well-defined belts running from Canada through the United States into Mexico, in Central America, and in the Andean chain of South America. There are many unique features and characteristics of these deposits. First, the rocks are roughly granitic in composition, and always porphyritic. Second, they are always parallel to subduction zones. Third, they occur in distinct age brackets, with most of the world's porphyry coppers found in Triassic to Tertiary rocks. These factors make exploration for porphyry coppers somewhat straightforward (though not necessarily easy). The search for porphyry coppers has also been attempted using satellite imagery as well. From photographs taken of the earth from great distances by satellites, and using special optical filters, areas that contain favorable rocks can be identified. This greatly reduces the often otherwise "blind" initial phases of exploration, and the U.S. Geological Survey estimates that several potential porphyry coppers have been identified in the arc system running through Iran and Iraq. Unfortunately, this was discovered in the late 1970s just before the violent conflict in that part of the world, hence data to document these findings are lacking.

Major porphyry coppers in the western hemisphere include those in Chile (Chuquicamata, El Teniente, Braden, El Salvador), the United States (Morenci, Ajo, Bagdad, Florence and others in Arizona; Santa Rita Tyrone and others in

New Mexico; Bingham Canyon, Utah; Butte, Montana; Ely, Nevada; and others), Peru, Mexico, and western Canada. In other parts of the world, porphyry coppers are found in New Guinea, Australia, southeastern Asia, Iran-Iraq, and elsewhere. Typically, porphyry coppers are large but with low-grade ore, so mining is by open pit methods. Figure 4–6 shows a view of one of the typical porphyry copper mines.

Another characteristic feature of porphyry coppers is that they usually possess well-defined zones of alteration around the main mineralization. This is shown in Figure 4–7. Typically, the ore is concentrated around a barren core of

FIGURE 4–6 The large open pit porphyry copper mine at Chino, New Mexico. The pit is more than one mile across. (Source: Brookins, 1981)

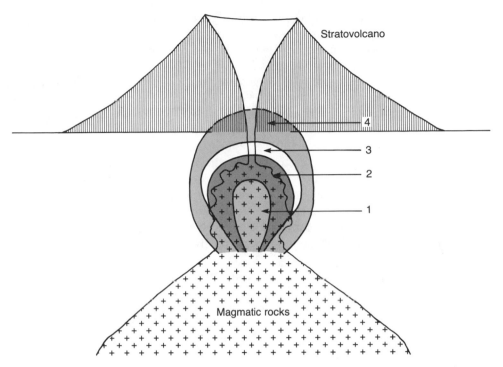

FIGURE 4–7   Cross section of a hypothetical porphyry copper. In the complex rocks above a magmatic intrusive body, and commonly below a stratovolcano, mineralization by copper-rich fluids is accompanied by different zones of alteration. 1, Zone of potassic alteration; 2, zone of sericitic alteration; 3, zone of argillic alteration; 4, zone of propylitic alteration. Most porphyry copper deposits contain one or more of these tell-tale alteration zones, and recognition of these zones can assist the exploration geologist in locating the copper mineralized rocks. (Modified from Sillitoe, (1973), The tops and bottoms of porphyry copper deposits. *Economic Geology, (68),* 799–815.

igneous rock, with surrounding shells or envelopes of pyritic, sericitic, argillic, and propylitic zones (named for their characteristic mineralogy). Each zone is characterized by a unique mineralogy and often distinctive trace-element abundances. The explorationist with knowledge of such alteration haloes can pinpoint where the porphyry copper might occur.

The second most abundant type of copper deposit is the sedimentary stratiform or Kupferschiefer type, named for the type occurrence of central Europe. In this type of deposit copper sulfides, often with or near sulfides of other elements, occur in shale enriched to fairly high grade. The layers are not unusually thick but they are laterally continuous and uniform in ore grade, thus making mining somewhat easier. The Zambian-Zaire copper belt in Africa is the largest of this type, accounting for roughly 15 percent of world reserves. The copper mineralization appears to occur in marine shale deposited over red beds, often in an overall evaporitic section, but the origin of this type of deposit is not well understood, with proponents of both syngenetic and epigenetic, plate tectonic related mineralization adamant in their stands. In the United States the White Pine deposit in Michigan is best known.

Massive sulfides in volcanic rocks are numerous. This third type of copper deposit includes the very large Kidd Creek deposit in Canada. Massive sulfides are also found in ophiolites (Cyprus), greenstone belts (Canada), Kuroko-type deposits in Japan and elsewhere, and many more. The source of the copper in many if not most of these deposits is submarine fumarolic exhalations.

Copper mineralization accompanies other metals in layered igneous rocks, such as the Sudbury irruptive in Canada and the Bushveld complex in South Africa, and elsewhere. Copper deposits are also associated occasionally with carbonatites and other alkalic rocks, and in numerous vein and replacement deposits, including metamorphic rocks. Not covered here is the copper mineralization in ocean floor nodules and massive sulfides, which is discussed later in this chapter.

Recovery of copper from most ores is a straightforward process, and one which involves the flotation process. The ore is crushed and ground to a fine size, then mixed with water and various organics in large flotation cells (Figure 4–8). The metal-sulfide particles are fixed on the organic matter, which rises to the top of the flotation cell as the organics separate from the water. The nonsulfide fraction, the bulk of the rock, settles to the bottom as waste. The froth is then

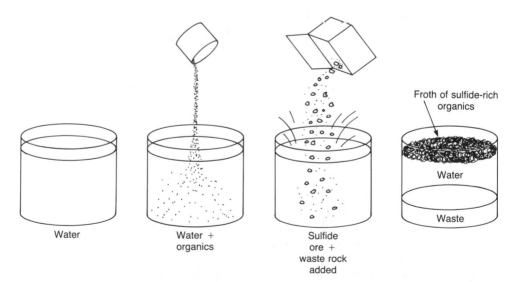

Water

Water +
organics

Sulfide
ore +
waste rock
added

Froth of sulfide-rich
organics

Water

Waste

FIGURE 4–8   Flotation. To a large vat of water is added some organic compounds, then crushed rock containing sulfide minerals. The sulfide minerals sorb onto the organics, which separate from the water and rise to the top (forming a froth rich in sulfide minerals) while the waste silicates and oxides sink to the bottom.

Typical flotation cell in the concentrator at Tyrone, New Mexico. Courtesy Phelps Dodge Corporation.

skimmed off. This material now contains several tens of percent copper as opposed to the 0.5 to 0.3 percent original copper ore. This concentrate is roasted to drive off water and carbon dioxide, and further smelted to drive off sulfur, arsenic, and other gaseous impurities. The iron present is oxidized during this smelting, so that when the copper melts (along with other metallic impurities such as zinc and lead), the iron stays in the solid slag left behind when the molten copper is poured off.

The copper at this stage is known as blister copper, and is usually refined further by heating under reducing conditions to drive off oxygen. It is then cast into large anode sheets and immersed in a bath of copper sulfate-sulfuric acid interspersed with thin copper cathode starter sheets. An electrical current is applied, and the anode copper, along with its impurities, dissolves and precipitates on the cathode starter sheets while the impurities remain in solution. This process is called electrolysis. The refined copper is better than 99.7 percent pure. The remaining solution is treated to remove the trace impurities, which are often a major source of the elements involved. These by-products and their percent of the total United States output are arsenic (100%), rhenium (100%), selenium (100%), palladium (100%), tellurium (100%), platinum (100%), silver (24%), molybdenum (35%), gold (16%), sulfur (6%), and lesser amounts of other elements. Clearly by-products for copper are extremely important.

Oxide ores and some sulfide ores of copper are usually processed by leaching. This process involves exposing crushed rock with dilute acid, which dissolves the copper minerals. The dissolved copper then is removed electrolytically on iron scrap. Leaching has been practiced for many years. In some areas rocks are blasted with explosives to fracture them and acid allowed to attack the ore. Iron scrap for electrolytic recovery is placed in excavations below the fractured rock.

Handling cathodes at Tyrone, New Mexico—solvent extraction/electrotwinning plant. Courtesy of Phelps Dodge Corporation.

Environmental problems for the copper industry are severe. The Clean Air Act of 1970, and as amended in 1977, restricts the amount of sulfur dioxide and other gaseous emissions from smelting and related activities. While a necessary step, this has nevertheless been extremely costly. In some areas the EPA has permitted smelting operations only during favorable weather conditions. This interim measure is unwise as it cuts into production, increases costs, and may locally cause pollution. Sulfuric acid is a major by-product of sulfide ore smelting, and unlike native sulfur, the acid cannot be stored. As it is not as pure as sulfuric acid made from salt-dome sulfur (see Chapter 5), its demand is less. This acid cannot be disposed of easily, and it often builds up in large amounts, creating additional

Close up of a tailings thickener, Tyrone, New Mexico. Courtesy of Phelps Dodge Corporation.

problems. All smelters had to comply with the 1970–1977 Clean Air Act by January 1, 1988, and many of the older smelters were closed.

The world copper picture presents an unfortunate situation for the domestic industry. In the United States, costs for mining, smelting, refining, transportation, and especially environmental controls have all increased markedly, as has the cost of providing the energy for the processes, while the price of foreign copper continues to drop. The Intergovernmental Council of Copper Exporting Countries (CIPEC), made up of Chile, Peru, Zambia, Zaire, and Indonesia, with associate members Australia, New Guinea, and Yugoslavia, has declined to reduce the overproduction which has led to the present glut on the market, so domestic copper faces an extremely difficult future. It is strange, for example, that the United States continues to import large amounts of Chilean copper when it further depresses the domestic copper market; and even stranger when it is remembered that the Chilean copper industry owes its existence to the U.S. companies that started the industry before being nationalized during the 1970s. In the case of Zaire, the United States is heavily dependent on this nation for cobalt, and the bulk of world cobalt reserves are held in Zaire, thus to ensure a supply of cobalt, the United States must also purchase Zairean copper.

World reserves of copper are adequate to meet demand for several decades, even with a projected growth of 2–3 percent per year consumption. Since foreign copper is produced more cheaply than in the United States, the domestic market looks poor. Growth will be very slow at best, and a decline is quite possible.

ZINC

The crustal abundance of zinc is only 70 ppm, yet zinc, like copper and some other trace metals, is often concentrated into economic deposits. Zinc has been widely used for more than 2000 years, mainly in alloys such as brass. It is the fourth most abundantly consumed metal behind iron, aluminum, and copper.

Zinc consumption in 1985 totaled 952 thousand metric tons as the metal and another 80 thousand as a nonmetal. Zinc coatings on iron and steel prevent corrosion, and there is no good substitute for it. In addition to its use as the metal in the construction, machinery, transportation, and other industries, much zinc is used in paint, rubber, and elsewhere in the chemical industry.

The zinc market is highly competitive, with 50 countries recording significant production in 1985, led by Canada, followed by the USSR, Australia, Peru, and the United States. The United States imports about 60 percent of the total zinc it consumes, produces about 30 percent from domestic mines, and obtains the remaining 10 percent from recycled zinc and from the national stockpile. Domestic production has dropped significantly in the last few years due to a variety of factors.

Zinc is often discussed with lead, since both elements are commonly found together in many metal deposits. Zinc as Zn^{2+} and lead as Pb^{2+} are commonly transported together, commonly segregated from other chalcophile elements, and commonly precipitated together but not in the same mineral. The ionic radius of Zn^{2+} is 0.8 angstroms whereas Pb^{2+} is 1.35 angstroms. So Zn^{2+} is incorporated into the mineral sphalerite (ZnS) and Pb^{2+} into galena (PbS), and many deposits contain both minerals for this reason.

Zinc deposits are quite varied. Massive sulfide deposits of zinc in stratified volcanic, sedimentary, and metamorphic rocks are known, with the source of the zinc from submarine fumarolic exhalative processes. In the United States much zinc is found in the Mississippi Valley-type deposits (described under "Lead") with lead in carbonate rocks. Zinc deposits in stratabound sandstones are common as are many kinds of vein deposits, and occasional vein deposits are found in metamorphic rocks. Supergene processes often cause oxidized zinc deposits to form over primary ore.

Massive stratabound deposits rich in zinc include the Broken Hill and Mount Isa locations in Australia; the Kidd Creek, Sullivan, and Bathurst occur-

TABLE 4–9   World zinc reserves

Continent/Country	Million Metric Tons
North America	
Canada	26
United States	22
Mexico	7
South America	
Peru	8
Brazil	2
Europe	
USSR	11
Spain	6
Ireland	5
Poland	3
Other	13
Africa	
South Africa	11
Zaire	7
Asia	
China	5
India	5
Iran	5
Japan	4
Australia	18

SOURCE: U.S. Bureau of Mines.

rences in Canada; and, in the United States, the Ducktown area in Tennessee. The reserves in the United States are adequate to meet domestic demand, but high prices, environmental controls, high cost of energy, and other factors have caused many U.S. operations to close. Much zinc is mined by underground methods. World reserves are shown in Table 4–9.

Recovery includes a two-step flotation following mining and grinding the ore to a powder. The first flotation step involves adding organics that effectively collect lead and copper sulfides, but not zinc or iron. After this step, new organics are added which selectively float off zinc and iron sulfides, and this material is the zinc concentrate sent to the smelter. The concentrate is roasted, driving off sulfur and other volatiles, and the zinc fixed in impure zinc oxide. This material is calcined (roasted without melting), then leached with sulfuric acid forming a dilute zinc sulfate solution from which the zinc is electrolytically recovered. Important by-products from zinc operations include (with amount of U.S. production in parentheses) thallium (100%), cadmium (100%), indium (100%), germanium (100%), mercury (small), and lead (small).

Environmental concerns include the usual emission of sulfur dioxide and other gaseous materials, sulfuric acid, and addition of zinc and other materials in solution into wastewater. The most polluting zinc retorts have been closed, and the remaining mills meet the EPA standards for gaseous and other concerns.

LEAD

Lead is a rare element in the crust of the earth, with an abundance of only 12.5 ppm. It occurs primarily in the mineral galena, PbS, in ore deposits. Use of lead dates back at least to 5000 BCE. It ranks fifth of metals consumed behind iron, aluminum, copper, and zinc. Its main use is for batteries in the transportation industry (about 770 thousand metric tons in 1984), with smaller but significant amounts (40 to 90 thousand metric tons) for construction, ammunition, chemicals, antiknock gasoline additive (which has decreased from 250 to 90 thousand metric tons between 1973 and 1984), and electrical uses. Lead is the substance of choice for shielding from X-rays and other radiation. Nonmetallic uses are dominated by protective coatings and paint. The Golden Gate Bridge in San Francisco is painted continuously with a lead-base paint to prevent weathering of the steel suspension cables. The United States consumed 1050 thousand metric tons of lead in 1985, with imports totaling 15 percent and recycled lead 45 percent.

Lead deposits are varied. Massive sulfide deposits rich in lead are found in stratiform volcanic, sedimentary, and some metamorphic rocks, and replacement deposits are found primarily in carbonate rocks. Well-known massive sulfide deposits rich in lead (and in other metals) include Kidd Creek, Sullivan, and Bathurst in Canada, while the replacement type in carbonate is best exemplified by the Mississippi Valley-type deposits in the United States. The Mississippi Valley-type lead-zinc ores are characterized by low-temperature galena and sphalerite in brecciated limestone and chert, and these deposits may be related to plate tectonic rifting in stable cratonic areas. The tri-state area of Oklahoma-Kansas-Missouri has been replaced by the Viburnum Trend lead-zinc deposits of southeastern Missouri as the leading lead producer in the United States, accounting for roughly 80 percent and 15 percent of domestic and world production, respectively. Vein lead deposits are common. The well-known Coeur d'Alene district of Idaho is a silver-rich galena vein deposit. World reserves for lead are given in Table 4–10.

Ores rich in galena and one or more other sulfides are processed by a two-stage flotation process, and often with a preliminary heavy-mineral separation. Galena can be concentrated by gravity separation and then separated from the other sulfides by selective flotation. The flotation concentrate is then roasted (sintered), smelted, and refined. The emitted sulfur dioxide is trapped and used to make sulfuric acid. Other volatiles are trapped and stored as well. Important by-products of lead production include (with the percent of U.S. total in parentheses) bismuth (100%), antimony (42%), zinc (22%), silver (5.6%), and others.

TABLE 4–10   World lead reserves

Continent/Country	Million Metric Tons
North America	
United States	21
Canada	12
Mexico and others	3.5
South America	
Peru	2
Others	0.5
Europe	
USSR	12
Yugoslavia	4
Bulgaria	3
Spain	2
Other	7.5
Africa	
South Africa	4
Morocco	2
Asia	
China	2
India	2
Others	1.5
Australia	16

SOURCE: U.S. Bureau of Mines.

Recycling works very well for lead. Domestic production of lead involves 45 percent from old scrap, and this figure is expected to increase to 55 percent by the year 2000. Scrap accounts for 40 percent of total world production.

Environmental concerns about lead are extreme. Poisoning resulting from lead use and processing is well known. The toxicity of lead requires that it be carefully monitored. Lead uptake by people near lead smelters, in cities with high automobile density, and in other areas is known. The phasing out of lead in gasolines is a step in the right direction, as is the cessation of lead-based paints in habitable buildings. Lead inhalation and adsorption may lead to anemia, neurological damage, and even death. Lead, a mutagenic element, may also cause birth defects. It is speculated that Nero, Emperor of ancient Rome, suffered severe neurological damage from drinking from lead containers. While a great deal of research has been, and continues to be, carried out on environmental health effects of lead, it is apparent that there is much more to learn. Lead is discussed in more detail in Chapter 8.

The outlook for domestic lead production is one of cautious optimism. Despite increased costs in environmental controls, energy for processing, and other costs, the recovery of lead continues to become more and more efficient. This offsets in part the competition from foreign sources, often subsidized by their respective governments, and cheap labor costs. The U.S. Bureau of Mines projects an increase in the consumption of lead in the United States to reach 1.6 million tons by the year 2000.

MANGANESE

Manganese is the twelfth most abundant element in the earth's crust. While abundant and widely distributed, it is not commonly concentrated into ore grade (35% or more Mn) over much of the continents. Manganese is an element capable of possessing many different valencies in natural materials (from $+2$ to $+7$) and thus has many valuable chemical properties.

Manganese is a critical element for the iron and steel industry. For every ton of steel or cast iron, 14 pounds (6.4 kilograms) of manganese is required. The manganese is used for deoxidizing and desulfurizing the feed to the steel and iron foundries, and some of it alloys into the steel and iron (about 0.4 percent). No element is as good as manganese for these purposes. The ore minerals of manganese (Table 4–5) are commonly oxides, although rhodochrosite is common in some deposits. Pyrolusite is the most common of a large number of oxide minerals. The ore deposits of manganese are predominantly either a chemical sedimentary type or products of secondary enrichment by manganese fixation in residual (commonly lateritic) deposits. The large deposits of South Africa, the USSR, Australia, and Mexico are of the chemical type, while the deposits of Gabon and Brazil are of the secondary enrichment type. While there are fairly large deposits in the western hemisphere, most of the world's manganese reserves are in Africa, the USSR, and Australia (Table 4–11). The United States imports 100 percent of its manganese.

The possiblility of recovering economic amounts of manganese from the ocean floor has received a great deal of attention over the last few decades. Seafloor nodules, commonly called manganese nodules due to their high manganese content, may some day be mined, and thus will decrease our dependence on imports. The subject is covered later in this chapter under "Seafloor Metal Deposits."

The 1984 consumption of manganese in the United States was about 630,000 short tons, of which about 0.5 percent was produced domestically (as a by-product of some iron ores). The consumption trends for manganese closely follow the combined steel and iron trend. Recycling does not play an important role for manganese.

Uses for manganese outside iron and steel are varied, with the chemical industry accounting for 10 percent. Environmental aspects of manganese are not severe since there is no domestic mining, and most upgrading of manganese used for steel and iron foundries is done outside the United States. Past emission problems of ferromanganese and manganese alloy plants have been corrected, and no ferromanganese plant has operated in the United States since 1977. Only 1000 workers were employed to specifically handle domestic manganese (grinding, processing, Mn chemicals, and so on) in 1984.

The outlook for manganese is the same as for steel and iron. Since there is a tendency now to import higher grade manganese concentrates, the bulk weight

TABLE 4–11 World manganese reserves

Continent/Country	Thousand Short Tons
North America	
Mexico	3,700
South America	
Brazil	20,900
Europe	
USSR	365,000
Africa	
South Africa	407,000
Gabon	110,000
Ghana	4,000
Australia	75,000
Asia	
India	20,000
China	14,000

SOURCE: U.S. Bureau of Mines.

of manganese imported, but not the manganese content, has decreased. If and when manganese from ocean floor nodules is available, the United States reliance on foreign manganese will be greatly reduced.

MAGNESIUM

Magnesium is the eighth most abundant element in the crust of the earth, and it is also the third most concentrated dissolved element in seawater. It is enriched in several minerals, including magnesite, dolomite, brucite, and others (Table 4–5). Production in the United States comes mainly from seawater, from evaporite wells and seeps, and from magnesite deposits.

Magnesium is used principally as a nonmetal, with some 625,000 tons of magnesium compounds (for 1983) dwarfing the 115,000 tons of magnesium metal produced. The United States produced roughly 11 percent of the world total. The magnesium compounds are used primarily in the steel and iron industries as refractories, as well as in chemical, rubber, and many other industries. Magnesium metal is used primarily in the transportation and equipment industries alloyed with iron and steel.

Of 2.8 billion tons of magnesium in minerals reported by the U.S. Bureau of Mines, only 40 million tons occurs in the United States. Most of the world reserves are in Europe, Asia, and South America. Magnesite is the main ore mineral of the United States, and occurs in veins and alteration of serpentine, in sedimentary beds, and as a replacement of limestone and dolomite. Domestic reserves of magnesite are small, but locally important in Nevada, Washington, and North Carolina. Larger reserves are found in brines from evaporites, such as those in Michigan.

The main supply of magnesium, however, is the ocean. Magnesium is recovered from seawater by precipitation of magnesium hydroxide after eliminating carbonates and sulfates. The hydroxide is calcined to produce magnesia (MgO), and may be further treated to produce magnesium metal.

The United States is a net exporter of both magnesium compounds and magnesium metals. Due to its light weight and strength, more and more research is being devoted to finding new uses for this element. By-products include those other materials recoverable from seawater, especially salt and bromine.

CHROMIUM

The earth's crust contains 100 ppm chromium, much of which is concentrated in the mineral chromite (see Table 4–5). Only chromite serves as an ore of chromium. Chromium is a lithophile element, and it substitutes for ferric iron, vanadium, and other similar metals in rocks and minerals. Because of the ease by which Cr^{3+} and Fe^{3+} may substitute for each other, many chromites are so iron rich that they are not good ores for chromium.

Chromium reserves in the United States are essentially zero, and the United States is totally dependent on foreign sources for this strategically important metal. The United States consumed 378 thousand short tons of chromium in 1985, of which 23 percent came from scrap.

There are two major types of chromium ore in the world, the large stratiform type, and the much smaller podiform type. The former are deposits in ultramafic rocks in layered intrusive complexes such as the Bushveld complex in South Africa and the Great Dyke in Zimbabwe. The subeconomic deposits of the Stillwater complex of Montana are also of this type. The podiform type are found in Alpine serpentinites parallel to subduction zones, and include those in Cuba, the USSR, Philippines, Turkey, Albania, and elsewhere. The podiform type tend to be more chromium rich than the stratiform type, mainly because the serpentinite hosts are low in Fe^{3+} relative to the stratiform deposits. The stratiform deposits, however, are also typically enriched in the platinum-group elements.

Chromium is used principally in stainless steels and in other steel chromium alloys. Stainless steels possess a wide range of mechanical properties and are very

TABLE 4–12 World chromium reserves

Continent/Country	Million Short Tons Ore
North America	0
South America	
Brazil	9
Cuba	3
Europe	
USSR	142
Finland	19
Albania	7
Africa	
South Africa	913
Zimbabwe	19
Madagascar	8
Sudan	2
Asia	
India	15
Philippines	15
Turkey	5
Others	6
Oceania	
New Caledonia	2

SOURCE: U.S. Bureau of Mines.

resistant to corrosion and oxidation. Stainless steels typically contain 12 to 36 percent chromium. Chromium (from 0.5 to 30%) is also used to improve the performance of cast iron. Some chromium is also used in the paint industry as a pigment and in the refractory industry for firebricks and mortar. No element works as well as chromium for these purposes, and chromium supply to the United States in the future is of great concern. Most of the world's reserves of chromium are in South Africa and the USSR, with major amounts in Zimbabwe and Finland (Table 4–12). The state-controlled chromium production facilities in the USSR, Cuba, Vietnam, and Albania accounted for 48 percent of total world output in 1984. Chromite ores are separated by gravity means after crushing, and ferrochromium is produced by a complex furnace treatment with flux-reductant-fuel conditions somewhat similar to the iron industry. Chromium is separated from iron by several refining techniques.

Environmental costs to the chromium industry in the United States amounted to 45 percent of new capital, mainly for equipment to ensure meeting the Clean Air Act (1970) and the Clean Water Act (1972) as amended. Chromium as Cr^{3+} is apparently not toxic, but Cr^{6+}, which is used in the chemical industry, is highly toxic.

Chromium use in the United States is expected to increase significantly in the next 15 years or so, and, except for some scrap, foreign sources will provide the chromium. South Africa has the largest reserves, but continued social and political unrest may affect this supply. The United States is considered extremely vulnerable for chromium, and attempts are being made to stockpile an adequate supply.

TITANIUM

Titanium is the ninth most abundant element in the earth's crust, and it is concentrated in several accessory rock-forming minerals, such as rutile and ilmenite (see Table 4–5). Titanium metal is light, strong, and resistant to chemical attack, hence it is desirable for many purposes such as the aerospace and defense industries, electric generating plants, and the chemical industry. A future use may be as a canister material for stored radioactive wastes. Despite its favorable characteristics, however, titanium metal is not used more widely primarily because it is expensive to process. Further, most titanium is consumed as the dioxide (TiO_2) as a white pigment in paints, plastics, and other media. Some 90 to 95 percent of total titanium consumption in the United States is due to this use, with metal making up the remainder.

Rutile is a common accessory mineral in many rocks, especially metamorphic rocks. Ilmenite is a common accessory mineral in basic to ultrabasic rocks. Both rutile and ilmenite are resistant to chemical attack during weathering, thus are concentrated in sands and placers. Much ilmenite is contained in the rock anorthosite, formed by magmatic segregation. The large deposits of Sanford Lake, New York, and Allard Lake, Quebec are of this type. Much of the world production of ilmenite and virtually all the rutile comes from nearshore sand deposits. Deposits in Australia, Malaysia (where it is a by-product of tin dredging operations), India, Japan, South Africa, and the United States account for most of the production. The U.S. Bureau of Mines lists 190 million short tons of titanium reserves (Table 4–13).

Titanium metal and pigment production in the United States together total about 550 thousand short tons per year, and will increase to about 750 thousand short tons per year in 2000 (U.S. Bureau of Mines). Domestic mining, mainly from sand off the shores of New Jersey and Florida, and from anorthositic rocks in New York, accounts for about one-third of the demand; the remainder is imported, mainly from Australia.

Environmental concerns stem largely from the recovery and purification of titanium metal and dioxide from ores. Pigment plants have traditionally used sulfate to react with ilmenite to form TiO_2 plus ferric sulfate, a waste product. About 3.5 tons of such waste is produced per ton of pigment. The chloride process requires chlorine reacted with rutile to form $TiCl_4$, which can then be reacted with oxygen to form the dioxide pigment or further reduced for metal production. The acid wastes from the sulfate method must be neutralized with lime and carefully

Table 4–13 World titanium reserves

| Continent/Country | Thousand Short Tons | |
	Ilmenite	Rutile, Anatase
North America		
United States	18,000	—
Canada	7,900	200
South America		
Brazil	1,100	37,000
Europe		
Norway	21,000	—
USSR	4,000	1,600
Finland	1,000	—
Africa		
South Africa	25,000	2,400
Sierra Leone	—	1,400
Asia		
China	20,000	—
India	20,000	2,900
Sri Lanka	2,500	500
Australia	15,000	5,700

SOURCE: U.S. Bureau of Mines.

monitored, and there is a trend toward increasing use of the chloride process and a phasing down of the sulfate process.

By-products of titanium are usually from sand deposits. Minerals concentrated with ilmenite or rutile in the sands include zircon, monazite, kyanite, and others.

NICKEL

The crustal abundance of nickel is only 75 ppm, but since nickel is a strongly chalcophile element, it is readily incorporated into sulfide minerals. Geochemically, Ni^{2+} is almost identical in properties to Co^{2+}, and both elements behave in nearly identical fashion in igneous rocks. They tend to be somewhat segregated from other chalcophile elements, and are often concentrated in sulfides in mafic to ultramafic rocks. Norite is often a host rock for nickeliferous sulfides, with a fairly predictable amount of cobalt as well. Some pyroxenites and peridotites contain economic concentrations of nickeliferous sulfides.

Nickel alloys with steel and iron account for most of its use, as nickel adds resistance to chemical attack over a wide range of temperatures. These alloys are widespread in the chemical, petroleum, electrical, construction, and aerospace industries, and no good substitute is readily available.

Layered intrusives of ultramafic to mafic composition contain most of the world's nickel reserves. The Sudbury irruptive in Canada is perhaps the best known of these, although occurrences in the USSR and Australia may be as large. The largest western hemisphere reserves of nickel, however, are in Cuba (see Table 4–14). In the United States, nickel occurrences are at present marginally subeconomic, and those at Riddle, Oregon, and the copper-nickel low-grade ores of the Duluth gabbro, Minnesota, worth note. Cobalt is an important coproduct of nickel sulfide ores. When the sulfur content of layered mafic intrusives is low, the nickel is fixed in rock-forming minerals such as olivine and pyroxene, from which it is not profitably extracted. Such rocks may, however, be sources for laterite nickel deposits (discussed below).

In recent years nickel laterites have become increasingly important as a source of nickel. During the lateritization of nickeliferous mafic to ultramafic rocks, nickel is only partly remobilized and, with silica, forms the nickel silicate garnierite (see Table 4–5). The cobalt is segregated from nickel here and fixed with the residual impure iron-aluminum-manganese oxides. The reason is that while Co^{2+} oxidizes to Co^{3+}, Ni^{2+} does not change valence, hence cobalt behaves like iron and nickel more like magnesium; but whereas magnesium is highly mobile, the nickel is only partly mobile. The largest nickel laterites occur in Cuba, New Caledonia, New Guinea, Indonesia, Brazil, Burundi, Greece, and the Philippines, with production from the New Caledonia site far greater than the others.

World production of nickel is led by the USSR (25%), Canada (18%), Australia (13%), New Caledonia (9%), Indonesia (7%), and Cuba (5%). The produc-

TABLE 4–14 World nickel reserves

Continent/Country	Thousand Short Tons
North America	
Canada	8,000
United States	30
Cuba	20,000
Dominican Republic	800
South America	
Brazil	900
Colombia	650
Europe	
USSR	7,300
Greece	2,600
Other	2,800
Asia	
Indonesia	4,300
Philippines	2,800
China	800
Africa	
South Africa	2,800
Botswana	450
Zimbabwe	200
Oceania	
Australia	2,300
New Caledonia	2,000

SOURCE: U.S. Bureau of Mines.

ers with government subsidies (USSR, Cuba) are likely to play an increasingly important role in the world market. The United States depends heavily on Canada for its nickel.

Nickel sulfides are processed by flotation followed by roasting and smelting prior to refining. The nickel laterite ores are smelted under reducing conditions, with nickel separated from other metal impurities.

The United States consumed 230 thousand short tons of nickel in 1985. Canada led the nations which supplied 70 percent of our domestic needs. Scrap accounted for about 25 percent of the total consumed.

Environmental problems involving nickel are primarily from outside the United States. The Copper Cliff smelter in Canada, for example, may provide by itself 1 percent of the total world sulfur dioxide emission (U.S. Bureau of Mines), much of which drifts over the United States. This problem is discussed more under "Coal" in Chapter 7. In the United States, older refining and smelting plants were a source of some nickel carcinogenesis among plant workers, but these have been phased out.

A future reserve of nickel (with other metals) may come from ocean floor nodules, many of which are typically rich in nickel. These nodules are discussed later in this chapter.

COBALT

Cobalt is even scarcer than nickel in the earth's crust, with an abundance of 22 ppm. It is usually camouflaged by nickel in sulfide and other minerals, but may be segregated from nickel during weathering (including lateritization). While cobalt has been used for several thousand years, initially as a blue pigment, it has now become an extremely important metal for the aerospace industries. Cobalt alloys of steel and other metals are superb in preserving magnetism and preventing corrosion, and they are remarkably durable under different temperature and stress conditions. No metal works as well as cobalt for these purposes. The United States consumption of cobalt in 1985 was approximately 8500 short tons of metal.

The largest cobalt reserves are those in the Zaire-Zambian copper belt (Table 4–15), where the sedimentary copper ores contain up to several percent cobalt. Zaire dominates the world cobalt market, as production from other countries is slight. Producers of cobalt metal include Belgium (using mainly Zairean and Zambian ore). In addition to the cobaltiferous sedimentary copper ores of Zaire and Zambia, cobalt is camouflaged by nickel in magmatic sulfide ores such as the Sudbury irruptive, Canada. The cobalt content of these sulfides is only 0.05 to 0.07 percent; however, the Zaire-Zambia ores typically run 0.1 to 2 percent. Laterites are a major source of reserves, many untapped now due to technological problems in processing the ores. Cobalt behaves like iron and is fixed in residual laterites, segregated from the more mobile nickel. Cobalt averages 0.1 percent in laterites with the largest found in Cuba, New Caledonia, the Philippines, Australia, and Indonesia. Small but rich deposits of cobalt with arsenic and other metals occur as hydrothermal veins in several parts of the world, including Canada and Morocco.

Seafloor nodules (discussed later in this chapter) contain appreciable amounts of cobalt, and may lessen our dependence on imports (now 100 percent) in the future.

Cobaltiferous sulfides are processed by mineral enrichment and flotation, with cobalt recovered as a by-product during the copper or nickel refining. Oxide or mixed oxide-sulfide ores are leached with sulfuric acid, roasted, and cobalt separated electrolytically. Technology (U.S. Bureau of Mines) is now being researched extensively to process cobaltiferous seafloor nodules.

TABLE 4–15   World cobalt
reserves

Continent/Country	Thousand Short Tons
North America	
Cuba	1150
Canada	50
Europe	
USSR	150
Finland	25
Greece	15
Yugoslavia	10
Africa	
Zaire	1500
Zambia	400
South Africa	20
Botswana	10
Asia	
Indonesia	200
Philippines	150
India	20
Oceania	
New Caledonia	250
Australia	25

SOURCE: U.S. Bureau of Mines.

While nickel can substitute for cobalt in some cases, the use of cobalt in jet engines, electrical purposes, cutting tools, and for alloys in the oil and gas industries is unique.

Environmental aspects of cobalt are restricted to processing and specialty plants in the United States. The EPA has set a standard of 0.1 milligram cobalt per cubic meter of air for plants where cobalt-bearing carbide tools and other objects are produced. Cobalt is toxic to animals in high concentrations, and causes anemia at low doses. No standard for cobalt in water has been set.

MOLYBDENUM

Molybdenum, with a crustal abundance of only 1.5 ppm, is nevertheless a readily available metal in the United States. This country contains the world's largest molybdenum deposit at Climax, Colorado, and the largest national reserves as well (see Table 4–16). Molybdenum is used principally for refractory metallic purposes, mainly alloyed with steel, cast iron, and superalloys to enhance tough-

TABLE 4–16 World
molybdenum reserves

Continent/Country	Thousand Short Tons
North America	
United States	3000
Canada	500
Other	100
South America	
Chile	1250
Peru	150
Europe	
USSR	500
Asia	
China	500
Iran	75

SOURCE: U.S. Bureau of Mines.

ness, strength, and resistance to corrosion. It has numerous chemical applications
as well, including pigments for paints and ceramics, as a catalyst, and for lubricants. Demand is anticipated to grow rapidly for molybdenum as research uncov-
ers more uses, based in large part on the fact of our large domestic reserves.

The main ore mineral of molybdenum is molybdenite (Table 4–5), although
30–35 percent of production is from by-product porphyry copper operations
(with only 0.02% molybdenum). There are several types of molybdenum deposits: hydrothermal deposits in porphyritic siliceous rocks, breccia pipes, and in contact-metasomatic rocks adjacent to siliceous intrusions. The molybdenum content
of hydrothermal deposits is 0.2 to 0.5 percent. These hydrothermal deposits constitute most of the world reserves. Quartz veins with molybdenite are also important, and lesser amounts are found in pegmatites and in bedded sedimentary
rocks. The large deposits at Climax and Henderson, Colorado, and at Questa, New
Mexico, are hydrothermal stockwork types. There is a strong spatial relationship
of molybdenum deposits to both porphyry coppers and subduction zones, although there is clearly a disproportionately large amount of molybdenum concentrated in the Americas relative to the rest of the world. The United States currently
provides 60 percent of world molybdenum, followed by Canada (17%). Also, the
Untied States is a net exporter of molybdenum, exporting 47 million pounds in
1984. That same year, U.S. consumption was 44 million pounds, of which imports provided 8.3 million pounds, mainly from China, Canada, and Mexico. Although these imports are small compared to total production, they nevertheless
hurt the domestic market. The molybdenum mining operations are in difficult
times due to a glut on the world market and low prices, and Chile and other
subsidizing countries refuse to cut back production of both copper and molybdenum, thus increasing the glut and lowering prices still further.

There are also important by-products associated with molybdenum. Because deposits are commonly associated with silicic igneous rocks, moderate amounts of tin and tungsten are often found with the molybdenum. Molybdenum, in turn, is a by-product of some tungsten operations, and also a by-product in small amounts from sandstone-type uranium deposits.

Environmental aspects of molybdenum operations are not extreme. The metal is not toxic, and the processing plants which convert molybdenite to metallurgical grade molybdic oxide, located mainly in the Midwest and East, are also small. Since molybdenite ores are low grade, large tonnages are involved, and waste dump material is the main problem.

TIN AND TUNGSTEN

Tin and tungsten are considered together as both elements are commonly enriched in the same rocks. The crustal abundances are low, with tin at 2.5 ppm and tungsten 1.2 ppm. Both elements are typically enriched in granitic rocks, often associated spatially with convergent plate boundaries. Tin is present commonly in the mineral cassiterite and tungsten in scheelite or wolframite (Table 4–5). Most of the tungsten mineralization, however, occurs in contact-metasomatic rocks, often in complex veins. Placers are the most important source of tin, as cassiterite is both heavy and resistant to weathering. Distinct metallogenic provinces are noted for both tin and tungsten. The belt from Korea through China to southeastern Asia is rich in both tin and tungsten, and the belt from Thailand through Indonesia is extremely rich in tin. In the Americas, tin is concentrated with copper in Bolivia-Peru while tungsten is more concentrated in Peru-Ecuador. Historically, these two elements were mined with copper from the granitic deposits of Cornwall, England. Prospecting for tin-tungsten using plate tectonic reconstruction helped bring in at least one tin operation in Nova Scotia in the last few years.

Both tin and tungsten are lithophile in most instances, although tin will form some sulfides. Tin is used primarily as solder (29%) and tinplate (27%), with additional uses in the transportation industries (12%), chemicals (13%), electrical industry (18%), and others. Tungsten is used as the carbide for a wide variety of cutting tools and related purposes (65%), in special mill products (25%), as special steel-tungsten alloy (8%), and in chemicals (1%). Reserves worldwide for tin and tungsten are good, but domestic reserves are low (Table 4–17).

Processing of tin ores is straightforward. Cassiterite is concentrated, then tin is formed by reduction by heating it with carbon at 1300°C. For tungsten, some scheelite is used directly with steel foundry feed for steel-tungsten alloys. Otherwise, the ore is leached with soda ash or hydrochloric acid to form tungstic acid, from which tungsten metal can be separated electrolytically.

The United States consumed 3700 metric tons of tin and about 4500 metric tons of tungsten in 1984. Domestic mines produced primary or by-product tungsten to account for 20 percent of the total, while only insignificant by-product tin

TABLE 4–17   World tin and
tungsten reserves

Continent/Country	Thousand Metric Tons	
	Tin	Tungsten
North America		
Canada	60	480
United States	20	150
Mexico	10	8
South America		
Bolivia	140	45
Brazil	70	20
Other	40	15
Europe		
United Kingdom	90	70
USSR	80	280
Portugal	—	40
France	—	20
Austria	—	15
Other	40	25
Africa		
Namibia	60	—
South Africa	30	—
Nigeria	20	—
Zaire	20	—
Zimbabwe	20	5
Rwanda	—	5
Other	10	10
Asia		
Malaysia	1100	17
Indonesia	680	—
Thailand	270	30
China	80	1200
Burma	10	15
Japan	10	—
North Korea	—	80
South Korea	—	58
Turkey	—	65
Other	100	5
Oceania		
Australia	180	130

SOURCE: U.S. Bureau of Mines.

was produced. Thus this nation relies very heavily on imports for both strategic metals. World tin output was led by Malaysia (20%), Indonesia (13%), Bolivia (12%), and Thailand in 1984, with the remainder from other market-economy countries and planned-economy countries. For tungsten, planned-economy countries (China and North Korea) may have led in total world production, with significant contributions from Canada, Bolivia, and Portugal. The United States imported over 50 percent of its tungsten from these last three countries.

Exploration for new tin and tungsten deposits, following the plate tectonic theory, is ongoing in the United States and in other western countries. There is some optimism for tungsten, but tin may be more problematic.

GOLD

Probably no metal has received as much literature and interest over time as gold. It is extremely rare, with a crustal abundance of only 3 ppb (0.0000003 weight percent). Gold is used as a long-term store of value, but it has many other uses. In addition to its use in jewelry, gold is important in aerospace, computers, electronic communications equipment, and in many other areas. Still, the U.S. Bureau of Mines estimates that 1.1 billion troy ounces (about 34 million kilograms), about a third of all gold ever mined, is in government vaults.

World production is dominated by South Africa, followed by the USSR, Canada and the United States, China, and Brazil. South Africa has actually turned out 40 percent of all gold mined. World gold reserves are given in Table 4–18. Gold deposits are extremely varied. The huge Witwatersrand deposit in South Africa is typical of deposits in metamorphic rocks, where the gold is present in a metamorphosed "fossil conglomerate." Many other occurrences in Archean through Early Proterozoic settings are in metamorphic rocks, including the large deposits in India, Brazil, and the USSR, and the Homestake mine in the United States.

There is no rule for many gold deposits. For example, many hydrothermal veins are found in very to fairly silicic igneous rocks or in their host rocks, whereas in other areas gold is found in more mafic rocks. In the United States, gold is an important by-product of porphyry copper milling, and smaller occurrences include breccia pipes, often associated with Tertiary stocks. Placer deposits are a small part of modern gold mining in the United States. New technology has made possible gold recovery from very low grade rocks. In such rocks, crushing followed by a cyanide leach releases the gold as a cyanide complex, and the gold is then fixed on charcoal, stripped by acid, and electroplated on foil. Due in no small part to such advances in technology, the United States now produces about 35 percent of its domestic need, with the balance imported from South Africa and Canada.

By-products from gold recovery operations include copper, silver, mercury, and some lead and zinc. In South Africa, a valuable coproduct is uranium, which was fixed in the same rocks as the gold during the Archean. Environmental prob-

TABLE 4–18   World gold reserves

Continent/Country	Million Troy Ounces
North America	
United States	80
Canada	42
Other	30
South America	
Brazil	23
Other	17
Europe	
USSR	200
Other	10
Africa	
South Africa	760
Zimbabwe	10
Asia	
Philippines	18
Japan	10
Other	15
Oceania	
Australia	23
Other	25

SOURCE: U.S. Bureau of Mines.

An overall view of the Homestake gold mine surface workings at Lead, South Dakota. Visible at top is the Yates shaft headframe, one of the two main surface shafts serving the mine. The stairstepped building at left center is the South Mill where the ore is ground prior to processing. Courtesy of Homestake Gold Mine, Lead, South Dakota.

lems from gold operations in the United States are not large, but are nevertheless monitored carefully. The cyanide leach used is kept from volatilizing by adding lime to keep the pH high, and the cyanide after being stripped of its gold content is recycled. Monitoring wells surround each property to ensure against pollution of water by the cyanide. Pyrite, and occasionally pyrrhotite, are waste sulfides which are stored as dump waste. This material must also be monitored as, with time, these iron sulfides will oxidize to form sulfuric acid (a problem discussed in Chapter 7 under "Coal").

The price of gold has fluctuated wildly since the Untied States went off the gold standard in the late 1970s, and many commercial operations are predicated on the assumption of a price of $300 per troy ounce or higher. Gold prices are not just influenced by supply and demand, but also by world turmoil, world and local economics, and anxiety. A further complication is that the gold output from South Africa will decrease by 1990 (U.S. Bureau of Mines prediction). In the United States, by-product gold from porphyry coppers will decrease as copper output diminishes. While the gold supply is adequate for the United States to 2000, it may not be in the early part of the next century.

SILVER

Silver, like gold, is an extremely rare element, with a crustal abundance of only 70 ppb. It occurs primarily in sulfides, although significant amounts of native silver are known. Historically, silver has long been a source of stored value, and this still accounts for its greatest usage. The prime uses of silver are in the photographic and electronics industries.

Roughly two-thirds of world silver production comes from by-products of copper, lead, and zinc smelting. Galena is a favorite host to silver as the ionic radius of Ag^+ (1.33 angstroms) is almost identical to that of Pb^{2+} (1.35 angstroms). All lead deposits contain some silver, but those forming hydrothermal veins near igneous rocks, such as the Coeur d'Alene district in Idaho, are especially rich. Mississippi Valley-type lead-zinc deposits are much less enriched in silver, probably reflecting the differences in source and transport mechanisms for these different kinds of deposits. The remaining one-third of world silver production is from gold operations (including gold-silver types) as a coproduct. Almost 80 percent of the reserves are in the market economy countries, with the largest reserves in South Africa, Canada, and Mexico (Table 4-19).

Consumption in the United States of silver in 1984 was approximately 93 million troy ounces, of which 44 million was produced domestically. The remainder was imported, primarily from Canada, Mexico, Peru, and Australia. World reserves are considered adequate to meet domestic demand to the year 2000.

Silver recovery from copper sulfide ores is by electrolysis. From lead ores, silver is separated by adding zinc to molten lead-silver; the zinc and silver combine and rise to the top where it is skimmed off and silver is separated from the zinc. For gold and gold-silver operations, both are leached from the ore by cyan-

TABLE 4–19 World silver reserves

Continent/Country	Million Troy Ounces
North America	
Mexico	1370
Canada	1160
United States	920
South America	
Peru	680
Other	240
Europe	
USSR	1400
Other	650
Africa	
All	300
Asia	
Japan	70
Other	170
Oceania	
Australia	780
Other	60

SOURCE: U.S. Bureau of Mines.

idation, and zinc dust is used to separate the gold and silver. Although silver processing plants are not very large, there is some environmental concern due to possible pollution of waters by materials used during flotation and cyanidation; adequate monitoring of pond storage and tailings are essential. The treatment of sulfide ores (discussed in connection with copper) involves sulfur dioxide, which must be trapped and safely used or disposed of.

The main special problem affecting silver is price. Since it is used for stored value, its price is very volatile, and has fluctuated widely in recent years.

MERCURY

Mercury is a rare element, constituting only 20 ppb of the crust of the earth. It is also unique among metals in being liquid at room temperature, which results in its being used for many special purposes for which there is no good substitute. Mercury is also a very toxic substance, and because of the importance of environmental concern over mercury, it is discussed in detail later in Chapter 8. Mercury can occur as the native metal, but most ore is in the form of the sulfide, cinnabar (see Table 4–5). Mercury deposits are mainly located in volcanic island-arc systems in subduction zone regimes, often concentrated in altered serpentinites or silica-carbonate rock. Ore deposits consist of highly faulted and fractured vein systems and replacement occurrences in the arc rocks, and as infiltrations in ash-rich lake beds such as in Nevada, where the McDermitt mine is the only important producer in the United States now operating. Because they are emplaced under surface to near-surface conditions, many deposits have been eroded away. Mercury is extremely volatile and is easy to extract from ore. Even small operations can successfully roast cinnabar, driving off sulfur dioxide and mercury vapor and trapping the latter under water. This same volatility, however, makes it environmentally problematic, as mercury, a trace constituent of other sulfide ores, may easily be emitted into the atmosphere.

Reserves of mercury in the United States are not large (Table 4–20). The world's largest reserves are in Spain, with fairly large amounts in the USSR, Yugoslavia, and China. The United States relies heavily on imports, amounting to 60 percent of the annual (based on 1984) consumption of 50,000 flasks, of which 48 percent was used in batteries, chlorine, and caustic soda production (16%), paints (12%), and miscellaneous uses (24%). Forty-one percent of the domestic consumption was made available from stocks held by industry and from government sources.

Due to environmental concerns, the use of mercury in pesticides, pharmaceuticals, amalgams, paper and pulp industries, and in antifouling paints has either been curtailed or greatly diminished over the last 15 years or so. There are, however, specific uses of mercury for which there are no good substitutes, thus mercury consumption is expected to increase slightly between now and the year 2000.

TABLE 4–20 World mercury reserves

Continent/Country	Thousand Flasks*
North America	
Mexico	150
United States	140
Europe	
Spain	2600
Yugoslavia	350
USSR	300
Africa	
Algeria	80
Asia	
China	300

*One flask = 76 pounds.
SOURCE: U.S. Bureau of Mines.

PLATINUM-GROUP ELEMENTS

The platinum-group elements (PGEs) consist of platinum, palladium, rhodium, ruthenium, iridium, and osmium. These elements are extremely rare in the crust, constituting only a few parts per billion. Only rarely are these elements concentrated to economic amounts in rocks.

The largest single occurrence of platinum-group elements is found in the Bushveld complex in South Africa. In a part of the complex known as the Merensky Reef, the PGEs are concentrated with chromite in the layered ultramafic intrusive. Similar occurrences are found in the USSR and in the Stillwater Complex of Montana in the United States. While most of the PGEs are in the native state due to their siderophile nature, they are in part found in sulfides as well. Some of the nickel-cobalt ores from layered intrusives such as the Sudbury irruptive in Canada contain significant amounts of trace PGEs in the sulfides, which is recovered as a by-product. Placers are important for some PGEs (Colombia). The PGEs are essentially chemically inert over a wide range of natural conditions, hence they are in demand for the petroleum, chemical, and other industries as catalysts and as corrosion-resistant materials. Except for a small amount of recycled scrap (5%), all platinum-group metals are imported, mainly from South Africa. Should the South African source be cut off, the United States will be hard pressed to meet its expected 40 to 50 percent increase in consumption by the year 2000. World PGE reserves are given in Table 4–21.

OTHER METALS

Many other metals are important, but space does not permit a thorough discussion of each. In this section, an attempt is made to cover just a few of the most significant.

TABLE 4–21 World platinum-group element reserves

Continent/Country	Million Troy Ounces
North America	
Canada	4
United States	Less than 1
Europe	
USSR	60
Africa	
South Africa	385
Zimbabwe	Less than 1

SOURCE: U.S. Bureau of Mines.

Niobium

A strategic metal, niobium (formerly called columbium) is currently being stock-piled in the United States because niobium alloys are presently the only materials capable of withstanding the superhigh temperatures to be reached in a nuclear fusion reactor. Niobium is also an important alloy for use in jet engines and for other purposes both in the aerospace and defense industries. The United States contains no identifiable niobium reserves, and 100 percent is imported, mainly from Canada, which has the world's largest niobium reserves. Niobium is a strongly lithophile element, and forms complex oxides and oxy-salts. It is concentrated in alkalic rocks and especially in carbonatites and pegmatites, which produce most of the world's supply.

Beryllium

Beryllium is also a strongly lithophile element used primarily in various industries as an alloy, mainly with copper. Beryllium is extremely light (atomic weight 9), strong, and becoming increasingly important in the United States. Historically, most beryllium was produced from beryl-bearing rocks such as pegmatites, but in the United States the main source of beryllium is from bertrandite (see Table 4–5). The mineral is found in altered tuffaceous rocks in the western United States, with those deposits at Spar Mountain, Utah the largest. Reserves are huge, with an identified supply to meet demand well into the next century. The technology to treat the bertrandite ores is new, however, and the United States still imports roughly 40 percent of its nceds (down from 90 percent ten years ago, and projected to be less than 10 percent by 2000).

Rare-Earth Elements

The REEs (lanthanum, cerium, praseodymium, neodymium, samarium, europium, gadolinium, terbium, dysprosium, holmium, erbium, thulium, ytterbium, and lutetium) are most commonly concentrated in carbonatites and pegmatites, mostly with alkalic rocks. Rare-earth minerals are mainly oxides, silicates, carbonates, and phosphates; and all are complex. Uses of the REEs are in their embryonic stage in the United States; these include semiconductors, additives to glasses, special alloys, petroleum cracking catalysts, magnets, video display tubes, and other electronics. The reserves in the United States are large, although the largest reserves are in China.

Arsenic

Arsenic is most abundantly consumed as arsenic compounds, especially arsenic trioxide used in herbicides, wood preservatives, and as an additive to many glasses. Domestic arsenic is a by-product of sulfide ores, mainly those of copper. Environmental concern over arsenic has curtailed many recovery operations. This will result in increased imports, which even now total 95 percent. Disposal of arsenic products and wastes will thus be more of a problem in the future. The United States consumes about 25,000 tons of arsenic metal and arsenous compounds per year. Increased use of solar energy in the United States will increase the demand for gallium arsenide, used in solar collector systems.

Antimony

Antimony alloys are used as hardening agents in lead batteries, and antimony oxide is a flame retardant. Worldwide, antimony is most commonly produced as a by-product from sulfide ores, although rich antimony ores are known. Stibnite, the main ore mineral of antimony, is common in many epithermal deposits, often associated with mercury, arsenic, and other metals such as gold; all are found in shallow to surface igneous-related deposits. Domestic production accounts for only 5 percent of our demand, with imports from Bolivia, South Africa, and China accounting for most of the rest. Recycled scrap accounts for roughly 10 percent.

Cadmium

A minor chalcophile element, cadmium is commonly concentrated in zinc and lead-zinc ores. Domestic zinc ores account for about 80 percent of the U.S. demand and recovery from imported zinc ores the remainder. Cadmium metal is used for a variety of applications, including coatings on iron, aluminum, and steel to prevent corrosion under alkaline conditions, on numerous electronic devices, and in several kinds of batteries, including those in space vehicles. Cadmium compounds are used as a pigment, as a stabilizing agent in plastics, and in solar photovoltaic cells. Cadmium is highly toxic, and is discussed in detail in Chapter 8.

Vanadium

Vanadium is an important alloying agent for steel. In the United States reserves from sandstone-type vanadium or uranium-vanadium deposits are adequate to meet demand for many years. It is also recovered from petroleum refining. Domestic production met 74 percent of the demand in 1984, with imports of vanadium ores and ferrovanadium from South Africa supplying most of the rest. The balance came from refining of imported crude oil. Vanadium is a minor environmental pollutant, but most of this comes from the burning of coal and residual fuel oil.

SEAFLOOR METAL DEPOSITS

Seafloor deposits include a wide variety of different materials, from mechanically accumulated placers, sand, and shells to hydrothermally precipitated polymetallic sulfides. The U.S. Bureau of Mines classifies seafloor deposits into several main types: oil and gas reserves, construction materials (sand and gravel), calcareous shell and sand, phosphorites, placer deposits, metalliferous oxides, and polymetallic sulfides. Only the last three of these are discussed here, and the others appear in later chapters where appropriate.

Exploration for seabed mineral deposits has been going on for many years, but received governmental support in 1983 when the U.S. Exclusive Economic Zone (EEZ) was established by Presidential proclamation. This action gives the United States the sole mineral rights to all inorganic and organic resources, dead or alive, in a zone 200 nautical miles (370 kilometers) from the shores of all U.S. properties. This zone, shown in Figure 4–9, adds over 13 million square kilometers of territory to the mineral jurisdiction of the United States. While this establishes a basis for exploiting the mineral resources of the EEZ, it does not address the thornier question of mineral resources on the ocean floor beyond the EEZ.

The four main metals of interest in seafloor deposits are manganese, cobalt, nickel, and copper. Of these the United States relies heavily on importing the first three, having sufficient reserves of copper. Seafloor metalliferous nodules, commonly called manganese nodules because of their 10–35 percent manganese content, are common on many parts of the ocean floor. They contain appreciable amounts of iron, zinc, lead, cerium (a rare-earth element), and other metals, all of which can be recovered as by-products if mining commences for them. The nodules consist of metalliferous oxides, and they form at depths of 3000 to 6000 meters in areas of very low sedimentation. Evidently, in zones of high rates of sedimentation, the metals are widely dispersed. Where the rates of sedimentation are low, however, and where there is nucleating material such as rock detritus or biogenic material, the metals will start to precipitate. While the source of the metals is still unclear, it is known that the nodules form slowly after initial nu-

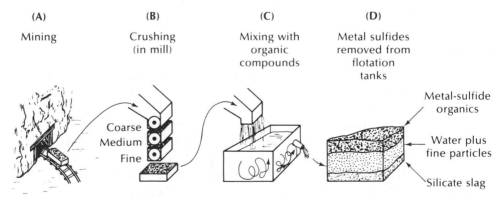

(A)	(B)	(C)	(D)
Mining	Crushing (in mill)	Mixing with organic compounds	Metal sulfides removed from flotation tanks

Coarse
Medium
Fine

Metal-sulfide organics

Water plus fine particles

Silicate slag

FIGURE 4–9   Metal sulfide recovery. Ore from the mine (A) is first crushed (B) and then mixed with a water solvent rich in organic compounds (C). The organic compounds affix themselves to the metal sulfides and, because of their low density, float to the surface (D) where the froth of metal sulfide and organic matter can be skimmed off. The metal is then removed from the sulfide by a combination of roasting and electrolysis. (Source: Brookins, 1981)

cleation, and take about 1 million years to form a 3-millimeter-thick layer. In many parts of the ocean floor, the cover of nodules is extremely dense (Figure 4–10), and economic amounts of manganese, nickel, cobalt, and copper accumulate in the nodules.

In the United States, EEZ nodule beds are known from the Blake Plateau off the coast of South Carolina and Georgia as well as off Johnston Island in the Pacific (Figures 4–11 and 4–12). One of the prime nodule provinces is the Clarion-Clipperton area in the northeast Pacific, but this location is outside of the EEZ.

On seamounts, seamount chains, and midocean ridges, metalliferous oxide crusts rich in cobalt and other metals are found. The role of bottom currents, which may prevent nodules from forming, may in turn be responsible for crust formation. Most crusts on seamounts are at depths of 1000 to 2000 meters, and water depth is probably a factor in cobalt availability (in deeper water nodules are less cobalt-rich). The thickness of the crust is a function of its age. Crusts up to 7 cm thick, containing 1 percent or more cobalt, have been reported. It is estimated by the Bureau of Mines that a typical seamount of a few hundred square kilometers could yield several million metric tons of ore rich in cobalt and in many cases manganese. The Pacific seamounts and chains are especially rich in these elements. Crusts are known in seamounts off the coasts of Hawaii, central California, Midway Island, Johnston Island, and Necker Island in the Pacific. All are on or close to seamounts or seamount chains (Figure 4–12).

Placer deposits of many minerals occur in the EEZ. Gold-bearing placers are present in several places off the coast of Alaska (Figure 4–13), while chromite occurs off the Oregon coast. Extensive titanium-mineral sands occur off the coast of South Carolina, Georgia, and Florida. Heavy-mineral sands total at least 1.3 billion cubic meters on the Pacific coast and 2 billion cubic meters on the Atlantic coast. The full potential of these resources and how to exploit them are being studied by various government agencies. Most interest in recent years, however, has focused on ocean floor polymetallic sulfide deposits, often rich in zinc, iron, copper, and other metals. These deposits owe their origin to exhalative fumarolic output from ocean ridges in response to plate tectonic activity. Sulfides deposited from these hydrothermal sources may form massive mounds, or thick deposits on the ocean floor which presumably penetrate to depth in fracture systems. The discovery of these deposits explains the massive sulfides found in a variety of rocks now exposed on the continents, including massive sulfides in ophiolites, stratified volcanics, and stratiform sedimentary and metamorphic rocks (see "Plate Tectonics and Metal Deposits" in this chapter). In or near ridge systems, rising mafic magma breaks and fractures the overlying rock and heats infiltrating seawater. A hydrothermal fluid, with water derived from both the magma and seawater, carries dissolved metals in it plus reduced sulfur. When these solutions are cooled away from their vents, they precipitate as metallic sulfides. The sulfides build stacks and thick deposits, sometimes tens of meters above the ocean floor and perhaps extending to depths of more than a kilometer (assuming the analogy between the seafloor deposits and known massive sulfides on land is correct) as

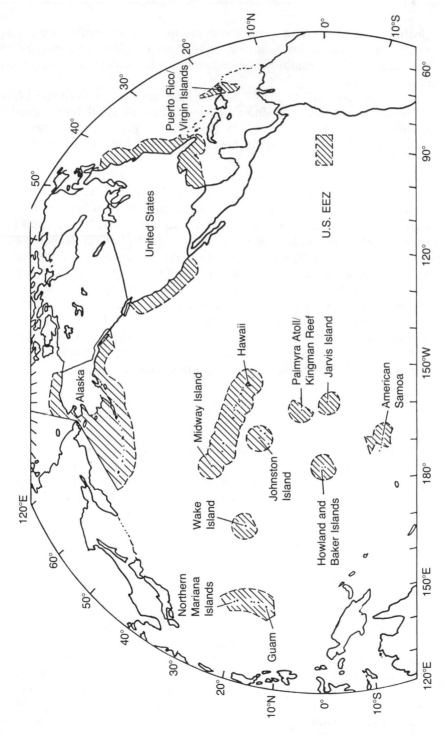

FIGURE 4-10 The Exclusive Economic Zone of the United States (pattern). (Source: U.S. Bureau of Mines)

FIGURE 4–11   Manganese nodules form on the floor of the ocean as chemical precipitates. Besides manganese, the nodules contain iron, copper, nickel, and cobalt. (Photo courtesy R. F. Weiss, O. H. Kirsten, and R. Ackerman, Scripps Institute of Oceanography, University of California, San Diego)

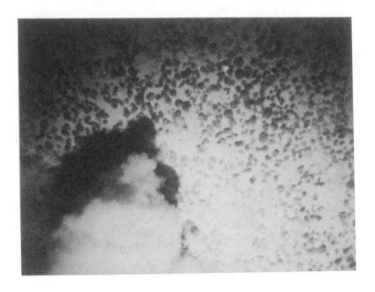

complex veins and replacement deposits. Many sulfides are deposited at depths of 2500 to 3000 meters under topographic conditions very different from those for ocean floor nodules and crusts. Whereas nodules are deposited on flat plains in the ocean deeps, and crusts on rolling topography and seamounts, the polymetallic sulfides are deposited in terrain consisting of steep cliffs flanking the narrow ridge center. Polymetallic sulfide occurrences now known in the EEZ include large deposits in the Gorda Ridge area off Oregon, the Juan de Fuca Ridge off Washington, the Aleutian Arc off Alaska, and probably off Hawaii (Figures 4–11, 4–12, 4–13). Research on and exploration for these deposits is truly only in the embryonic stage, and certainly more will be found.

The resources from metalliferous seafloor deposits are huge. The U.S. Geological Survey has estimated billions to tens of billions of mineable oxide nodules. The U.S. Bureau of Mines has estimated 800 million tons of cobalt-rich crust covering seamounts and surrounding seafloor terrain in the United States EEZ, and the hydrothermal metal sulfides in the Juan de Fuca Ridge area alone may contain 250 million tons zinc with high silver content. But the key to mining this wealth from the ocean floor may not be technological but political and legal.

For any successful deep-sea mining venture, there must be some good assurance that exclusive mining rights are available. In the 200 nautical miles off the coast of the United States and its territories, there are several laws, acts, and proclamations that assure rights. These include the Outer Continental Shelf Lands Act, as well as the 1958 Geneva Convention on the Continental Shelf, the Deep Seabed Hard Minerals Resources Act, and the recent Presidential proclamation on the Exclusive Economic Zone. Since most nations with ocean shorelines have such regulations, there should be little dispute over these properties. But what about the bulk of the deposits, which lie well away from land masses? Who has the right to mine these, and who should reap the profits?

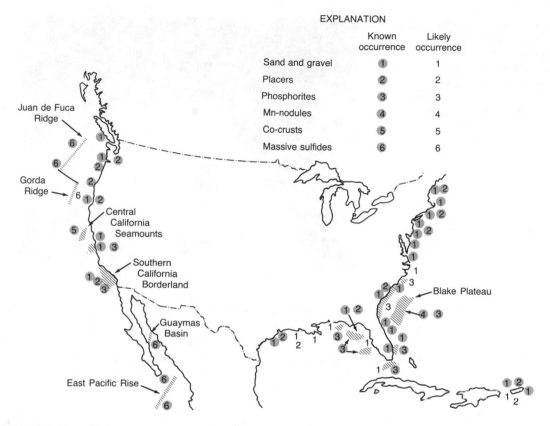

FIGURE 4–12   Offshore nonenergy mineral occurrences of the conterminous United States. (Source: U.S. Bureau of Mines)

This is a thorny issue. The United Nations in 1970 passed a resolution (No. 2749) which states in part that deep-ocean resources are "the common heritage of mankind" and thus should be available and shared by all nations. The United States signed this resolution. An earlier UN resolution in 1969 (No. 2574D) called for a moratorium on mining of deep-sea nodules until a new Law of the Sea treaty (LOS) could be passed; the United States did not sign this one. In 1982 a new LOS was passed by the United Nations by a vote of 130 to 4 with 17 abstentions. The United States voted against the LOS, although it had negotiated the treaty, and most of the world's industrialized countries abstained. The parts of the 1982 LOS objectionable to the United States were these provisions:

Protection of land-based producers on seafloor mineral mining is limited.

Decisions concerning all seafloor mining would be by a newly created international mining company called The Enterprise, with each nation having a voting member on the Board of The Enterprise. This body would compete with private mining firms.

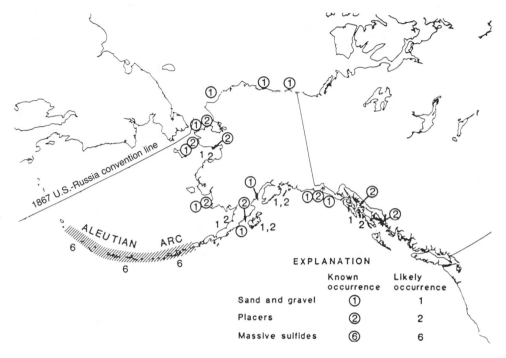

FIGURE 4–13   The Hawaiian Exclusive Economic Zone. (Source: U.S. Bureau of Mines)

Mining companies working on seabed recovery would be required to sell their technology to The Enterprise.

The Enterprise, after 20 years and without approval of individual nations affected, could revise the legal-regulatory LOS framework in any manner.

The U.S. administration stated that these provisions are unacceptable to a market economy nation, and that a large, unwieldy body such as The Enterprise could not easily manage such a complex task. Further, all the preliminary technology and expertise would be forfeit: There is no patent protection for the United States or other industrialized nations in the LOS. In 1984 the United States and seven other Western nations signed a provisional understanding essentially in opposition to the LOS and the developing nations that supported it. The ramifications of these actions, both short- and long-range, are unknown. Further confusion comes from the fact that Japan and France, two nations with ocean floor mining interests and activities, signed the LOS.

At present, many nations are involved in pursuing ways to exploit the ocean floor's mineral wealth. These include several companies from the United States, Belgium, the United Kingdom, France, Italy, Canada, West Germany, Japan, and the Netherlands. The USSR is also continuing its long-standing sole investigatory role.

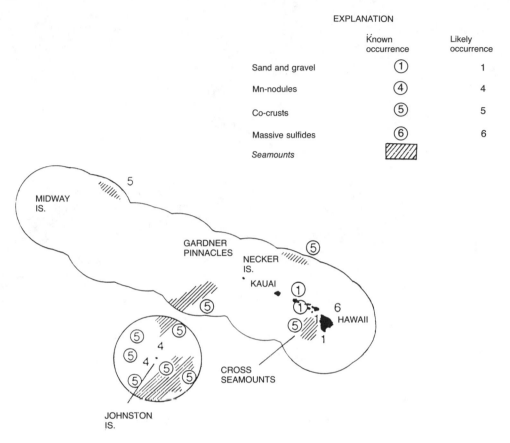

FIGURE 4–14   Nonenergy mineral occurrences in offshore Alaska. (Source: U.S. Bureau of Mines)

Environmental concerns are of two main types: The effect on the ocean floor of the physical mining activities, and the processing effects. The Deep Ocean Mining Environmental Studies (DOMES), sponsored by the National Oceanic and Atmospheric Administration, concluded that the effects of mining the nodules and crusts from the deep ocean will be of little or no consequence. Processing, on the other hand, may result in large problems. The early metal recovery will probably be on land, but it is probable that later, larger scale metal recovery will be done at sea. One plan calls for an Ocean Thermal Energy Conversion (OTEC) plant to provide the energy for the metal extraction, and this is of environmental concern by itself (see Chapter 7). Water pollution from dumped toxic mine wastes, toxic chemicals used in processing, and trace elements released by the processing is likely, but at present the magnitude of the problem is unknown. Since most life in the ocean is present in the upper tens to hundreds of meters of the ocean, the effect of such dumping will have pronounced effects on organisms and on the marine ecosystem.

The technology to harvest nodules from the ocean floor is now available, and scraping devices to harvest crusts and sulfides are probably also available, at least on pilot scales. Not resolved, however, are the very difficult problems of international agreement and regulation of seafloor mining acceptable to all nations and the environmental aspects of metal ore processing at sea.

FURTHER READINGS

Amstutz, G. C., and Bernard, A. J. 1973. *Ores in Sediments.* International Union of Geological Sciences, Series A, No. 3. Berlin: Springer-Verlag.

Barnes, H. L., ed. 1967. *Geochemistry of Hydrothermal Ore Deposits.* New York: Holt-Rinehart-Winston, 670 p.

Barnes, H. L., ed. 1979, *Geochemistry of Hydrothermal Ore Deposits. II.* New York: Wiley-Interscience, 798 p.

Berger, B. R., and Bethke, P. M. 1985. *Geology and Geochemistry of Epithermal Systems:* Reviews in Economic Geology Z. Golden, Colo.: Society of Economic Geologists.

Burchfiel, B. C., Foster, R. J., Keller, E. A., Melhorn, W. N., Brookins, D. G., Mintz, L. W., and Thurman, H. V., 1982, *Physical Geology.* Columbus, Ohio: Merrill Publishing Co.

Dixon, C. J. 1979. *Atlas of Economic Mineral Deposits.* Ithaca, N.Y.: Cornell University Press.

Douglas, R. J. W., ed. 1970. *Geology and Economic Minerals of Canada.* Economic Geology Report No. 1, Geological Survey of Canada.

Henley, R. W., Truesdell, A. H., Barton, P. B., and Whitney, J. A. 1984. *Fluid–Mineral Equilibria in Hydrothermal Systems:* Reviews in Economic Geology 1. Golden, Colo.: Society of Economic Geologists.

Hutchinson, C. S. 1983. *Economic Deposits and their Tectonic Setting.* New York: Springer-Verlag.

Jensen, M. L., and Bateman, A. M. 1979. *Economic Mineral Deposits,* 3d ed. New York: Wiley.

Levinson, A. A., ed. 1981. *Precious Metals in the Northern Cordillera.* Calgary: Association of Exploration Geochemists, 214 p.

Manheim, F. T. 1972. Mineral resources off the northeastern coast of the United States. U.S. Geological Survey Circular 669.

Maynard, J. B. 1983. *Geochemistry of Sedimentary Ore Deposits.* New York: John Wiley and Sons.

Meyer, C. 1981. Ore-forming processes in geologic history. In *Economic Geology—Seventy-Fifth Anniversary Volume,* 6–41. El Paso, Tex.: Economic Geology Publishing Co.

Noble, J.A. 1955. The classification of ore deposits. *Economic Geology* (special issue): 155–69.

Oceanus. 1984. Deep-sea hot springs and cold seeps. Special issue of *Oceanus,* vol. 27, no. 3.

Park, C. F., Jr., and MacDiarmid, R. A. 1970. *Ore Deposits,* 2d ed. San Francisco: W. H. Freeman.

Ridge, J. D., ed. 1968. *Ore Deposits of the United States, 1933/67.* New York: American Institute of Mining, Metallurgical and Petroleum Engineers.

Routhier, P. 1963. *Les gisements metallifères.* Paris: Masson.

Sawkins F. J. 1984. Metal Deposits in Relation to Plate Tectonics. New York: Springer-Verlag.

Stanton, R. L. 1972. *Ore petrology.* New York: McGraw-Hill.

Titley, S. R., ed. 1982 *Advances in Geology of the Porphyry Copper Deposits, Southwestern North America.* Tucson: University of Arizona Press, 560 p.

United Nations. 1982. *The Development Potential of Precambrian Mineral Deposits.* Oxford: Pergamon Press, 432 p.

U.S. Bureau of Mines. 1985. *Mineral Facts and Problems.*

U.S. Geological Survey, 1973. *U.S. Mineral Resources.*

Verwoerd, W. J., ed., 1978. *Mineralization in Metamorphic Terranes.* Pretoria: J. L. Van Schaik, 552 p.

Warren, K. 1973. *Mineral Resources.* New York: John Wiley & Sons.

5
Elements for the Agricultural and Chemical Industries

INTRODUCTION

This chapter deals with the extensively consumed elements essential to our agricultural industry: potassium (as potash), nitrogen, phosphorus, and sulfur. In addition, elements and their compounds important to the chemical and related industries are discussed, including soda ash, sodium sulfate, boron, salt, lime, fluorine, and bromine. Except for iron, all the most abundantly consumed elements in the world, in excess of 100 million tons per year, are nonmetals. Carbon, from fossil fuels, is first, followed by iron, sodium, nitrogen, oxygen, sulfur, potassium, and calcium with phosphorus not far behind.

OVERVIEW

The consumption figures and projected demand for the elements of interest in this chapter are given in Table 5–1. Together, potash, nitrogen, and phosphorus consumption in 2000 will be about 75 million metric tons per year, sulfur another 19 million, and the other elements and compounds for the chemical industries roughly 95 million short tons per year. All of these materials are essential to the United States' role in modern society.

Table 5–2 shows the reserve and demand figures for the United States and the world for the period 1983–2000. It is noteworthy that the U.S. demand can be met by domestic reserves for most of the commodities listed (and lime and sodium sulfate, which are not shown), as well as nitrogen, which is recovered from the atmosphere. For potash and sulfur, however, demand will exceed supply. Worldwide potash reserves are adequate to meet demand to 2000, but sulfur

TABLE 5-1   United States demand for elements for the agricultural and chemical industries, 1975-2000.

Commodity	Units	1975	1980	1981	1982	1983	2000
Nitrogen (as ammonia)	Mmt	12.0	16.0	14.9	12.8	12.4	18.0
Phosphorus (phosphate)	Mmt	31.0	40.8	35.1	28.8	34.8	47.0
Potassium (as K_2O)	Mmt	4.6	6.3	6.2	5.1	5.7	7.6
Sulfur	Mmt	10.8	13.7	12.8	10.1	11.0	18.8
Soda ash	tt	6785	7134	7112	6667	6868	8400
Boron (boron oxide)	tt	273	384	373	266	381	580
Bromine	tt	171	149	163	181	159	190
Fluorspar	tons	1245	977	933	531	564	890
Salt	Mt	42.9	44.8	42.2	42.3	40.1	58.0
Lime	Mt	19.2	19.0	18.9	14.1	14.9	27.5

SOURCE: U.S. Bureau of Mines.

Abbreviations: Mmt, thousand or million metric tons; tt or Mt, thousand or million tons.

reserves are marginally adequate. These matters are discussed in detail later in this chapter.

Many of the elements discussed in this chapter are found in evaporites or in brines or in the oceans, and the oceans are responsible for marine evaporites and most brines. Nonmarine evaporites are important for soda ash and boron, as well as some other elements. It is necessary then to briefly discuss the chemistry of the oceans, and then marine evaporites, nonmarine evaporites, and brines. The various minerals involved, and their formulas, are listed in Table 5-3.

Composition of the Oceans

Seawater contains about 3.5 percent dissolved solids by weight. Chloride ion is the most abundant dissolved ion, followed by sodium ion (Table 5-4). These elements are more concentrated in the oceans relative to continental rocks for several reasons. Sodium (as Na^+) and chlorine (as Cl^-) are both highly soluble, and remain in solution for very long times. Sodium, for example, is released by rocks during weathering and transported to the oceans where it resides. Submarine volcanism and hydrothermal activity also add sodium, and especially chlorine, to the oceans as well as large amounts of sulfur (quickly oxidized to SO_4^{2-}), magnesium, and other elements. This must be the case, for composition of stream waters includes large amounts of silica, bicarbonate ion, calcium, and other ele-

TABLE 5–2   United States and world reserve/demand data for industrial material, 1983–2000.

Commodity	U.S. Reserve / U.S. Cumulative Demand	World Reserve / World Cumulative Demand
Boron	120 Mt / 7.8 Mt	360 Mt / 22 Mt
Bromine	12.5 Mt / 3 Mt	Adequate / 7.5 Mt
Fluorspar	36 Mt / 12 Mt	850 Mt / 110 Mt
Phosphate	1400 Mmt / 700 Mmt	14 Gmt / 3.2 Gmt
Potash (as K_2O)	95 Mmt / 110 Mmt	9.1 Gmt / 590 Mmt
Salt	Adequate / 833 Mt	Adequate / 4200 Mt
Soda ash	26 Gt / 130 Mt	26 Gt / 720 Mt
Sulfur	160 Mmt / 250 Mmt	1.3 Gmt / 1.3 Gmt

SOURCE: U.S. Bureau of Mines.

Abbreviations: Mt or Gt, million or billion tons; Mmt or Gmt, million or billion metric tons.

ments relative to the proportions in seawater. This is explained by the fact that much of the calcium and bicarbonate are removed in the oceans by shell-secreting organisms in the nearshore environment, but especially in deep-sea sediments. Additional bicarbonate formed by calcium carbonate dissolution is in large part taken up by organic life and ultimately released to the atmosphere as carbon dioxide. These mechanisms prevent calcium and bicarbonate from building up with time in the oceans. Similarly, silica is removed by incorporation into the shells of silica-secreting organisms such as diatoms, and by reactions involving clay minerals.

Clay mineral reactions are very important in affecting the composition of the oceans. As streams meet the oceans, a series of complex reactions involving clay minerals takes place. For example, potassium is preferentially removed by clay minerals relative to sodium, so that the amount of sodium that stays in solution is much greater than that in streams: The Na/K ratio in streams is roughly 3, yet in the oceans this ratio is closer to 30. Some silica is removed by the clay minerals as well. As mentioned above, magnesium and sulfur balances are dependent not only on weathered materials brought to the oceans but on submarine

TABLE 5–3   Minerals and their formulas for selected nonmetals.

Mineral	Formula
Potash	
Sylvite	KCl
Carnallite	$KMgCl_3 \cdot 6H_2O$
Polyhalite	$K_2Ca_2Mg(SO_4)_4 \cdot 2H_2O$
Anhydrite	$CaSO_4$
Gypsum	$CaSO_4 \cdot 2H_2O$
Halite	NaCl
Langbeinite	$K_2MG_2(SO_4)_3$
Phosphorus	
Apatite	$Ca_5(PO_4)_3(F,OH)$
Soda ash, sodium sulfate	
Trona	$Na_3(CO_3)(HCO_3) \cdot 2H_2O$
Nahcolite	$NaHCO_3$
Thenardite	Na_2SO_4
Mirabilite	$Na_2SO_4 \cdot 10H_2O$
Glauberite	$Na_2Ca(SO_4)_2$
Boron	
Tincalconite	$Na_2B_4O_7 \cdot 5H_2O$
Ulexite	$NaCaB_5O_4 \cdot 8H_2O$
Colemanite	$Ca_2B_6O_{11} \cdot 5H_2O$
Fluorine	
Fluorite (fluorspar)	CaF_2

TABLE 5–4   Selected elements in seawater and streams.

Element (Ion)	Seawater (ppm)	Streams (ppm)
Sodium (Na^+)	10,770	6.3
Chlorine (Cl^-)	18,800	7.8
Sulfur (SO_4^{2-})	905	11.2
Magnesium (Mg^{2+})	1,290	4.1
Calcium (Ca^{2+})	412	15
Potassium (K^+)	380	2.3
Silica (SiO_2)	2	13.1
Carbon (CO_3^{2-})	28	58.4
Phosphorus ($PO_4)^{3-}$	0.06	
Bromine (Br^-)	67	
Strontium (Sr^{2+})	8	

volcanic activity as well. All of the elements in the ocean are in a delicate balance; sodium and chlorine, for example, must have stayed in the same relative proportions over much of geologic time. Otherwise there would be evidence for far more acidic conditions in the Precambrian geologic rock column, and this is not the case. Hence chlorine added by submarine volcanism and hydrothermal activity must keep pace with sodium additions by continental weathering, and their removal must also be in balance. Many trace elements are highly depleted in seawater, the result of chemical precipitation due to chemical reactions under reducing conditions in deep oceans among other factors. For this reason, ocean floor materials such as manganese nodules and cobalt-rich crusts (see Chapter 4) are enriched in many elements, and seafloor metallic sulfide deposits are enriched in others. The advantage of this lack of trace-element impurities in the oceans is discussed later in this chapter.

More important for the purposes of this book, however, are the types of deposits that form when seawater is trapped and evaporates.

Marine Evaporites

The solubility of minerals varies widely. Thus halite (NaCl) is more soluble than gypsum ($CaSO_4 \cdot 2H_2O$), and gypsum is more soluble than calcite ($CaCO_3$). These three minerals are used as examples here because of their importance in marine evaporites: those evaporites formed by evaporation of seawater. If one evaporates a beaker of seawater under an inert atmosphere, an assemblage of minerals in the following proportions results: NaCl 78%, $MgCl_2$ 9%, $MgSO_4$ 6.5%, $CaSO_4$ 3.5%, KCl 2%, and small amounts of $CaCO_3$, $MgBr_2$, and $SrSO_4$. Yet naturally occurring marine evaporites consist predominantly of interlayered halite, gypsum and anhydrite, and calcite (limestone). To attempt to answer this discrepancy, J. Usiglio, in the 1840s, conducted a series of experiments to see what minerals precipitated during evaporation of seawater under atmospheric conditions. He found that calcite precipitated first when 30–40 percent of the seawater remained, followed by $CaSO_4$ (gypsum in his experiments) at 20 percent, and then halite at 10 percent (Figure 5–1). Only when 2 to 4 percent of original volume remained did potassium and magnesium salts form. He then noted that in the thick salt deposits of Germany, limestones were abundant on the edges of the sequence, gradually grading and interbedded with $CaSO_4$, and then halite as the center of the basin was reached. Usiglio concluded that his experiments were entirely consistent with this assemblage. His work is an excellent example of the scientific method.

Yet there are some problems. While the early precipitation of calcite is easily explained as due to its relative insolubility and the addition of carbon dioxide from the atmosphere to allow $CaCO_3$ to precipitate, and the occurrence likewise of thick accumulations of gypsum or anhydrite and halite, two other questions were problematic: the absence of sylvite in most marine evaporites and the absence of much of the seawater magnesium.

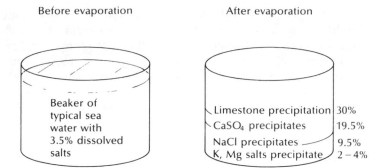

Before evaporation

After evaporation

Beaker of typical sea water with 3.5% dissolved salts

Limestone precipitation	30%
CaSO$_4$ precipitates	19.5%
NaCl precipitates	9.5%
K, Mg salts precipitate	2 – 4%

FIGURE 5–1   Usiglio's evaporation experiments. The Italian scientist J. Usiglio conducted a series of studies on seawater evaporation in the late 1840s. He found that normal evaporation led to some calcium carbonate (i.e., limestone) formation after about 70 percent of the seawater had been evaporated. Gypsum formed after 80 percent had evaporated, and common table salt, halite (NaCl), formed after 90.5 percent evaporation. Not until 96 to 98 percent had evaporated did potassium and magnesium salts appear. In nature rocks formed by evaporation follow a typical sequence of thick layers of limestone and gypsum (or anhydrite) followed by minor amounts of halite and rarely sylvite (KCl). (Source: Brookins, Merrill, 1981)

What happens is that as the point is reached where potassium and magne-sium salts might form, only a few percent of the original volume remains. Any inflow of new water will easily dissolve potassium or magnesium salts or prevent them from forming. Only when conditions are extremely static for fairly long times is sylvite able to form, often including appreciable but not all of the mag-nesium. The remaining Mg^{2+}, it turns out, is attempting to return to the sea by moving through the entrapping reef as shown in Figure 5–2. The Mg^{2+} that en-ters the reef reacts with limestone to form dolomite and releases Ca^{2+}. This pro-cess can be seen working today in the Bahamas, for example. When sylvite can-not form in the restricted marine basins, it commonly winds up trapped in trace amounts in halite disseminated over a large volume, and thus is not economic. This is the rule rather than the exception, and most marine evaporites do not contain appreciable potassium-salt concentrations. Figure 5–3 a map of the dis-tribution of large marine evaporites in the United States, shows that only in a small part of the Permian Basin and in the Paradox Basin are commercial amounts of potassium salts found. They occur in the Michigan and Gulf basin evaporites, but only in subeconomic quantities.

The simplest models for marine evaporites to form are shown in Figure 5–2. The most common features a barrier reef (or other barrier), which after a long period of growth starts to effectively block off seawater from reaching the trapped water. The Great Barrier Reef of Australia will eventually do this. When the water can no longer be in communication with the open ocean, evaporation begins, and the sequence of salts predicted by Usiglio results. Geologists have long recognized that this model cannot explain all marine evaporites, and propose that

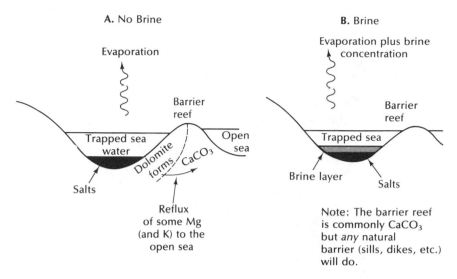

FIGURE 5–2   Marine evaporation formation. Under natural conditions evaporites may form from either condition **A** (simple evaporation) or **B** (evaporation plus formation of a brine layer). Thick layers of limestone (at 70 percent evaporation) can accumulate before gypsum (at 80 percent evaporation) starts to form. Flooding at about the time gypsum is starting to form will precipitate more limestone. Hence alternating layers of limestone and gypsum are the rule. Similarly, at about 90 percent evaporation, alternating layers of halite and gypsum are noted; and, near total evaporation, sylvite and magnesium salts are interlayered with halite. Some of the magnesium is not incorporated into the evaporite and causes dolomitization of the barrier reef, as shown in condition **A**. (Source: Brookins, Merrill, 1981)

a deep marine basin model, as shown in Figure 5–2B, must be necessary to explain some marine evaporites.

Salt Domes

Salt domes (or salt diapirs) form when buried halite-bearing evaporites are overlain by thick sediments in a tectonically sinking environment. Thus off the Gulf of Mexico coast there exists a deeply buried and sinking sedimentary section which includes evaporites. Salt, because it is plastic, will flow upward through fractures in overlying sediments, as shown in Figure 5–4. The flowing halite pushes up original overlying anhydrite and limestone, which serve as an effective cap to the dome, and it is highly fractured. Such fractured cap rocks may be suitable hosts for migrating petroleum or gas, and are also where abundant sulfur resources are found. Petroliferous material that migrates into the fractured salt-dome cap rocks reacts with anhydrite and reduces the sulfur to hydrogen sulfide, that is, from S^{6+} (anhydrite) to S^{2-} (H_2S). This H_2S is then oxidized to elemental sulfur, which fills cracks and voids in the fractured cap rock. The sulfur in such domes is recovered by the Frasch process (see "Sulfur" in this chapter).

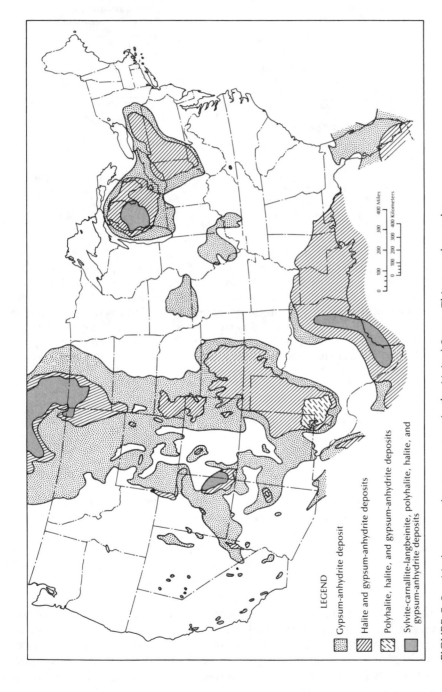

FIGURE 5–3   Marine evaporite occurrences in the United States. This map shows the abundance of marine evaporites and concentrations of potash-rich (sylvite) areas. The potash deposits of New Mexico are the largest producers in the United States. However, the potash deposits of the Williston Basin (which extend northward from the Montana–North Dakota border into Canada) are just now reaching full capacity. Since evaporites are also considered for storage of radioactive waste, this map shows where logical sites might be. (Modified from U.S. Department of Energy)

LEGEND

Gypsum-anhydrite deposit

Halite and gypsum-anhydrite deposits

Polyhalite, halite, and gypsum-anhydrite deposits

Sylvite-carnallite-langbeinite, polyhalite, halite, and gypsum-anhydrite deposits

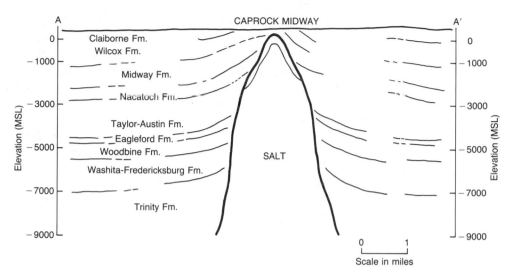

FIGURE 5–4 Generalized cross section of Keechi dome. (Modified from U.S. Department of Energy)

Nonmarine Evaporites

We have already seen how trapped seawater can evaporate to form thick marine evaporites, all of which have a more or less predictable mineralogy and chemistry because the starting water is chemically homogeneous. But when lakes or inland seas are impounded in closed hydrologic basins, they evaporate to form nonmarine evaporites (Figure 5–5). These rocks possess a mineralogy and chemistry dependent on the chemistry of their waters, in turn dependent on the chemistry of source materials. Thus a closed hydrologic basin surrounded entirely by limestone, and with limestone-dominated springs, will form, upon evaporation, de-

FIGURE 5–5 Formation of nonmarine evaporites. Nonmarine evaporites may form anywhere a restricted (closed) drainage basin is found surrounded by continental rocks. Nonmarine evaporites, unlike marine evaporites, possess very complex chemistries due to the varied composition of rocks being weathered around the closed basin. (Source: Brookins, Merrill, 1981)

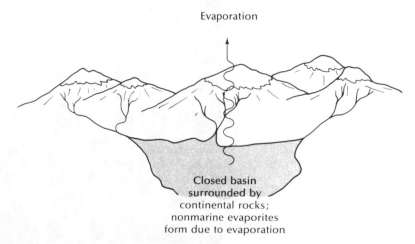

posits of calcium carbonate, without gypsum and halite and other typical marine evaporite minerals. If granitic rocks surround such a basin, an entirely different type of evaporite will result, one which will reflect the granite chemistry. If the source rocks, including springs, have a complex chemistry including many rare elements, then these rare elements will be found in the nonmarine evaporite.

In some areas of the western United States abundant sodium is available from weathering volcanic rocks, as well as substantial carbonates. When such rocks feed a closed basin, many minerals form, but sodium carbonates are very common (see "Soda Ash" in this chapter). Conspicuous by its absence here is chlorine, the most abundant ion in seawater. In other places, boron may be rich in weathering materials, and natural borate minerals occur in the nonmarine evaporite (see "Boron" in this chapter). Hence nonmarine evaporites not only possess individually different chemistries and mineralogies, but often are enriched in economic amounts of important minerals.

Brines

Halite and other evaporite minerals commonly contain small inclusions of gas plus liquid, called fluid inclusions as they result from originally trapped fluid in the minerals that couldn't crystallize. The gas separates from the fluid due to

Mg-K-sulfate stalactites forming from brine seeps in an underground potash mine, near Carlsbad, New Mexico. The source of the brine is from the evaporite and not from surface waters.

changes in temperature and pressure. If pockets of such fluid were originally present in the rocks crystallizing, or if they form by partial dissolution, so-called brines or brine pockets may result. These range from tiny cavities of a few microliters to very large pockets containing hundreds of millions of liters. If the brines represent original trapped seawater, they usually have highly concentrated dissolved salts to the point that they are economic. Their bromine and magnesium contents, for instance, are higher than in seawater, thus they are a better and cheaper source of these elements (see "Bromine" in this chapter and "Magnesium" in Chapter 4). Similarly, lithium and other elements are enriched in these brines. If brines accompany nonmarine evaporites, their chemistries may be very complex, and they may contain high concentrations of many elements, including some valuable metals such as tungsten and tin. Ways to exploit these brines are an active area for research today.

POTASSIUM (POTASH)

World potassium (as potash) deposits of economic importance are all found in marine evaporites. Potash is a generic term used here to describe all potassium ores, whether sylvite, polyhalite, carnallite, or others. Potash is also used to refer to any water-soluble potassium additive to crops. Most of the world's potash is used as fertilizer, with the remainder used in the chemical industry.

During World War I, Germany attempted to cause mass starvation in the world by cutting off supplies of potash salts from the Stassfurt area of Germany, then the world's main supplier. Had World War I lasted a few more years, this could possibly have had disastrous effects. This action, however, prompted widespread exploration for potash deposits which eventually led to the discovery of the large deposits in the United States in the 1920s.

Reserves of potash in the United States are large (Table 5–5), yet the United States imports roughly 80 percent of its potash. However, 25 percent of U.S. production is exported. In both cases, transportation is the key. Potash from huge reserves in Saskatchewan, Canada can be more cheaply transported into the Great Plains farm belt than from southeastern New Mexico where most of the U.S. reserves are found. The latter source is more cheaply sent to Central and South America.

Potash ores in the United States and in most of the other world deposits are sylvite layers (Table 5–3 has formulas for sylvite and other evaporite minerals). Although such layers are rare, as explained earlier, when they are found they are laterally continuous for great distances, thus making them economic. Room and pillar mining is used. A shaft is sunk to the ore zone and tunnels (drifts) excavated away from the shaft. Pillars of ore are left in place to provide support. When a mine is near the end of its productive lifetime, pillars from the mine are removed, starting with those farthest from the shaft, and the mine is allowed to cave. Since potash mines in the United States are typically at depths of over 300 meters, and all are located in lands of very little surface value, this subsidence caused by the eventual caving will not be problematic. However, that date is far

TABLE 5–5   World potash reserves.

Continent/Country	Potash as K_2O (Million Metric Tons)
North America	
Canada	4400
United States	95
South America	
Brazil	60
Europe	
USSR	3000
Germany (FRG and GDR)	1300
France	25
Others	60
Africa	Small
Asia	
China	200

SOURCE: U.S. Bureau of Mines.

off as most mines in the United States have closed due to the impact of foreign imports. Lower output is predicted. Ores rich in carnallite or polyhalite are less potassic, and very subeconomic.

The processing of sylvite ores is straightforward. The individual layers of potash are quite pure (Figure 5–6), but are always mixed with halite, some clay minerals, oxide impurities, and other evaporite salts. A process of froth flotation on crushed ore floats the sylvite and halite away from the other minerals, and the sylvite is then floated away from the halite using an aqueous solution rich in both

Small electric car hauler, IMC mine, New Mexico. Note whitish seams of potash ore, sylvite, on wall. Courtesy of International Minerals Corporation.

FIGURE 5–6   Underground view of pure sylvite (KCl, or potash ore) from a depth of 640 meters near Carlsbad, New Mexico. The chisel is 20 centimeters long. This bed, and others like it, extend laterally for thousands of square kilometers. (Source: Brookins, Merrill, 1981)

sodium and potassium chlorides. Very pure sylvite is produced, and this material can be, and usually is, prepared as a direct additive to crops. Occasionally the potash is reacted with other ingredients to make potassium-ammonium salts as well.

There are no substitutes for potash, an essential ingredient in the plant and animal diet. About 90 percent of the potash added to any area is taken up by the crops, hence a fresh application of potash must be made for each harvest. Manure is a common source of nutrients, but contains less than half the potash of sylvite. Further, manure is less efficient in releasing potash to the growing plants. Sewage, another source of potash, contains undesirable bacteria, worms, heavy metals, and a variety of carcinogens. Brines can also be treated for potash removal, as can seawater. Although unimportant in the United States, brines are locally important in other countries, and Israel has developed technology to recover potash from the seawater being used for recovery of other elements. A future source of some potash may come from alunite $KAl_3(SO_4)_2(OH)_{67}$ if this mineral is ever used as a source of aluminum. It contains several percent potassium as well as aluminum, and potash would be a by-product of aluminum production.

The future of domestic potash production is not bright. The very large deposits of sylvite in Saskatchewan will continue to dominate the U.S. market for

FIGURE 5–7   Trends in apparent consumption, production, imports, and exports for potash, 1955– 1983. (Source: U.S. Bureau of Mines)

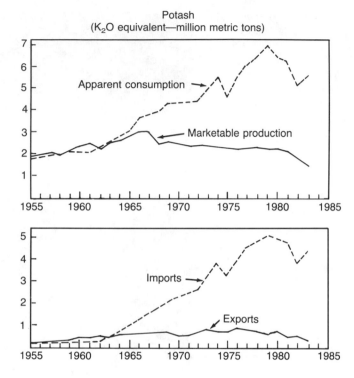

many years. Figure 5–7 shows the decline in both production and exports of domestic potash, along with increased imports to meet increased demand. Domestic production has been declining in New Mexico for some years, and is at a virtual standstill in the Paradox region of Utah (where there are smaller deposits of potash-bearing evaporites). There is some hope that refined solution mining of some Michigan potash ores may be economic in the not too distant future. Domestic reserves of about 95 million metric tons (see Table 5–2) are, in any event, short of the projected 100 million metric tons demand for the United States to the year 2000 (U.S. Bureau of Mines); hence imports are inevitable.

NITROGEN

Although nitrogen constitutes 78 percent of the atmosphere, mainly as elemental N_2 gas but also as fixed nitrogen in compounds (ammonia), it is not easily fixed in terrestrial materials. Only in the Atacama Desert deposits of Chile are nitrate minerals abundant, for example. These deposits are large, but could not possibly meet world demand. Other scattered nitrate deposits are too small to be of any consequence.

Plants require fixed nitrogen as a nutrient, but few plants are able to fix the nitrogen gas from the atmosphere, so fixed nitrogen is used in the form of any number of crop additives, chiefly ammonium compounds. Nitrogen and ammonia are also used in explosives, fibers, resins, and plastics.

The nitrogen used by the agricultural industry today is the result of the Haber-Bosch process by which ammonia is produced by combining nitrogen with hydrogen at high temperature, according to the equation

$$N_2 + 3\,H_2 = 2\,NH_3 \text{ (ammonia)}$$

The hydrogen is obtained from natural gas. Pure nitrogen can also be taken directly from the atmosphere by cryogenic processes, where the air is cooled to ultracold temperatures and oxygen and nitrogen condensed out as coproducts. The liquid nitrogen obtained is used for a variety of vacuum, industrial, and other uses. Fertilizers account for 80 percent of ammonia consumption in the United States; the rest is urea, ammonium sulfate, ammonium phosphate, and ammonium nitrate. Processing plants are usually built near areas with natural gas, or near phosphate, potash, or native sulfur operations. This minimizes some costs and allows in some cases direct formation of ammonium salts of potassium or phosphorus. Nonfertilizer uses of ammonium nitrate as an explosive, urea for resins and plastics, and the production of nitric acid are well known. The consumption of ammonia in the United States is striking. In 1950, 1.5 million tons were produced, and this amount increased to over 16 million tons by 1980 and fell to a low of 12–13 million tons in 1983. The lowered production in 1980 and 1983 was due to higher costs for deregulated natural gas and foreign competition. The cost for natural gas, for example, was 73 percent of total production cost of ammonia in 1983, and several U.S. companies are relocating efforts in countries where cheaper natural gas is available. The trend now is on the upswing (Table 5–1) and is projected to increase to nearly 20 million tons by the year 2000.

Despite the unlimited availability of nitrogen from the atmosphere, the United States imports roughly 15 percent of the ammonia it needs for the fertilizer industry, mainly from the USSR, Canada, Mexico, and Tobago and Trinidad. The USSR is the largest single producer of ammonia in the world, with 21 percent of the total production, followed closely by China. The United States ranks third.

Environmental problems with ammonia production are few. Modern plants remove natural gas's sulfur content, which is stored as native sulfur, and remove other gases (CO, CO_2) as well. There is little waste in the manufacture of ammonium salts.

PHOSPHORUS

Phosphorus is another of the essential elements for the agricultural industry, and the United States is the leader in production of phosphorus and phosphate rock, followed by the USSR, Morocco, and China, with significant amounts from Israel, Jordan, Tunisia, Togo, Senegal, and South Africa. Domestically, most of the need

TABLE 5–6 World
phosphorus reserves.

Continent/Country	Million Metric Tons
North America	
United States	1400
Europe	
USSR	1300
Africa	
Morocco	6900
South Africa	2600
Western Sahara	850
Others	250
Asia	
China	210
Others	170

SOURCE: U.S. Bureau of Mines.

NOTE: Only countries with 100 and greater million metric tons reserves considered.

is met from operations in Florida and North Carolina (84 percent) with 13 percent from Montana, Idaho, and Utah; 3 percent is from Tennessee.

Historically, guano deposits were an extremely important source of phosphorus for the world until early in this century. These deposits still account for roughly 8 percent of world output, coming mainly from islands in the South Pacific. Guano has been replaced by large deposits of the mineral apatite concentrated in various rocks. Apatite (see Table 5–3) contains 18 percent phosphorus and is the only readily available naturally occurring mineral that serves as a phosphorus ore.

Apatite is a common accessory mineral in igneous rocks, and is especially concentrated in alkaline rocks. Alkaline rock complexes on the Kola Peninsula, USSR, in the Palabora complex in Africa, and the Araxa and Jacipuranga complexes in Brazil are good examples. The deposits on the Kola Peninsula serve as an important source of phosphorus for the USSR, part of an overall multisystem mining effort where the nepheline and alkalic feldspar serve as a source for aluminum, the accessory magnetite yields iron, and other accessory minerals yield uranium, thorium, and the rare-earth elements. The ready availability of hydroelectric power helps lessen costs there. Igneous apatite accounts for 17 percent of world apatite production.

The largest reserves of apatite are in marine shales, located in the United States on the coastal plains of Florida and North Carolina and ancient "fossil" marine coastal plains in Tennessee. These deposits account for most of the production of the United States.

Even larger phosphorus reserves and resources occur in the nonmarine shales of the Phosphoria Formation in the western United States, which crops out over a million-square-kilometer area in parts of Idaho, Montana, Wyoming, and

Utah. This formation contains a 1.2-meter-thick horizon very rich in apatite, and contains at least 6 billion metric tons of identified reserves (of 24% or greater phosphorus) and nearly another 9 billion metric tons of resources. These figures far outweigh the Florida–North Carolina–Tennessee reserves of 4.4 billion metric tons. However, the southeastern deposits are situated in areas where water is available and ore processing is not a problem, unlike the Phosphoria Formation in the West where water is very scarce. Thus it is expected that the southeastern states will continue to meet most domestic demand well into the next century. World phosphorus reserves are given in Table 5–6.

Recovery of the phosphorus from marine apatites are, with silica sand and dolomite, separated from the shale's clay minerals by gentle grinding and flotation. The clays are discarded. By addition of anionic reagents, further flotation separates the phosphate-rich rock from the quartz and dolomite. This phosphate rock is very rich in apatite, but apatite does not dissolve readily enough to make it useful as a direct plant additive. When treated with sulfuric acid, however, phosphoric acid (52% phosphoric oxide) is produced, which is then used commonly to make ammonium phosphate fertilizers. Gypsum is produced in large quantities, some of which is packaged and sold but much of which is dumped, creating a disposal problem.

Marine apatites typically contain 3 to 4 percent fluorine and 50 to 200 ppm uranium. About 20 percent of the fluorine is recovered as a by-product, and the rest is waste. Two mills built for uranium recovery remove uranium from much of the phosphate rock, which is stored as uranium oxide.

Environmental problems of the phosphate industry are severe. The huge amounts of clay and gypsiferous wastes clog waterways and destroy plant and animal life where dumped, and the fluorine content of these waters is also very high. Fluorine in small amounts is very beneficial to human health, preventing tooth decay, for instance, but in large amounts is toxic. The groundwaters adjacent to the disposal ponds are gradually becoming polluted with fluorine around Tampa Bay and elsewhere. In other areas, trace elements such as arsenic may be locally concentrated. Yet the demand for phosphate to meet U.S. and world demands is great. Phosphate mine wastes are covered further in Chapter 8.

A future resource of phosphate may be the large phosphoric deposits off the coast of the United States in the Exclusive Economic Zone (EEZ). These deposits are especially large off the southeastern coast on the Portales Terrace, a bench off the Florida Keys, and on the Blake Plateau off the coast of South Carolina and Georgia. These deposits form in shallow seas due to vigorous biogenic activity, and gradually become coated with manganese oxides in deeper waters. Phosporites also occur off the coast of California and Baja California, and are extremely rich in phosphate. Many billion tons of phosphate ore is present in the EEZ of the United States.

It is projected (Figure 5–8) that there will be an overall decrease in U.S. phosphate production, due primarily to competition from foreign sources (Middle East and north Africa), and supply will not meet demand after about 1995. To compensate for this, the phosphate exported will probably be reduced. The United

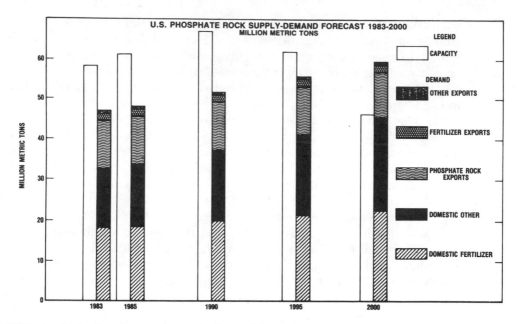

FIGURE 5–8   U.S. phosphate rock supply-demand forecast, 1985–2000. (Source: U.S. Bureau of Mines)

States has been a major exporter of phosphate, as shown in Figure 5–9 along with Israel, Jordan, Morocco, and others, but that situation will probably change in the late 1990s.

SULFUR

Sulfur is used for many purposes, but its use as sulfuric acid dominates and accounts for 80 percent of domestic consumption. About 65 percent is used for agricultural purposes. The remainder has a wide variety of industrial applications, including acid leaching of some copper and uranium ores, paper and plastic goods, petroleum refining, and chemicals.

Large deposits of sulfur are found in many parts of the world. Identified reserves are nearly 1300 million metric tons (Table 5–7), with the largest deposits in the USSR, the United States, Canada, Iraq, and Poland.

In the United States, sulfur from salt domes is the largest source of elemental sulfur. This native sulfur occurs as fracture fillings in the anhydrite-limestone cap rocks of the salt domes. The sulfur was presumably formed by bacterial reduction of sulfur in anhydrite due to petroliferous hydrocarbons seeping into the fractured dome rocks. It is recovered by the Frasch process, which takes advantage of the fact that sulfur melts at only 120°C and is very mobile up to temperatures near 180°C, above which it becomes viscous. A three-ringed pipe is used as illustrated

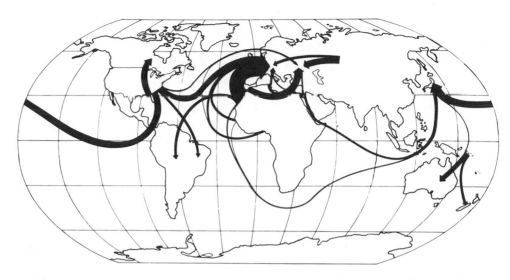

FIGURE 5–9   International trade in phosphate rock, 1983. (Source: U.S. Bureau of Mines)

in Figure 5–10. Hot water is pumped into the fractured dome rock from the outer ring. The innermost ring permits superheated steam to be injected into the rising sulfur to keep it molten as it is pumped up the middle ring. The Frasch process is also used to recover elemental bedded sulfur from the USSR, Poland, and Iraq. One major advantage of these elemental sulfur deposits in evaporites, both in

TABLE 5–7   World sulfur reserves.

Continent/Country	Million Metric Tons
North America	
United States	155
Canada	150
Latin America	
Mexico	80
Other	30
Europe	
USSR	350
Poland	150
Other	95
Africa	Small
Asia	
Iraq	155
China	25
Other	95
Oceania	Small

SOURCE: U.S. Bureau of Mines.

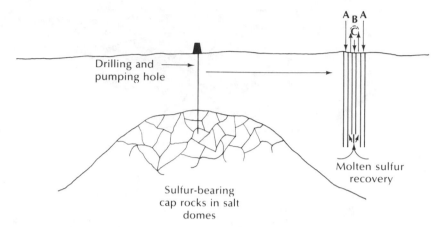

FIGURE 5–10   The Frasch process. A concentric pipe is inserted into sulfur-rich cap rocks found in many salt domes off the coasts of Louisiana, Texas, and Mississippi. Hot water, injected through the outer part of the pipe (A), melts the sulfur in the fractured rock. Because of its lower density, the molten sulfur moves to the recovery part of the pipe (B), where it is mixed with superheated air from the innermost part of the pipe (C) to keep the sulfur molten until it is recovered at the surface. (Source: Brookins, Merrill, 1981)

domes and bedded, is that they are very pure. Since the original source of the sulfur was anhydrite, which forms by evaporation of seawater, and because most trace elements associated with sulfur in pyrite of igneous rocks are absent in sea-water, the native sulfur pumped to the surface by the Frasch process is over 99.8 percent pure. This makes it directly applicable for a wide variety of agricultural and other purposes without further purification. Frasch process sulfur accounts for 34 percent of U.S. production. Another source of elemental sulfur is from volcanic related hot-springs areas. While not commercial in the United States, these deposits provide sulfur for several developing countries.

Sulfur is also recovered from sulfide minerals. Deposits of pyritiferous rock are sometimes mined exclusively for their sulfur content, usually to make sulfuric acid. In the United States, much of the sulfur released by the treatment of copper, lead, and zinc sulfide ore is converted to sulfuric acid. This acid is fairly impure, containing numerous trace elements (arsenic, selenium, lead, zinc, mercury, and so on), and must be used only in applications where impurity is not a problem (discussed in detail in Chapter 8). Sulfides yield 12 percent of the total U.S. demand for sulfur.

Natural gas normally contains some hydrogen sulfide, which must be removed prior to use as a fuel. Many petroleums also contain abundant sulfurous organic compounds, which must also be removed. The sulfur is removed from gas and petroleum as hydrogen sulfide and converted to elemental sulfur by the Clauss process, which involves hydrogenation of the gas, forming some sulfur dioxide which reacts with the hydrogen sulfide to form elemental sulfur plus water. The recovered elemental sulfur accounts for about 53 percent of domestic consumption.

Sulfur is also present in large amounts in tar sands, in coal, and in abundant sulfate minerals such as gypsum, but these sources are only resources at the present time.

Until 1976 the United States was a net exporter of sulfur, but since then demand has outstripped supply. The problems are several fold. Reserves of Frasch-recoverable sulfur have diminished, and costs of heating the water to melt the sulfur, mainly by natural gas, have increased dramatically, as have transportation costs. The decline of Frasch sulfur is shown in Figure 5–11. Recovered elemental sulfur from gas and petroleum refining has increased, and now together they have surpassed Frasch sulfur. Even with this increase, however, the total U.S. demand cannot be met domestically. In fact the U.S. Bureau of Mines estimates that there may be inadequate sulfur worldwide to meet demand by 2000. Further, it is likely that sulfur from sulfide mining operations will decrease in the near future due to escalating costs of environmental controls, cheap foreign sources of metals, and energy and labor costs. There is also no guarantee of a solid gas and petroleum industry (see discussion in Chapter 7), which would further reduce domestic supplies.

Sulfur is a good example of how reserve estimates can be off. In 1973 the U.S. Geological Survey published MACD (minimum anticipated cumulative de-

FIGURE 5–11   Trends in the consumption of sulfur in the United States, 1960–1983. (Source: U.S. Bureau of Mines)

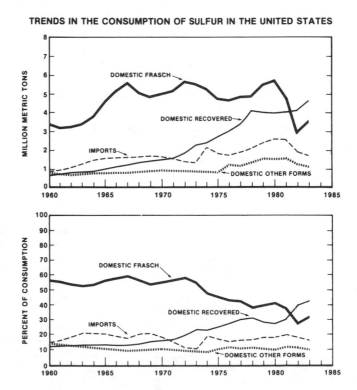

TRENDS IN THE CONSUMPTION OF SULFUR IN THE UNITED STATES

mand) figures of I and I for sulfur, meaning that reserves were more than 10 times that necessary to meet demand to the year 2000. Yet, about a dozen years later, the supply is inadequate. Clearly technology must be found to recover sulfur from new sources, perhaps from sulfide mine wastes, from tar sands, and especially from sulfur-bearing coals. Certainly if the sulfur dioxide released to the atmosphere from coal burning could be trapped and converted to sulfuric acid, the United States would be a long way toward solving two major problems— sulfur demand and health effects of sulfur dioxide emission.

SODA ASH AND SODIUM SULFATE

Both soda ash (trona is the main mineral; see Table 5–3) and sodium sulfate (mainly glauberite, thenardite, and mirabilite; see Table 5–3) both form under nonmarine evaporitic conditions. Soda ash forms in lakes fed by material weathered from sodic volcanics and carbonate rocks, with subsurface contributions from carbonate-rich springs, combined with abundant carbon dioxide from decay of plant and aquatic animal remains. The largest single deposit is in southwestern Wyoming, where ancient Lake Gosiute contained conditions favorable for trona formation. The deposits there are adequate to meet U.S. demand for over 3000 years and world needs for at least 700 years. Smaller nonmarine evaporitic lakes in California, some fed by sodium carbonate-rich brines (Searles Lake, Owens Lake, others), also contain appreciable trona. Another sodium carbonate, nahcolite (see Table 5–3) constitutes possibly 20 billion tons of soda ash resources, and dawsonite (Table 4–5) may also provide substantial soda ash if oil shales are mined. World reserves of soda ash are given in Table 5–8.

Sodium sulfate deposits are most common in postglacial settings. Closed nonmarine basins become traps for dissolved sodium and sulfate when source rocks are rich in sodium-rich volcanics or bentonitic shales along with sulfide minerals. As evaporation proceeds, the sodium sulfate content of the waters increases until mirabilite starts to precipitate. Thenardite forms from mirabilite after more evaporation. World reserves of sodium sulfate (Table 5–9) are largest in the USSR, but abundant also in North America.

TABLE 5–8   World soda ash reserves.

Continent/Country	Tons
North America	
United States	5,000
Mexico	Small
Africa	
Several	500

SOURCE: U.S. Bureau of Mines.

TABLE 5–9   World sodium
sulfate reserves.

Continent/Country	Million Tons
North America	
United States	945
Mexico	182
Canada	92
Europe	
USSR	2000
Spain	200
Africa	
Botswana	207

SOURCE: U.S. Bureau of Mines.

About 6900 thousand tons of soda ash is consumed in the United States each year (see Table 5–1) in glass and related industries (50%), sodium-based inorganic chemicals (20%), soap and detergents (9%), industrial water treatment (4%), paper and pulp industries (3%), and miscellaneous uses (14%). The United States consumes about 1100 thousand tons of sodium sulfate, most of which is used for the paper and pulp industries (47%), soaps and detergents (40%), and various chemical uses.

Soda ash demand will increase at about 1 percent per year to 2000 (Table 5–1), and sodium sulfate will show a slightly smaller increase over the same period. The choice of consumers for aluminum and especially plastic containers over glass has had, and will continue to have, an adverse effect on soda ash production. The use of substitutes for sodium sulfate in the paper and pulp industries will hurt this commodity's growth as well.

Environmental problems associated with soda ash are few. The amount of sodium chloride and calcium chloride that may be released to the surface area is regulated, and trona operators avoid this by pumping the wastes to large settling ponds with a pH near 10 due to the alkaline nature of the solutions. These ponds are of concern to migratory fowl, however. Sodium sulfate environmental problems are even fewer; large evaporating ponds present the only minor problem with land use.

BORON

Boron is another element concentrated into economic deposits only in nonmarine environments. Scarce rocks, such as certain volcanics in the western United States, have fairly high boron contents, which is released during weathering and transported to playa or dry-lake environments. Alternately, geothermal springs may feed dissolved boron into lake sediments. The large Kramer deposit at Boron,

The open pit mine of U.S. Borax at Boron, California, is the world's principal source of borates. To expose the ore, which lies approximately 300 to 600 feet below the surface, the overburden is removed by using explosives, electric shovels, and gigantic trucks. Shown here is a shovel which scoops up 19 cubic yards of earth in a single bite and loads it into a truck which can carry up to 150 tons.

California, consists of a thick accumulation of boron minerals (mainly tincalconite, overlain by colemanite and ulexite; see Table 5–3), 15 to 75 meters thick, occurring over an area of about 2 1/2 square kilometers. Other large deposits of this type are at Searles Lake and Furnace Creek in Death Valley, both in California. Smaller deposits are known in Nevada (historically, the first to be exploited). Even larger deposits are found in Turkey, and other large deposits are known in the USSR, Argentina, Bolivia, China, and Chile (see Table 5–10).

The open pit mined borate minerals are dissolved and washed to remove impurities, thickened, and crystallized. Different borate salts can be produced by differing the crystallization process to meet specific needs.

Boron is a very widely used element. Uses by the glass industry total 55 percent including fibers for insulation and borosilicate glasses; fire retardants and soap and detergents each total about 10 percent. Other uses are for metallurgy, agriculture, nuclear applications, and miscellaneous. The reserves of the United States are adequate to meet demand to the year 2000 with ease, even with a

Kernite, a rich sodium borate mineral, was discovered in 1925 in a large borate deposit in the Mojave Desert, some 130 miles northeast of Los Angeles. First called rasorite by U.S. Borax, the mineral was later called kernite, after Kern County. Courtesy United States Borax & Chemical Corporation.

projected growth of 3.2 percent per year (see Table 5–1). Boron compounds are also used in the agricultural industry as herbicides.

World demand for boron will continue to be met by exports from the United States, although exports from Turkey will become increasingly important in the next few decades.

TABLE 5–10   World boron reserves.

Continent/Country	Million Tons of Boron Oxide
North America United States	115
South America	30
Europe USSR	60
Asia Turkey China	120 30

SOURCE: U.S. Bureau of Mines.

Environmental problems with boron are few. Although some boron compounds, such as boranes, are extremely toxic, these are no longer produced. Boron in small amounts is an essential ingredient in plant and animal life.

SALT

Salt, or sodium chloride (the mineral halite), is a widespread and abundantly consumed material in the United States and in the world. The United States is the leading producer of salt, and has extensive deposits. Salt is used domestically for the chemical industries, and especially the chloralkali industry, makers of chlorine, caustic soda (sodium hydroxide) and soda ash. These uses accounted for some 50 percent of total consumption in the United States. Another 17 percent was for deicing roads, mainly in the northeastern states. Foods and food processing requires another 6 percent, and about 4 percent is used in the agricultural industries. The remainder is used for a wide variety of items in the petroleum, aluminum, and paper and pulp industries.

There are several sources of salt. The largest, of course, is the ocean. Trapped seawater has been used for evaporative recovery of salt for a very long time. Brines from evaporites are often very concentrated in salt, and thus can be pumped to solar evaporating ponds for salt recovery. Some inland lakes and seas are also very much enriched in salt, such as the Dead Sea in Israel and Jordan and the Great Salt Lake in Utah, from which water is impounded for solar evaporation. Bedded and dome halite from marine evaporites is the largest terrestrial source of salt, however. The United States contains major salt deposits in many areas. The four largest basins are the Gulf coast, which takes in the bedded and dome salts of Florida, Alabama, Mississippi, Louisiana, Arkansas, and east Texas; the Permian Basin, which takes in parts of New Mexico, Colorado, western Texas, Kansas, and Oklahoma; the Williston Basin, which includes parts of North Dakota, South Dakota, Montana, and Wyoming; and the Salina Basin, which takes in part of Michigan, Ohio, Pennsylvania, West Virginia, and New York. Salt resources of the United States total at least 61 trillion tons.

Despite these huge reserves, the United States imports 14 percent of its salt demand, mainly from Canada, Mexico, and the Bahamas. This reflects mainly the much cheaper production costs outside the United States. Brines account for 55 percent of U.S. production, rock salt 29 percent, and evaporated salt 15 percent. World reserves are more than adequate to meet demand for many decades.

Salt deposits are also used for storage due to their stability, impermeability, low water content, and other factors. The United States has stored a strategic reserve of some 450 million barrels of petroleum in salt domes in Louisiana and Texas, and transuranic radioactive wastes will be stored at the Waste Isolation Pilot Program (WIPP) site in southeastern New Mexico in the 1990s (see Chapter 7 for details). While salt is an excellent deicer, it dissolves and enters the water supply, increasing both sodium and chloride contents. The effects of this are discussed in Chapter 8.

LIME AND OTHER CALCIUM MATERIALS

When limestone is calcinated, lime (CaO, calcium oxide) is produced. As used here, lime refers not only to calcium oxide but also to calcium hydroxide, and magnesian varieties of these. All developed and most developing nations have lime production facilities, and the world supply is virtually inexhaustible. Major uses in the United States are in the chemical and steel industries, which consume about 89 percent of the total, mostly for steelmaking, water purification, stack gas treatment (to remove sulfur), paper and pulp industries, soil and road stabilization, and sewage treatment; while 8 percent is used in transportation, and refractory uses and agricultural uses account for the balance.

Although sufficient quantities exist domestically, the United States imports a small amount of lime from Canada and other countries because transportation and cost factors make it more economic to import in some northern areas. Uses for lime are increasing, and the industry is apparently very healthy well into the future. The U.S. demand for lime from 1975 to 1983 and projected for the year 2000 is shown in Table 5–1.

Calcium chloride use in the United States is modest but steady. It is used primarily as a pavement deicer (35%), for dust control (20%), in coal treatment (19%), by the oil and gas industry (12%), and for other miscellaneous uses. It is produced from brines in California and Michigan and made synthetically in New York, Louisiana, and Washington. As a deicer, it is usually mixed with salt for maximum efficiency in covering large areas.

FLUORINE

Fluorine is an important element to the United States, with major uses in the chemical industry (42%), treatment of aluminum ores (21%), steel production (31%), and other uses. The only important source of fluorine is from the mineral fluorite (CaF_2), a common accessory mineral in hydrothermal deposits. While fluorite (commonly referred to as fluorspar) is widespread in the United States, it is not concentrated in economic quantities. Table 5–11 shows world fluorite reserves, most of which are outside the western hemisphere.

For chemical industry purposes, fluorite is reacted with sulfuric acid to yield hydrogen fluoride and calcium sulfate. Fluorocarbons are chemically stable, low toxicity materials accounting for much of this hydrogen fluoride use.

Hydrogen fluoride is also used to make aluminum fluoride (AlF_3) or cryolite (Na_3AlF_6) for production of aluminum metal from bauxite ore, since natural cryolite deposits are insufficient to meet world demands.

The iron and steel industries use fluorite as an additive in foundries to promote desulfurization and dephosphorization of the feed, and to keep temperatures stable and to promote effective slag formation. About 5 pounds (2.3 kilograms) of fluorite per ton of steel produced is required.

Hydrogen fluoride is also used to make uranium hexafluoride for isotopic separation of uranium-235 from uranium-238 (see Chapter 7), while various inorganic fluorides, such as sodium fluoride used in water fluoridation, are also prepared from hydrogen fluoride.

Fluoride is commonly found in fissure veins, often with associated minerals of barium, silica, iron, lead, and zinc; and as a stratiform replacement mineral in limestone and other carbonate rocks, such as in southern Illinois, South Africa, and Mexico. Other occurrences include contact replacement deposits in carbonate rocks, shear-zone and breccia-zone stockworks (common in the western United States), in carbonates, and as gangue to base-metal deposits. United States depos-

TABLE 5–11   World fluorspar reserves.

Continent/Country	Million Tons Contained CaF_2
North America	
Mexico	46
South America	Small
Europe	
USSR	80
Spain	19
France	15
Italy	15
Other	8
Africa	
South Africa	83
Kenya	21
Other	5
Asia	
China	56
Mongolia	50
Other	4

SOURCE: U.S. Bureau of Mines.

its are small, poor grade, and cannot compete with foreign sources. Domestic production of fluorine is mainly as a by-product of phosphate rock processing (see "Phosphorus" in this chapter). Roughly 85 percent of the U.S. demand for fluorspar is imported, mainly from Mexico (45%), South Africa (29%), Italy (10%), and smaller amounts from Spain, Morocco, and China.

The demand for fluorspar has tapered off since peaking in the early 1970s (see Table 5–1), due mainly to decreased iron and steel production. Demand is expected to increase to nearly 900 tons per year in 2000, most of which will be met by imports.

Environmental aspects of fluorine are difficult to assess. High fluorine content in animals causes bone damage and other symptoms, but small amounts prevent tooth decay. Natural fluorite is not toxic, but if fluorocarbons burn, toxic fluorophosgene is produced, as well as gaseous fluorocarbons which can cause suffocation. Fluorine gas, hydrofluoric acid (used to dissolve rocks in laboratory experiments), and fluosilicic acid are all extremely dangerous chemicals and must be shipped, stored, and used with great precaution. Further, chlorofluorocarbons (CFCs) have been identified as a class of substance that can affect the ozone layer of the upper atmosphere (see Chapter 8).

BROMINE

Bromine is a rare element, present in seawater in small amounts and in many evaporitic brines. Technology was developed in the 1920s to recover bromine from seawater. Bromine is present in solution as bromide ion (Br^-), and this is oxidized to bromine gas (Br_2) by addition of chlorine. The bromine is removed from solution and either condensed or, preferably, reacted with various chemicals to form salts, which are easier to handle. Purification is the final step. The United States recovered bromine from seawater for well over 40 years, but since about 1970 has been recovering bromine from evaporitic brines because the bromine content is higher. World reserves, shown in Table 5–12, are based on the bromine reserves in brines only. The much larger bromine reserves of the oceans (100 trillion tons) are not included. The largest reserves are in the United States, France, the USSR, and Japan.

Bromine is used in lead-bearing gasolines as EDB (ethylene dibromide), where it scavenges lead from precipitating on cylinders of engines. EDB is also used in the dye and pharmaceutical industries. Methyl bromide is a fumigant, and other bromine compounds are used for fire retardants, dyes, pharmaceuticals, photography, and plastics. Uses in 1983 were divided among fire retardants (31%), EDB in gasoline (27%), drilling fluids (27%), and lesser uses.

The U.S. Bureau of Mines indicates that bromine reserves are adequate to meet U.S. demand well into the next century, and that other leading producers will be Israel, the USSR, France, West Germany, and Japan.

Environmental problems exist with bromine. It is highly toxic and corrosive as bromine gas, and is not allowed to enter the atmosphere. Bromine contents of

TABLE 5–12   World bromine reserves. Does not include bromine recovered from solar salts.

Continent/Country	Thousand Tons Bromine Content
North America	
United States	12,500
South America	Small
Europe	
France	1,750
USSR	1,500
Germany (FRG)	40
Middle East	
Israel	Not available
Asia	
Japan	900

SOURCE: U.S. Bureau of Mines.

only 500 ppm can cause fatalities in under an hour, and any concentration in air over 1 ppm is hazardous. Workers manufacturing EDB may only be exposed to 130 ppb bromine. Bromine as a fire retardant is also poisonous.

The outlook for bromine is favorable, although an expected drop in EDB use is predicted as leaded gasolines are phased out. Expanded use in pharmaceuticals, sanitary uses, drilling fluids, and fire retardants will offset the EDB loss. The demand for the United States will be met by reserves from Arkansas (80%) and Michigan (20%).

FURTHER READINGS

Austin, G. S. 1978. Geology and mineral deposits of Ochoan rocks in Delaware Basin and adjacent areas. New Mexico Bureau of Mines and Mineral Research Circular 159, 88 p.

Bates, R. L. 1960. *Geology of the Industrial Rocks and Minerals.* New York: Harper & Row.

Borchert, H., and Muir, R. O. 1964. *Salt Deposits—The Origin, Metamorphism, and Deformation of Evaporites.* Princeton, N.J.: Van Nostrand.

Dean, W. E., and Schreiber, B. C. 1978. Marine evaporites. Oklahoma City: Society of Economy, Paleontologists and Mineralogists Short Course No. 4.

Ellison, S. P. 1971. *Sulfur in Texas.* Texas Bureau of Economy Geology Handbook 2.

Gillson, J. L. et al. 1960. *Industrial minerals and rocks* 3d ed. New York: American Institute of Mining, Metallurgical and Petroleum Engineers.

Lamey, C. A. 1966. *Metallic and industrial mineral deposits.* New York: McGraw-Hill.

Multhauf, J. P. 1978. *Neptune's gift: A history of common salt.* Baltimore: Johns Hopkins University Press.

Nissenbaum, A., ed. 1980. *Hypersaline brines and evaporitic environments.* New York: Elsevier, 270 pp.

Skinner, B. J. et al., eds. 1979. Special issue devoted to phosphate, potash, and sulfur. *Economic Geology,* v. 74.

6
Industrial Minerals

INTRODUCTION

In dollar value, industrial materials are second only to fossil fuels among the world's important natural resources. In total volume of material processed, industrial materials are unsurpassed by any other resource except water. Industrial materials include sand and gravel, dimension and crushed stone, cement, various clays, pumice and perlite and other lightweight aggregate; industrial diamonds, mica, vermiculite, asbestos, gypsum, and plaster; abrasives, barite, feldspars, graphite, diatomite, and gemstones. For the more abundantly consumed industrial materials, such as sand and gravel, stone, and cement, supplies are adequate to meet demand well into the next century. For some of the others, such as barite and industrial diamonds, reserves are in very short supply and the United States is highly dependent upon imports.

This chapter focuses on the location of these resources, and the many factors that affect them. Thus transportation is a major factor for sand, gravel, and stone due to the weight involved, while dust and noise pollution may affect others. Industrial materials lack the romance of metals, except the small market for gemstones, yet they are of immense value in the United States. Delaware is the only state that does not have a flourishing industrial materials industry.

Asbestos is covered in this chapter in terms of its occurrence, uses, and reserves. The health aspects of asbestos are covered in some detail in Chapter 8.

The U.S. demand for industrial materials is given in Table 6–1, and the U.S. and world reserve and demand data appear in Table 6–2.

SAND AND GRAVEL

Sand and gravel constitute the second largest nonfuel mineral crop by tonnage after crushed stone. Ninety-six percent of the sand and gravel used in the United States is for construction purposes, with use in the concrete industry accounting

TABLE 6–1   United States demand for industrial materials, 1975–2000.

Material	Units	1975	1980	1981	1982	1983	2000
Barite	tt	1830	3770	4716	4060	2745	4200
Asbestos	tmt	551	359	349	247	217	470
Sand and gravel	Mt	761	763	689	592	654	1000
Silica sand	Mt	24.5	27.5	28.2	25.6	25.1	38
Crushed stone	Mt	901	984	873	790	863	1300
Dimension stone	tt	1067	974	988	1154	1078	1600
Perlite	tt	490	603	566	502	493	850
Pumice	tt	934	734	589	536	632	950
Cement	Mt	72	80	76	67	74	100
Clays	Mt	47.1	45.9	14.2	32.9	38.5	63
Diamond							
Powder	Mct	17.3	33.5	36.9	33	39.2	86
Stone	Mct	2.9	3.4	3.2	2.9	3.2	3.9
Emery	tons	2200	3800	2500	2000	2200	4500
Garnet	tt	14.2	22.9	22.2	23.7	27	38
Corundum	tons	1100	1500	1100	650	725	900
Diatomite	tt	430	516	525	472	473	750
Feldspar	tt	660	697	651	604	701	800
Gypsum	Mt	15.6	19.5	19.0	17.4	21.8	34
Graphite	tt	54	53	59	47	43	63
Vermiculite	tt	305	333	316	315	287	450
Sheet mica	tons	2685	2272	2277	1369	1104	650
Flake and							
scrap mica	tt	113	106	117	93	127	145

SOURCE: U.S. Bureau of Mines.

Abbreviations: Mt or tt, million or thousand tons; tmt, thousand metric tons; Mct, million carats.

for 45 percent of that. The remaining 4 percent of total consumption is for industrial sand, most of which is silica sand.

Sand and gravel deposits in the United States originate due to stream, glacial, lake, marine, and residual processes. Stream gravel is the most widely used, but lake and marine deposits are the best sorted while glacial deposits are the most poorly sorted. The sand and gravel operations are located close to the point of use to minimize transportation costs. The sand and gravel industry offers examples of both good and poor planning as to location. There are still good sand and gravel deposits located within metropolitan Chicago, for example, but they have been buried as the city expanded. Now it is necessary to bring sand and

TABLE 6–2 United States and world reserve/demand data for industrial materials, 1983–2000.

Material	U.S. Reserve U.S. Demand	World Reserve World Demand
Barite	11 million t 60 million t	160 million t 150 million t
Diamonds (stone)	0 60 million ct	600 million ct 460 million ct
Diatomite	250 milion t 10 million t	800 million t 29 million t
Feldspar	Adequate 13 million t	Adequate 72 million t
Garnet	5 million t 55,000 t	8.1 million t 860,000 t
Graphite	0 400,000 t	15 million t 7.8 million t
Gypsum	800 million t 480 million t	2.6 billion t 1.9 billion t
Perlite	50 million t 11 million t	700 million t 35 million t
Pumice	Adequate 13 million t	Adequate 260 million t
Sand and gravel	Adequate 14 billion t	Adequate 14 billion t
Stone	Adequate 18 billion t	Adequate Adequate
Vermiculite	25 million t 6.3 million t	50 million t 11 million t

SOURCE: U.S. Bureau of Mines.

Abbreviations: t, short tons; ct, carats.

gravel in from 60 to 100 miles away. In Austin, Texas and Albuquerque, New Mexico, however, sand and gravel operations have been planned so that the most economic lifetime of an operation is near completion as the city growth reaches it, and then switched to an operation further removed. In this way, the problems of dust, noise pollution, truck flow, and other necessary parts of the operation do not impact on the highly urbanized area.

In the next century, it is anticipated that offshore sand and gravel deposits along the Atlantic and perhaps part of the Pacific coast will be dredged to meet increasing demand, especially where land-based operations are limited due to environmental concerns (noise, dust, and traffic). These deposits are in the United States EEZ, and locations are given in Figure 4–13. The U.S. Bureau of Mines estimates that the United States will consume 1 billion tons of sand and gravel by the year 2000 (Table 6–1).

Industrial or silica sand is a very diversified material, used for many applications. Very pure quartz sand is termed glass sand because of its use in the manufacture of glass. Quartzite sand is used most commonly as refractory sand in brick making. Quartz sand is also used for abrasive sand, but is not as pure as glass sand. Foundry sand, used for molding in the steel and cast iron industries, has a mixed mineralogy, but must be able to be useful for making molds. Some sands are used as hydraulic fracturing sands to help oil and gas drilling. Silica flour is finely ground sand used for fillers in paint, ceramics, plastic, and other uses.

Geologically, the source rocks for these sands are pure quartz arenite or quartzite and quartzite sandstone. The Ridgeley Formation (or Oriskany Sandstone) of Early Devonian age is the main source rock in the eastern United States, extending from New York to Virginia. The St. Peter Sandstone (Middle Ordovician) is the leading source material in the Midwest, occurring in parts of Wisconsin, Minnesota, Iowa, and Illinois. The Ione Formation (Eocene) of California and the Tejon (Eocene) and Silverado Formations (Paleocene) are the most important in the West.

Use of silica sand is expected to increase to 38 million tons by the year 2000. As shown in Table 6–1, silica sand, because of its diversity of uses, did not suffer as great a decline during the recession of the early 1980s as many other commodities.

Supplies of silica sand arc adequate to meet the demands of the United States well into the next century.

CRUSHED STONE

Crushed stone is the largest nonfuel mineral tonnage crop in the United States, with consumption in 1983 at 863 million tons. Crushed stone means rock that, after quarrying, is crushed, broken, or ground to specific sizes. Crushed stone is used primarily in construction work on highways, railroads, and foundations for fill.

The data in Table 6–1 show that the United States demand has stayed between 800 and 900 million tons from 1975 to 1983, and is expected to reach 1.3 billion tons in 2000. Of this, about 65 percent will be for the construction industries, 12 percent for cement and lime, and the balance for agricultural, metallurgical, and other industries. Every state except North Dakota and Delaware produces crushed stone.

Crushed stone may be produced from a great variety of native stone. The site is usually located as close as possible to the intended use to keep transportation costs down. Granite, gneiss, limestone, dolostone, sandstone, trap (basalt, diabase, or gabbro), and some others may be used as crushed stone.

Environmental concerns are those associated with any large surface operation near urban areas, namely noise, dust, and discharge of water used in processing; visual appearance is an issue in some areas. Five percent of crushed stone is produced from underground workings, in part due to the above factors.

Only very small amounts of crushed stone are exported. Supplies of crushed stone are essentially inexhaustible.

DIMENSION STONE

Dimension stone refers to rock quarried and then shaped into slabs, blocks, or other shapes, usually to specific measurements. Most dimension stone is used for construction, but some is used for monuments and curbing. The United States consumed about 1.8 million tons of dimension stone in 1983, one-third of which was imported.

The most abundantly consumed dimension stone is granite (52%), followed by limestone (21%), sandstone (15%), marble (12%), and slate. Granite is quarried mainly in New England and Georgia, while Indiana, Minnesota, Texas, and Wisconsin produce mostly limestone. Ohio, Pennsylvania, and New York produce sandstone; Georgia, Vermont, and Idaho quarry marble, and slate is produced in Vermont and Virginia. The dimension stone industries in the United States are expected to show an increase in consumption to the year 2000 (Table 6–1), but imports will play an increasing role. The industries have been coerced into a position of modernizing their operations in order to stay competitive with foreign sources using the latest equipment. Limestone use in construction continues to be phased down, due in part to its poor resistance to chemical attack, which is being augmented by acid rain in many parts of the country. Alternative building materials continue to substitute for dimension stone.

CEMENT

Cement is a fired and crushed mixture of carbonate rocks and clay minerals (or equivalent) that has strong binding properties when wetted with water. Mixed with sand, gravel, or aggregate, cement yields numerous kinds of concrete.

Portland cement is the type accounting for 90 percent of cement used in the United States; the name is derived from Portland, England. Cement ranks third in terms of dollar value of finished product, after iron and steel and aluminum.

While specific cements have defined chemistries, a generic Portland cement can be made as follows: 2500 lb magnesian limestone plus 700 lb aluminosilicate material, fired in a kiln at 1500°C, yields one ton of Portland cement clinker. When crushed, and with the addition of small amounts of gypsum, one ton of Portland cement results. The aluminosilicate can be clay, shale, or fly ash from coal-fired power plants, or equivalent material (see Figure 6–1 for detail).

Limestone capping granite, Sandia Mountains, New Mexico. The limestone contains enough shale to make it valuable for nearby cement manufacturing.

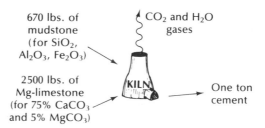

670 lbs. of mudstone (for SiO_2, Al_2O_3, Fe_2O_3)

2500 lbs. of Mg-limestone (for 75% $CaCO_3$ and 5% $MgCO_3$)

CO_2 and H_2O gases

KILN

One ton cement

FIGURE 6–1 Portland cement process. The ingredients and proportions shown here must be fairly constant for high-quality cement. Argillaceous limestones sometimes can be used directly as feed; more likely, however, fairly pure limestone must be mixed with mudstone (clay or shale) to provide the proper mix. This crushed material is roasted in a kiln at a temperature just under 1500°C. This drives off the carbon dioxide and water as gases and leaves behind a partially molten *clinker* of silicate-oxide material. When crushed to a fine size, Portland cement is the result. To produce concrete from this, sand and gravel are added. (Source: Brookins, Merrill, 1981)

The USSR is the world's leading producer of cement (141 million tons in 1983), followed by China (119 Mt), Japan (89 Mt), and the United States (74 Mt). The growth in the United States will be 1.6 percent per year to 2000 (see Table 6–1) whereas the world growth will be twice that at 3.3 percent per year. The United States possesses ample raw materials to meet domestic demand into the next century, although a small amount is imported yearly. World cement production figures are given in Table 6–3.

Environmental problems include the usual surface factors such as dust, noise, traffic, and unsightliness, none of which are serious, and occasional pollution of streams. Competition for land for raw materials is also problematic. The

TABLE 6–3   World cement production, 1983, and clinker capacity, 1983, 1984, and 1990 (million tons).

	Production 1983	Clinker Capacity		
		1983	1984	1990[1]
North America				
United States	71	89	90	92
Canada	9	17	17	18
Mexico	19	29	31	32
Other	8	15	15	17
Total	107	150	153	159
South America				
Argentina	6	11	11	12
Brazil	26	33	39	41
Venezuela	5	7	7	9
Other	29	35	36	40
Total	61	79	86	93
Europe				
Belgium	6	8	8	10
Czechoslovakia	12	12	12	14
France	27	30	30	33
Germany (DRG)	13	14	14	15
Germany (FRG)	34	45	46	48
Greece	9	12	12	14
Italy	43	55	55	58
Poland	18	20	20	22
Portugal	7	8	8	10
Romania	17	20	20	22
Spain	34	39	39	41
USSR	141	154	154	160
United Kingdom	15	20	20	23
Yugoslavia	10	16	15	16
Other	38	56	57	67
Total	424	509	510	553

TABLE 6–3   Continued

	Production 1983	Clinker Capacity		
		1983	1984	1990[1]
Africa				
Egypt	7	8	14	18
South Africa	9	12	15	17
Other	28	40	44	40
Total	44	60	73	75
Asia				
China	119	125	125	128
India	28	33	36	38
Indonesia	9	11	13	15
Iran	11	14	14	15
Japan	89	120	121	125
North Korea	9	10	10	11
South Korea	23	24	25	25
Philippines	5	10	11	12
Taiwan	16	17	17	18
Thailand	8	8	8	10
Turkey	15	22	24	27
Other	53	70	81	91
Total	376	464	485	515
Oceania				
Australia	7	8	8	10
Other	1	2	2	4
Total	8	10	10	14
World total	1,020	1,272	1,317	1,409

[1]Forecast.

SOURCE: U.S. Bureau of Mines.

cement plants spew out large amounts of water and carbon dioxide, which, while of little actual harm, tend to alarm individuals living near these operations. The cement industry has spent well over $1.5 billion to meet the EPA control standards for water and air pollution from their operations, apparently with good success.

CLAYS

Clay is the generic name for a wide variety of rocks rich in one or more clay minerals. Three important clay minerals mined in the United States are kaolinite, montmorillonite, and illite. Their formulas are given in Table 6–4, and the differ-

TABLE 6–4   Clay mineral formulas.

Clay Mineral	Approximate Formula
Kaolinite	$Al_2Si_2O_5(OH)_4$
Illite	$K_{1-x}(Mg,Fe,Al)_2(Al,Si)_4(O)_{10}(OH)_2$
Montmorillonite (smectite)	$(Na,Ca)_{0.33}(Al,Mg)_2Si_4O_{10}(OH)_2 \cdot nH_2O$
Chlorite	$(Mg,Fe,Al)_6(Al,Si)_4O_{10}(OH)_8$

ent industrial types of clays described in Table 6–5. The United States contains abundant supplies of shale (often rich in montmorillonite and illite) and fire clay, but smaller amounts of ball clay, kaolin, fuller's earth, and bentonite. The United States not only contains excellent deposits of these clays, but has developed the world's best processing and technology to deal with them.

Clay minerals are common in many sedimentary rocks: shale, argillaceous (clayey) limestone and arkose and graywacke, and as residual deposits in the weathering of rocks: kaolinite from granitic or volcanic rocks, montmorillonite from volcanic rocks, and others. Kaolinite deposits in the southeastern United States are predominantly sedimentary, although residual kaolinites are found in North Carolina. Ball clays from Tennessee and Kentucky are sedimentary, as is the fire clay found in Missouri. Many volcanic ash or tuff deposits alter to ben-

TABLE 6–5   Classification of clays.

Kaolin ("china clay"): Essentially pure kaolinite ideal for ceramic products.

Fire clay: Impure kaolinite with variable amounts of organic carbon associated with it; ideal for refractory material which must be fired at high temperatures without deformation.

Bentonite and fuller's earth: Varieties of montmorillonite: Bentonite is derived from volcanic ash and contains abundant Mg; in addition, Ca is common, K less so, and Na is usually scarce. Fuller's earth, on the other hand, is typically a Na-montmorillonite. Both bentonite and fuller's earth are widely used as drilling muds although in very different areas.

Ball clay: Kaolinite which contains some illite or montmorillonite; its name is derived from its plasticity, which allows it to be rolled into 30- to 50-pound (14- to 23-kilogram) balls. After firing, it is used in tiles and refractories.

Miscellaneous clay: Illite-rich clay and shale; most consumption as fired-clay component of tiles, piping, and cement additives.

SOURCE: Brookins, *Earth Resources, Energy, and The Environment,* Merrill Publishing Co., 1981.

tonite, fuller's earth, or other montmorillonites. Many units (such as the Morrison Formation of the western United States, an argillaceous sandstone) contain all three major clay minerals (see Figure 6–2). Some fuller's earth found in the southeastern states is of sedimentary origin.

Kaolin is widely used in industry, because of its chemical inertness over a wide range of pH and its white color, as a filler and binding agent. The paper industries consume abundant kaolin. Ball clay is used in refractories, such as white-wares. Fire clay is used in firebrick and other similar materials. Fuller's earth and sodic bentonites are used as a prime ingredient of drilling mud in areas where the waters contain abundant sodium, as in the Gulf of Mexico or the Gulf coastal states. In the western states, where the waters are more rich in calcium and magnesium, calcic-magnesium bentonites are used for the same purposes. In both cases, advantage is made of the fact that the clays will not exchange cations with the surrounding waters if their exchangeable cations are identical; if the wrong mud is used, the ions will exchange and the mud will bind up so that the hole closes. Fuller's earth and other bentonites are also widely used in purifying and decolorizing various animal, vegetable, and mineral oils. Illites are the most common clay minerals in miscellaneous clays, although other clay minerals may be present. Miscellaneous clays are used in industry for aggregates, the cement industry, the iron and steel industry (for processing of taconite ores), construction, and other applications.

The United States consumed 38.6 million tons of clay in 1983, and is expected to consume 63 million by the year 2000. Supplies are adequate to meet demand well into the next century. Worldwide, clays of most kinds are available to most communities.

FIGURE 6–2   Scanning electron micrograph showing three types of important clay minerals: the pseudohexagonal booklets are kaolinite; the honeycomb material is montmorillonite; and the rosette material is chlorite. All three clays are common in the Morrison Formation of the western United States. (Source: Brookins, Merrill, 1981)

ASBESTOS

Asbestos is a widely used industrial material. It is an important ingredient in floorings, certain cement pipes, friction products (such as brake drums), coatings, roofing, and compounds. No other industrial material has received as much attention over the last decade due to environmental concerns, which are discussed in depth in Chapter 8. Asbestos is a necessary ingredient with no viable alternatives for many of the purposes listed above. Production in the United States has decreased steadily over the last decade due to cost and environmental factors, but worldwide asbestos use is growing rapidly.

The largest deposits of asbestos in the world are in the USSR (see Table 6–6), while other large deposits are in Canada. These two countries should monop-

TABLE 6–6 World asbestos reserves.

Continent/Country	Thousand metric tons
Chrysotile	
North America	
Canada	829
United States	70
South America	
Brazil	135
Others	7
Europe	
USSR	2250
Italy	120
Greece	93
Other	11
Africa	
Zimbabwe	190
South Africa	92
Swaziland	11
Asia	
China	110
India	25
Others	40
Australia	20
Crocidolite	
South Africa	87
Amosite	
South Africa	41

SOURCE: U.S. Bureau of Mines.

olize production in the near future. The United States imports 70 percent of its need, mainly from Canada and South Africa.

Asbestos, the name for fibrous minerals like hydrated magnesium silicate, $Mg_3Si_2O_5(OH)_4$, occurs in nature as three common polytypes: chrysotile (white asbestos), crocidolite (blue asbestos), and amosite (brown asbestos). Crocidolite is found in minable amounts only in South Africa and Australia, while amosite is found only in South Africa. The most common variety is chrysotile, which is common in many parts of the world. Chrysotile is found in veins in serpentine, whereas crocidolite and amosite are found in banded ironstones. Only crocidolite has been linked to cancer of the lung lining (see Chapter 8), but concern about this has been passed on to all asbestos and asbestos-like minerals (such as the common rock-forming mineral group, the amphiboles), with the result that many taxpayer dollars are being spent to remove all asbestos (most of it the harmless white chrysotile) from schools, offices, and other buildings. Two asbestos mines in the United States have closed due to environmental concerns. The EPA's decision to consider "asbestos-like" materials as identical to asbestos is a prime example of enviroscare (see Chapter 8).

BARITE

Barite ($BaSO_4$) is an important weighting agent in drilling muds, primarily used by the oil and gas drilling industries. Over 95 percent of its use is in this category. As an additive to drilling mud, barite serves many purposes. It lubricates and cools the drill bit, serves as a transporting agent for cuttings, maintains formation pressures, keeps the drill hole stable, and protects (oil and gas) producing zones.

There are large reserves of barite in the United States, but not sufficient to meet projected demand. Some 96 percent of domestic barite is imported, mainly from China, Morocco, Chile, and Peru (see Table 6–7). Substitutes for barite will have to be thoroughly investigated in the near future; these include barite-hematite mixtures and some iron ores.

Barite occurs in bedded deposits of uncertain origin, along with chert and siltstone; in epithermal veins in many places; as replacement bodies in limestone, and in some other residual deposits such as the bedded deposits of Nevada, Arkansas, and California. Nevada produces about 60 percent of the U.S. total.

The only environmental concern of barite is the waste rock from open pit barite mines, but this is a minor problem. Some barium compounds are highly poisonous if ingested, but the amounts are small and easy to monitor.

Barite consumption closely follows oil and gas activities. In peak periods of drilling, barite consumption is very high. The U.S. Bureau of Mines predicts that roughly 4200 thousand tons of barite will be needed in 2000, of which over 75 percent will be met by imports. While world reserves of barite are only marginally able to meet demand by 2000, there are indications that new deposits in North America, China, and India may be large.

TABLE 6–7   World barite reserves.

Continent/Country	Million tons
North America	
United States	30
Other	10
South America	
Several	6
Europe	
USSR	10
Other	18
Africa	
Morocco	10
Other	3
Asia	
China	40
India	34
Thailand	9
Other	12
Oceania	1

SOURCE: U.S. Bureau of Mines.

LIGHTWEIGHT AGGREGATE

Perlite

Perlite is glassy to partly devitrified volcanic material found in volcanic domes, welded ash-flow tuffs, and in contact zones to intrusive plugs and dikes. It is an important insulating material, and has the advantage that, when heated to fairly high temperature, it expands to roughly 20 times its original volume. Thus it can be shipped in small volume, then heated and expanded prior to use at its destination. The expanded material may be highly porous, fluffy material or glazed glassy material with low porosity, depending on the nature of the starting material.

Perlite is used for many things. In the construction industry it is used as insulation, acoustical tile, and plaster, as aggregate for concrete, and in wallboard. It is a common filler in paints and plastics. In agricultural industries it is used as a fertilizer dilutant and carrier for insecticides. Breweries use perlite to filter out impurities. It is also used as a sorbent for oil or paint spills.

Domestic reserves are adequate to well into the next century. The only environmental effect of perlite production is the generation of very fine dust, which can be handled by installing closed collection systems to meet emission standards.

The United States consumed 493 thousand tons of perlite in 1983, .and this is expected to rise to 850 thousand tons by 2000.

Pumice

Pumice (including so-called pumicite) is a porous, volcanic material formed by explosive acid igneous activity, especially where significant horizontal flow occurs. Deposits can be subaerial or subaqueous, and are often reworked. Because these are easily weathered and altered in other ways, only Tertiary to recent deposits are of interest. The United States has abundant supplies to last well into the next century. Anticipated demand in 2000 will be 950 thousand tons (Table 6–1).

Pumice is used mainly in construction as insulation, concrete aggregate, building blocks, landscaping, and industrial abrasives. The uses are very similar to those for perlite.

ABRASIVES

Diamond

Industrial diamonds are primarily those that for reasons of color or impurities are not suitable as gemstones. Diamond is especially valuable for drilling, cutting, and similar applications because it is the hardest known naturally occurring substance. Rock bits account for roughly 75 percent of domestic use, of 3.2 million carats consumed in 1983. The United States has the technology to produce artificial diamonds and uses it to produce diamond powder. It exports diamond powder, but imports all coarser material as there are no economic deposits of diamond in the United States.

Diamond larger than 16 to 20 mesh is referred to as stone, from 16–20 to 325–400 mesh as bort, and below 325–400 mesh as grit or powder. The United States imports 100 percent of the diamond stone used in the drilling, cutting, and other industries, but produces 100 percent of the diamond powder from artificially grown diamonds. For example, in 1983, 39.2 million carats of diamond powder was consumed in the United States, while production was 69 million carats.

Diamonds are found in an ultramafic rock known as kimberlite and rarely in lamproite, and in numerous alluvial occurrences. Kimberlites and lamproites form in the earth's mantle, perhaps as deep as several hundred kilometers, where diamond can form, along with other high-pressure minerals such as pyrope, chromian spinel, and Mg-ilmenite. Prior to the last decade, African countries dominated the world picture, but now Australia possesses the largest reserves. Diamonds have been found in the United States in Arkansas (lamporite), along

Kimberlite boulder on top of kimberlite-rich regolith. Murfreesboro, Arkansas. Diamonds have been recovered from this area since the early 1900s, although the area is now a state park.

the Wyoming-Colorado state border (kimberlite), in Kansas (kimberlite), and in alluvial occurrences in most of the conterminous 48 states and Alaska. No economic source of diamonds has been found, however.

World reserves of diamond stone are adequate to meet demand to 2000, and there is an abundance of diamond powder from industrial sources. The opening of the large Australian diamond fields has taken away some of the world's concern over possible loss of industrial diamonds from African countries.

Corundum and Emery

Much of the world's corundum, aluminum oxide (Al_2O_3), is from metamorphic corundum granulites formed in Archean greenstone belts in Africa, India, South America, and in the USSR. Because of its resistance to weathering, some corundum is recovered from stream and other placer deposits. World reserves are adequate to meet demand into the next century, although it is not present in sufficient deposits in the United States; thus 100 percent of corundum used in abrasives is imported, mainly from African and European countries.

Artificially produced alumina, as well as fused silica and diamond grit, have replaced much of the corundum. It is used for optical grinding, and for grinding and finishing metals. The United States will consume about 900 tons of corundum in 2000.

Emery is a natural mixture of magnetite and corundum. In the United States, fairly large deposits are found near Peekskill, New York, which are being worked. Inactive sites are in Virginia, Maryland, and North Carolina. The largest emery deposits in the world are located in Turkey, followed by Greece and the USSR. The New York deposits are part of the metamorphic Cortland Complex, a regionally metamorphosed gneiss.

Emery is used primarily in wear-resistant floors and pavement, in abrasives, and in tumbling-polishing apparatus. The U.S. demand for emery is expected to increase to 4500 tons by 2000, compared to 2200 tons consumed in 1983. The United States will import roughly one-third of this amount. Emery is faced with possible phase-down since some of the operations are located in rural residential areas, where concerns over noise and dust have forced operational cutbacks.

Garnet

Garnet is widely used as an industrial material, much of it the iron-aluminum variety almandine. The largest reserves of garnet are in the United States, which is the leading producer of industrial garnets in the world (75%), and also the largest consumer (70%). Domestic production is entirely adequate to meet demand to 2000 and beyond.

View of part of the Hale pegmatite quarry, Connecticut. First mined for gem tourmaline and beryl and later for uraninite, the abundant quartz-feldspar rock is now mined for the nearby glass manufacturing industry in Portland-Middletown, Connecticut.

FELDSPAR

Feldspars, as used in industry, are predominantly albite and potassium feldspar, and to a much lesser extent calcic plagioclase. Feldspars are used in the glass industry primarily, but also in making pottery and in other parts of the ceramic industry. The best feldspar ores are from granitic pegmatites, which are common in many places in the United States. Production comes primarily from pegmatites in North Carolina, Connecticut, Georgia, and South Dakota. In Oklahoma and California, feldspar is recovered from high-feldspar, low-iron arkosic sandstones. Feldspar use in filler applications is increasing.

Domestic reserves are very large, and adequate to meet demand well into the next century. The same is true for much of the rest of the world. The U.S. demand is expected to reach 800 thousand tons by 2000, up 14 percent from 1983 (Table 6–1).

DIATOMITE

Diatomite, or diatomaceous earth, forms from accumulation of tiny marine organisms known as diatoms which secrete silica shells. These deposits are restricted to the Tertiary and Quaternary and are most common in lake deposits, although marine deposits are much larger. The United States contains the major diatomite reserves of the world, about 30 percent of those known. Collectively, European deposits are larger.

Most of the consumption of diatomite around the world is for filtering of water and other liquids. This accounts for over 60 percent of total consumption, and includes standard water treatment plants, purification of beverages, and waste-water treatment. Other uses include fillers for the paint, plastic, and related industries. The U.S. reserves are more than adequate to meet local demand to 2000 and also allow for export. Demand is expected to grow to 750 thousand tons in 2000, up nearly 60 percent from 473 thousand tons in 1983. Diatomite has a very low density, hence the cost of transportation is high. There are no significant environmental problems associated with diatomite production in the United States.

GYPSUM

Gypsum, $CaSO_4 \cdot 2H_2O$, is a common constituent of both marine and nonmarine evaporites, and forms on many surfaces by evaporation as well. Natural occurrences of these are very large, and the United States leads the world in gypsum reserves of some 800 million tons. Canadian reserves are also large, as are those in France and England.

Gypsum is a common and important building material, used especially in ceilings, wallboard, and partitions. When calcined, it forms plaster of Paris

FIGURE 6–3 World gypsum production, 1963–83 and projected to 2000. (Source: U.S. Bureau of Mines)

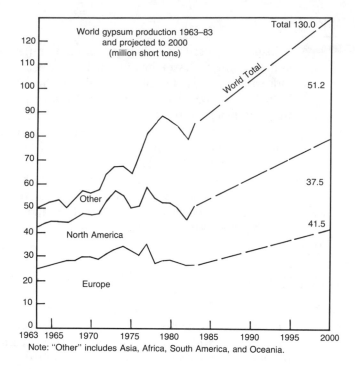

World gypsum production 1963–83 and projected to 2000 (million short tons)

Total 130.0

World Total

51.2

37.5

Other

41.5

North America

Europe

Note: "Other" includes Asia, Africa, South America, and Oceania.

$\left(CaSO_4 \cdot \frac{1}{2}H_2O \right)$ which, when wetted, reverts to gypsum and hardens. Gypsum is also used for fillers in paint and plastics, as a cement additive to retard setting, and as an additive to soils to neutralize saline and alkaline conditions. Despite a decline in production over the period 1975–1983, the U.S. demand is expected to increase significantly to 34 million tons by 2000 (Figure 6–3). The United States is a net exporter of gypsum and gypsum products.

GRAPHITE

Graphite occurs as fissure vein fillings, and as flakes, in metamorphic rocks. So-called amorphous graphite is the product of metamorphism of anthracite coal. The United States contains no workable deposits of graphite, and relies on imports from Sri Lanka, southern African countries, and others.

Graphite is used in the iron and steel industries, primarily as foundry facings and refractories. Other uses include lubricants, batteries, pencils, brake linings, and a wide variety of miscellaneous uses. The U.S. demand for graphite is expected to increase by 33 percent from 1983 to 2000, with 63 thousand tons predicted by the U.S. Bureau of Mines for 2000.

VERMICULITE

Vermiculite forms from other ferromagnesian silicates, mainly biotite. It has a very expandable structure in response to chemical changes, water sorption, or heating. As expanded vermiculite, it has wide use in several industries. It is used as a lightweight aggregate, as well as for thermal insulation in construction operations. It is used as a soil conditioner and as a carrier for fertilizer in agriculture. It is also used as a filler in the paint, plastic, and rubber industries and for a wide variety of other applications.

The largest domestic reserve (of approximately 25 million tons total for the United States) is at Libby, Montana, where a Cretaceous biotite pyroxenite cuts Precambrian sedimentary rocks. The core is very pure biotite, and it is surrounded by biotite-rich pyroxenite; much of the surface and near-surface biotite is altered to vermiculite. Ultramafic igneous and some metamorphic rocks elsewhere in the United States contain small vermiculite deposits as well. Outside the United States, the Palabara complex in South Africa contains some 20 million tons of vermiculite reserves. Together, the United States and the Republic of South Africa account for 90 percent of total world vermiculite reserves. United States production is expected to increase almost 60 percent between 1983 and 2000, as shown in Table 6–1. Supplies are more than adequate to the next century (Table 6–2).

GEMSTONES

Gemstones originate predominantly outside the United States. Diamonds are from South Africa, Botswana, Angola, Namibia, Zaire, the USSR, Australia, Venezuela, and Brazil. More than 80 percent of the diamonds sold as gems are controlled by the DeBeers Corporation of South Africa. Other gems and their main countries of reserves are as follows: ruby: Burma, Sri Lanka, Brazil; emerald: Colombia and Zambia; beryl: Afghanistan, Brazil, Burma, Sri Lanka, Madagascar; sapphire: Australia, Brazil, Burma, Madagascar, Sri Lanka; topaz: Brazil, Burma, Sri Lanka, Mexico; opal: Australia, Mexico. Detailed discussion of semiprecious stones can be found in the references at the end of this chapter.

The United States imported about 6200 thousand carats of diamonds, 2150 thousand carats of emeralds, and 6200 thousand carats of sapphires and rubies in 1983. The value of all these is roughly $2.3 billion. The supply of foreign gems appears to be adequate, as long as people are willing to invest in or buy gems. It is probable that gem prices, at least for diamond, are highly inflated and carefully controlled by the Diamond Trading Company, which not only sets diamond prices but also is a wholly owned subsidiary of DeBeers.

MICA

Mica as an industrial material refers mainly to muscovite ($KAl_3SI_3O_{10}(OH)_2$) but sometimes to phlogopite ($KMg_3AlSi_3O_{10}(OH)_2$). Mica is described as scrap, flake, or sheet varieties for commercial purposes. Sheet mica, mainly muscovite, is obtained from coarse-grained pegmatites. While sheet mica is fairly common in the United States, its recovery is expensive, and 100 percent of sheet mica is imported from India, Brazil, and other countries where labor costs are low. Flake mica is recovered as a by-product of feldspar mining from pegmatites in North Carolina and elsewhere, and it is commonly recovered from muscovite schists as well. Scrap mica is predominantly excess sheet mica recovered after trimming.

Sheet mica is important to the electrical and electronic industries, and varieties of sheet mica are used in vacuum tubes, heater plates, insulators, and for many other purposes. Substitutes for sheet mica as well as the phasing out of vacuum tubes have caused marked decrease in sheet mica consumption, and this trend is expected to continue (Table 6–1).

Flake and scrap mica are used in plasterboard, in paints, drilling muds, rubber, roofing, asphalt, and in other areas. Demand for flake and scrap mica is expected to increase somewhat for the United States (Table 6–1) as well as the world for the period 1983–2000.

FURTHER READINGS

Bates, R. L. 1969. *Geology of the industrial rocks and minerals.* New York: Dover, 178 pp.

Bates, R. L., and Jackson, J. A. 1982. *Our modern stone age.* Palo Alto: William Kaufmann.

Boynton, R. S. 1980. *Chemistry and technology of lime and limestone,* 2d ed. New York: John Wiley & Sons.

Flawn, P. T. 1970. *Environmental geology.* New York: Harper and Row.

Gillson, J. L., et al. 1960. *Industrial minerals and rocks,* 3d ed. New York: American Institute of Mining, Metallurgical and Petroleum Engineers.

Griggs, G. B., and Gilchrist, J. A. 1977. *The earth and land use planning.* North Scituate, Mass.: Duxbury Press.

Harben, P. W., ed. 1977. *Raw materials for the glass industry.* London: Metal Bulletin Ltd.

Knill, J. L., ed. 1978. *Industrial geology.* London: Oxford University Press.

Lefond, S. J., ed. 1975. *Industrial minerals and rocks,* 4th ed. New York: American Institute of Mining and Metalurgical Engineers.

Legget, R. F. 1973. *Cities and geology.* New York: McGraw-Hill.

Taylor, G. C. 1981. California's diatomite industry. California Geology v. 34.

Utgard, R. O., McKenzie, G. D., and Foley, D. 1979. *Geology in the urban environment.* Minneapolis, MN: Burgess Pub. Co.

Winkler, E. M. 1973. Stone: Properties, durability in man's environment. New York: Springer-Verlag Pub.

Yundt, S. E., and Augaitis, B. E. S., 1979. From pits to playgrounds. Toronto: Ontario Ministry of Nat. Res.

7
Energy

INTRODUCTION

This chapter deals with many aspects of energy from consumption to alternate energy sources. Energy is a very complex topic, and this chapter's length reflects the difficult task of attempting to cover many aspects of energy applicable to today's society. The focus is on the occurrence, location, exploitation, and availability of the various energy resources. In the United States there is a very great reliance on petroleum, mainly oil and gas, yet domestic production has peaked. World production may peak in the early part of the twenty-first century, and then alternatives to oil and gas will be forced upon us.

Of necessity, a good deal of this chapter is devoted to a comparison of the major energy options in terms of environmental aspects, public health, and related topics. Availability of resources goes hand in hand with many parts of this discussion, as do topics such as disposal of wastes from all energy options and risks involved from all aspects of the energy cycle. This chapter also focuses on some aspects of risk assessment, public perception of risk, and factors that affect this perception.

Because of the public's concern over nuclear power, I have included an expanded treatment of the nuclear fuel cycle and a discussion of the natural radiation background. Of perhaps special interest, the growing concern over naturally occurring radon gas and indoor radon problems are covered in some detail.

Additional topics include alternative energy sources such as geothermal, nuclear fusion, biomass, and oil shale.

Energy Projections by Sector

Residential energy needs are expected to increase very slightly through 1995–2000, according to the U.S. Department of Energy's *Annual Energy Outlook*. The reduced energy use for the residential sector is due primarily to more energy ef-

193

ficient appliances, more insulation and sealing of structures, and conservation measures. Further, part of the energy decline and tapering off reflects the shift in population to the warmer regions of the West and South. The slight increase in residential-sector energy use will be due to increased use of electricity, mainly for air conditioning and refrigeration. Natural gas use in the residential sector will stay roughly constant through about 1990, then decline somewhat due to an anticipated price increase. Similarly, petroleum for space heating and other uses will also decline.

In the commercial sector most energy is consumed by structures, including warehouses, offices, and public buildings. The energy use in this sector is proportional to floor space, and as floor space is projected to increase about 2 to 2.5 percent annually from 1985 to 1995, then energy consumption in this sector will probably rise similarly. A combination of increased energy prices and more efficient building designs will keep the energy growth down in this sector. There will, however, be an increased reliance on electricity from 39 to 44 percent between 1985 and 1995.

The industrial sector is the largest consumer of energy in the United States, and this sector is especially sensitive to energy prices, supply, and other considerations. Energy growth, estimated to be 2.7 percent per year between 1985 and 1990, will be 1.3 percent from 1990 to 1995. The decline in the period 1990–1995 will be a combination of higher energy prices plus replacement of equipment with new energy-efficient models.

Electricity provides 26 percent of the total energy demand in the United States. After the Arab oil embargo of 1973, utility companies set out to reduce reliance on oil and gas for generating electricity. As shown in Figure 7–1 and Table 7–1, coal will provide about 55 percent of total electricity generation in 1990 (roughly the same as in 1985), nuclear power will increase from 14 percent (1985) to 20 percent (1990), and oil and gas will provide only 15 percent in 1990 compared to about 18 percent in 1986. Hydroelectric power will provide about 10 percent in 1990 compared to 14 percent in 1986.

Demand for electricity is growing at about 3 percent per year. Generation of electricity is crucial to the future of the United States: Not only are petroleum resources on the decline, but it is probable that coal use will decrease for a variety of environmental reasons, as discussed in Chapters 3 and 8. Hydroelectric power requires water, but because most available water for such facilities has already been tapped, its potential for growth is limited. Geothermal will continue as a small contributor, along with solar, wind, and other minor sources in the national picture. If demand must continue to rise, then only nuclear power can provide the increased electricity demand with the least disruption or long-term effects on the environment.

Interestingly, petroleum use in industry will increase, reaching 48 percent of total use in this sector by 1990. This reflects in part the growth in plastics manufacturing and in other industries using petrochemical feedstocks. The principal industries, which account for over 80 percent of manufacturing energy for heat and power (exclusive of petroleum refining), are the chemical and rubber indus-

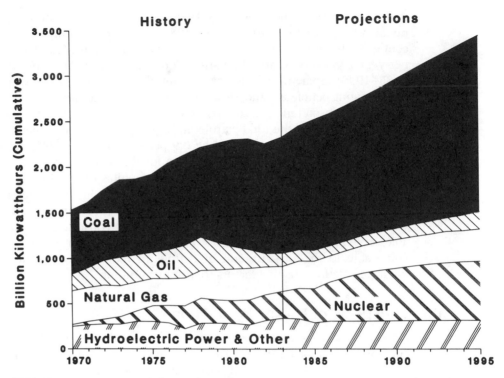

FIGURE 7–1   Sources of electrical supply, 1970–1995. (Source: Energy Information Administration, *Annual Energy Outlook 1984)*

TABLE 7–1   Electricity generation by type, 1973–1995.

Energy	\multicolumn{7}{c}{Gigawatts at Year's End}
	1973	1978	1983	1984	1985	1990	1995
Coal	184	234	286	296	304	329	360
Nuclear	21	53	64	69	81	110	117
Oil and Gas							
Steam	135	161	157	157	157	157	157
Turbine	37	50	51	51	51	53	59
Hydroelectric	64	76	79	81	83	88	88
Other (mainly geothermal)	—	—	3	4	4	4	4

SOURCE: U.S. Department of Energy, *Annual Energy Outlook 1985.*

tries (4.8 percent growth per year 1985–1990), steel, aluminum, and other base metals (4.5 percent growth per year 1985–1990), paper-pulp industries (3.0 percent growth per year 1985–1990), stone-clay-glass industries (4.1 percent growth per year 1985–1990), and the various food industries (2 percent growth per year 1985–1990). Some industries are attempting to convert coal power for their operations, but petroleum and electricity will continue to dominate.

The transportation sector relies very heavily on petroleum, accounting for about 60 percent of the total petroleum consumption in the United States. Annual energy growth in this sector will be 0.9 percent in 1990–1995, according to the Department of Energy. Part of the reason for this slight increase is that more and more vehicles are energy-efficient, and automobiles in 1995 may be, on the average, 20 percent more fuel-efficient than in 1985. In 1986 the speed limit on U.S. highways was raised from 55 miles per hour to 65 mph on nonurban interstate highways. The effect of this increase on fuel consumption will be considerable, as will the increase of traffic fatalities (see Chapter 8). No sector is more vulnerable to the Organization of Petroleum Exporting Countries (OPEC) and other petroleum-sensitive factors than transportation.

International Aspects of Energy Production

Primary energy production is now nearly 300 quadrillion Btu (roughly 80 trillion kilowatt-hours), of which the United States produces about one third. In Figure 7–2 and Table 7–2 it is noted, however, that the increase for the United States and other countries in the western hemisphere between 1973 and 1983 was only 5 quadrillion Btu (quads) while significantly greater increases were noted in western Europe (12 quads), eastern Europe–USSR (18 quads), and the Far East (15 quads). A pronounced drop occurred in the Middle East (− 14 quads), and African energy production stayed about the same (about 1 quad). Any future use of energy requires an estimation of reserves for the major energy sources, and this will be covered below.

The estimated world oil and gas proven reserves are shown in Figure 7–3 and Table 7–3. Note that of the world total 698.7 billion barrels of oil, only 27.3 billion is accounted for in the United States. Similarly, for gas, the total for the United States is only 198 of the world total 3,402 trillion cubic feet. In all, only 17 percent of world petroleum occurs in the western hemisphere, and only 13 percent of the natural gas reserves. Very simply, the United States and other countries in the western hemisphere must become increasingly dependent on oil and gas imports from the eastern countries. The United States, for example, consumed almost 26 percent of world petroleum and gas in 1985 (Figure 7–3). Crude oil flow for 1985 is shown in Figure 7–4, and the staggering importance of the Middle East is obvious. As shown in Figure 7–5, the situation for natural gas is presently heavily dependent on supplies from the USSR for most of Europe, while the United States relies on domestic and Canadian supplies.

The picture for coal on an international scale is much more favorable to the United States, if coal should be continued as a major energy source. The United

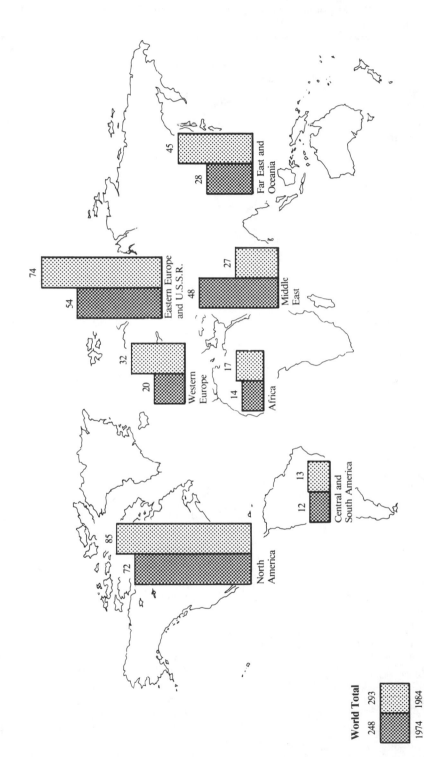

FIGURE 7–2 International primary energy production, 1973 and 1983, in quadrillion Btu. (Source: Energy Information Administration, *Annual Energy Review 1984*)

TABLE 7–2   World primary energy production by area and country, 1976 – 1986 (quadrillion Btu).

Area and Country	1976	1981	1986
North, Central, and South America			
Canada	8.68	9.81	11.98
Mexico	2.74	6.78	7.10
United States	59.81	64.30	64.09
Venezuela	5.79	5.58	5.13
Other	4.88	6.56	8.96
Total	81.89	93.03	97.26
Western Europe			
France	1.71	2.65	3.87
Netherlands	3.44	3.10	2.74
Norway	1.47	3.09	4.00
United Kingdom	5.39	8.71	10.53
West Germany	5.14	5.57	5.81
Other	5.19	6.63	8.67
Total	22.34	29.76	35.61
Eastern Europe and U.S.S.R.			
East Germany	2.25	2.53	2.96
Poland	5.00	4.54	5.69
Romania	2.39	2.54	2.74
U.S.S.R.	47.46	56.71	66.48
Other	3.13	3.35	3.82
Total	60.23	69.67	81.69
Middle East			
Iran	13.43	3.28	4.74
Iraq	5.25	2.16	3.63
Kuwait	4.84	2.70	3.33
Saudi Arabia	18.88	22.57	12.24
United Arab Emirates	4.20	3.45	3.80
Other	2.54	2.49	3.00
Total	49.13	36.65	30.75
Africa			
Algeria	2.65	2.85	3.26
Libya	4.26	2.57	2.42
Nigeria	4.51	3.81	3.29
South Africa, Republic of	1.84	3.09	4.29
Other	2.23	3.31	4.56
Total	15.49	15.00	17.83
Far East and Oceania			
Australia	3.35	3.89	5.68
China	15.35	18.10	24.96
India	2.88	4.09	5.85
Indonesia	3.38	4.27	4.35
Japan	1.86	2.27	2.99
Other	3.78	4.81	7.12
Total	30.59	37.42	50.96
World Total	259.67	281.54	314.09

SOURCE: Energy Information Administration, *Annual Energy Review 1987*.

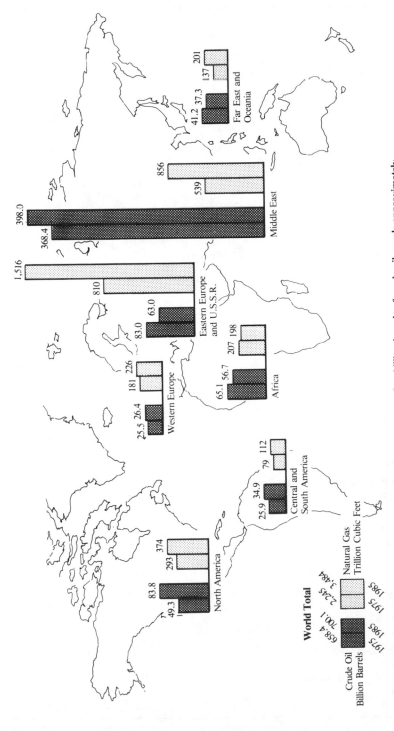

Note: Bars are scaled in proportion to the btu content of the reserves. One billion barrels of crude oil equals approximately 5.3 trillion cubic feet of wet natural gas.

FIGURE 7–3   Estimated international crude oil and natural gas proved reserves, December 31, 1984. (Source: Energy Information Administration, *Annual Energy Review 1984*)

TABLE 7–3   World crude oil and natural gas proved reserves, end of year 1977 and 1987.

Area and Country	Crude Oil (billion barrels)		Natural Gas (trillion cubic feet)	
	1977	1987	1977	1987
North America				
Canada	6.0	6.8	58	98
Mexico	14.0	48.6	30	77
United States	31.8	26.9	207	192
Total	51.8	82.3	295	366
Central and South America				
Argentina	2.5	2.3	8	24
Bolivia	0.4	0.2	5	5
Brazil	0.9	2.3	1	4
Chile	0.4	0.3	2	4
Colombia	1.0	1.6	6	4
Ecuador	1.6	1.6	5	4
Peru	0.7	0.5	1	1
Trinidad and Tobago	0.7	0.6	9	10
Venezuela	18.2	56.3	41	95
Other	([2])	0.1	0	([3])
Total	26.4	65.7	78	150
Western Europe				
Denmark	0.1	0.4	2	4
Italy	0.6	0.7	8	10
Netherlands	0.1	0.2	60	64
Norway	6.0	14.8	20	106
United Kingdom	19.0	5.2	29	22
West Germany	0.3	0.3	7	6
Other	0.8	0.8	10	5
Total	26.9	22.4	137	218
Eastern Europe and U.S.S.R.				
U.S.S.R.	75.0	59.0	920	1,450
Other[4]	3.3	1.8	11	29
Total	78.3	60.8	931	1,479
Middle East				
Bahrain	0.3	0.1	3	7
Iran	62.0	92.9	500	489
Iraq	34.5	100.0	28	26
Kuwait[1]	70.1	94.5	34	43
Oman	5.7	4.0	2	10
Qatar	5.6	3.2	40	157
Saudi Arabia[1]	153.1	169.6	88	146
Syria	2.2	1.8	3	2
United Arab Emirates	32.4	98.1	22	204
Other	([2])	0.6	([3])	2
Total	365.8	564.7	719	1,084

TABLE 7–3 *Continued*

Area and Country	Crude Oil (billion barrels)		Natural Gas (trillion cubic feet)	
	1977	1987	1977	1987
Africa				
Algeria	6.6	8.5	125	106
Angola	1.2	1.1	2	2
Cameroon	0.1	0.5	0	4
Congo	0.4	0.7	(^{3})	2
Egypt	2.5	4.3	3	10
Gabon	2.1	0.6	2	1
Libya	25.0	21.0	26	26
Nigeria	18.7	16.0	43	84
Tunisia	2.7	1.8	6	3
Other	0.2	0.6	(^{3})	11
Total	59.2	55.2	207	249
Far East and Oceania				
Australia	2.0	1.7	32	19
Bangladesh	0.0	(^{2})	8	13
Brunei	1.6	1.4	8	7
China	20.0	18.4	25	31
India	3.0	4.3	4	18
Indonesia	10.0	8.4	24	73
Malaysia	2.5	2.9	17	52
New Zealand	0.1	0.2	6	5
Pakistan	0.3	0.1	16	22
Thailand	(^{2})	0.1	5	4
Other	0.3	0.3	4	12
Total	39.7	37.8	148	256
World Total	648.1	888.9	2,515	3,802

[1] Includes one-half of the Partitioned Zone (formerly called Neutral Zone).

[2] Less than 0.05 billion barrels.

[3] Less than 0.5 trillion cubic feet.

[4] Includes also Cuba, Mongolia, North Korea, and Vietnam.

Note: All reserve figures except those for the U.S.S.R. and natural gas reserves in Canada are proved reserves recoverable with present technology and prices. U.S.S.R. figures are "explored reserves," which include proved, probable, and some possible. The Canadian natural gas figure includes proved and some probable. The latest Energy Information Administration data for the United States are for December 31, 1986.

SOURCE: Energy Information Administration, *Annual Energy Review 1987.*

States contains 28.7 percent of world coal reserves (Figure 7–6 and Table 7–4), although a significant amount of this coal is of the high-sulfur variety, which is harder to process. Only the USSR, with 26.8 percent of the total, is close. The USSR is now the leading coal producer (Figure 7–7 and Table 7–5) with the

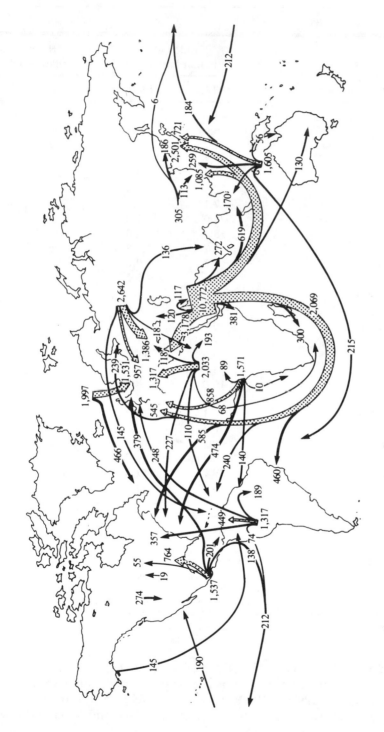

Arrows Indicate Origins and Destinations but Not Necessarily Specific Routes. Several Minor Routes and Quantities Are Not Displayed.

FIGURE 7-4   International crude oil flow, 1982 in thousand barrels per day. (Source: Energy Information Administration, *Annual Energy Review 1984*)

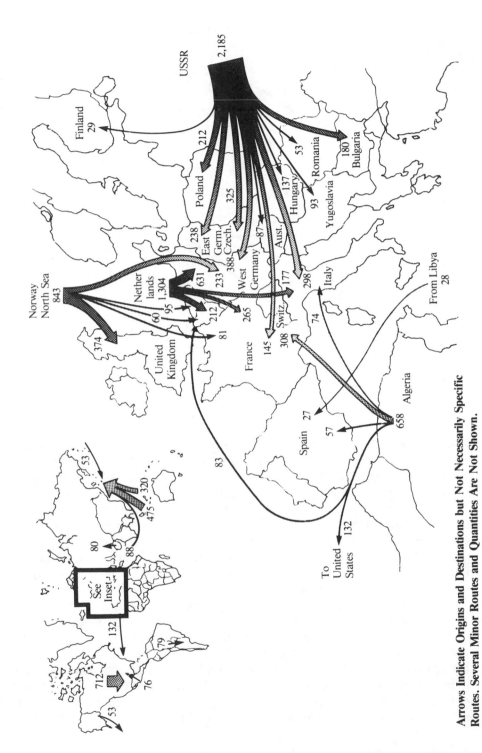

Arrows Indicate Origins and Destinations but Not Necessarily Specific Routes. Several Minor Routes and Quantities Are Not Shown.

FIGURE 7–5 International natural gas flow, 1982, billion cubic feet. (Source: Energy Information Administration, *Annual Energy Review 1984*)

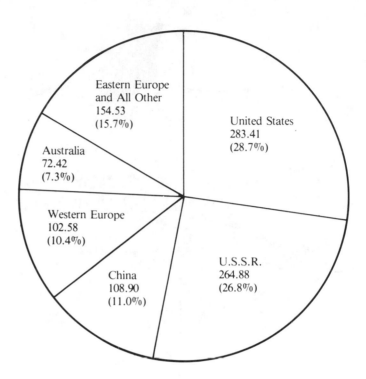

World Total: 986.72

Eastern Europe
and All Other
154.53
(15.7%)

Australia
72.42
(7.3%)

United States
283.41
(28.7%)

Western Europe
102.58
(10.4%)

China
108.90
(11.0%)

U.S.S.R.
264.88
(26.8%)

United States a close second, while China is rapidly closing in at third. It is also noteworthy that coal reserves and production are meager in the southern hemisphere, in part because most of the world's land is in the northern hemisphere, and in part because of lack of favorable geology.

Hydroelectric power generation for the world is shown in Figure 7–8 and data given in Table 7–6. The United States continues to lead the world in hydroelectric power generation, although its share of the total has dropped from 21 percent (1973) to 17 percent (1985). This trend must continue as world hydropower resources are, in many areas, as yet undeveloped whereas these resources in the United States are already strained.

Nuclear electricity generation for non-communist countries is shown in Figure 7–9 and in Table 7–7. The United States still leads with about one third of the world total, but other nations are making greater strides. This trend will also continue as the nuclear energy effort in the United States will increase, but at a slower pace than for the world. The American Nuclear Society has reported that there will be 548 nuclear power plants in operation worldwide by 1995, representing 35 countries, with a total electricity generating capacity of 423 billion watts (of which the United States will provide 28 percent from 132 power plants).

Figure 7–10 shows the distribution of nuclear power plants in the United States as of 1987.

OIL AND GAS

Oil and gas are discussed together because of their many similarities. Both consist of hydrocarbons, and they form in similar geologic settings. Briefly, oil and gas represent remobilized, fluid hydrocarbons formed in sedimentary rocks from decaying animal and some plant matter. These petroliferous materials form basically from animal remains that were deposited under swamplike conditions and accumulated over millions of years without destruction or dissemination. The process for oil and gas formation is moderately well known, and the key parts of the story are told below.

It is a well-known fact that hydrocarbons, those simple organic compounds consisting entirely or almost entirely of hydrogen and carbon, are extremely common in surface rocks and in pores of near-surface rocks. They may, however, be very nonuniformly distributed in rocks. Organic carbon is an even more common constituent of rocks, occurring nearly everywhere on the earth's surface, and accumulating readily in shale, sandstone, limestone, soil, and other media. Only a small amount of the organic carbon is hydrocarbon, however. While organic carbon is found in sedimentary rocks dating back beyond 3.2 billion years, hydrocarbons are extremely rare until after 1 billion years ago, and petroliferous hydrocarbons only became important after animal life began to flourish in the Devonian Period. Where this life existed in marginal marine to nearshore marine basins, the remains of the abundant animal life accumulated. As burial of the remains increased, the organic matter was altered by the compression and compaction due to the overlying sediment, and chemical changes took place internally. At some depth, possibly as little as 450 meters, the organic material may be broken down ("cracked") into simpler hydrocarbons that may be fluid or gaseous. This requires temperature as the main force, but pressure and pore fluid also play a role. As the sediments are buried, several processes go on hand in hand: The rocks are physically compacted and compressed, the temperature increases with depth, the complex hydrocarbons are cracked to simpler ones, and the expulsed simple hydrocarbons tend to move upward in the rock.

If the route to the surface contains only rocks similar to those in which the petroliferous material formed, then the mobile hydrocarbons are likely to reach the surface and dissipate. However, if their path toward the surface is blocked by some kind of impervious layer, then these petroliferous materials may be blocked in place and start to accumulate. The migration of the hydrocarbons can take place when pressure and temperature are sufficient for mobilization, probably at a depth of 500 to 600 meters or so, although more migration will occur at greater depths. As migration takes place from the source rocks into coarser grained rocks, the mobility of the fluid hydrocarbons increases. "Reservoir rocks" are those in

TABLE 7–4 World recoverable reserves of coal, 1984 (billion tons).

Area and Country	Anthracite and Bituminous Coal[1]			Lignite		Total Recoverable
	Recoverable	Portion Surface Minable	Portion Coking Quality	Recoverable	Portion Surface Minable	
North, Central, and South America						
Canada	4.88	4.10	2.58	2.67	2.67	7.55
United States	254.78	64.37	NA	36.06	36.06	290.84
Other	7.77	3.02	1.77	0.02	0.00	7.79
Total	267.43	[2]71.49	[2]4.35	38.75	38.73	306.18
Western Europe						
Turkey	0.10	0.00	0.09	5.25	3.68	5.35
United Kingdom	5.07	0.00	1.98	0.00	0.00	5.07
West Germany	26.37	NA	15.82	38.75	38.75	65.12
Yugoslavia	1.73	0.58	0.00	16.50	13.20	18.23
Other	2.22	[2]0.31	[2]0.61	3.69	0.36	5.91
Total	35.49	[2]0.89	[2]18.50	64.19	[2]55.99	99.68
Eastern Europe and U.S.S.R.						
Bulgaria	0.03	NA	0.02	4.00	2.60	4.03
Czechoslovakia	3.00	NA	NA	3.15	0.00	6.15
Hungary	1.74	0.00	0.17	3.18	3.18	4.92
Poland	31.20	0.00	10.00	15.88	14.33	47.08
U.S.S.R.	165.56	39.11	60.00	104.20	96.88	269.76
Other	0.00	0.00	0.00	23.95	23.90	23.95
Total	201.53	[2]39.11	[2]70.19	154.36	[2]140.89	355.89

Africa						
Botswana	3.85	0.00	0.00	0.00	0.00	3.85
South Africa, Republic of	64.38	NA	NA	0.00	0.00	64.38
Swaziland	1.00	0.00	0.00	0.00	0.00	1.00
Other	2.06	0.56	0.55	0.00	0.00	2.06
Total	71.29	[2]0.56	[2]0.55	0.00	0.00	71.29
Middle East, Far East, and Oceania						
Australia	32.53	7.26	12.10	39.90	39.90	72.43
China	108.90	14.16	32.67	0.00	0.00	108.90
India	NA	NA	NA	1.74	1.65	1.74
Other	2.30	[2]0.07	[2]0.71	0.92	0.70	3.22
Total	143.73	[2]21.49	[2]45.48	42.56	[2]42.25	186.29
World Total	719.47	[2]133.54	[2]139.07	299.86	[2]277.86	1,019.33

[1]Includes subbituminous coal.

[2]Not all countries in this group reported under this category.

NA-Not Available.

SOURCE: Energy Information Administration, *Annual Energy Review 1987*.

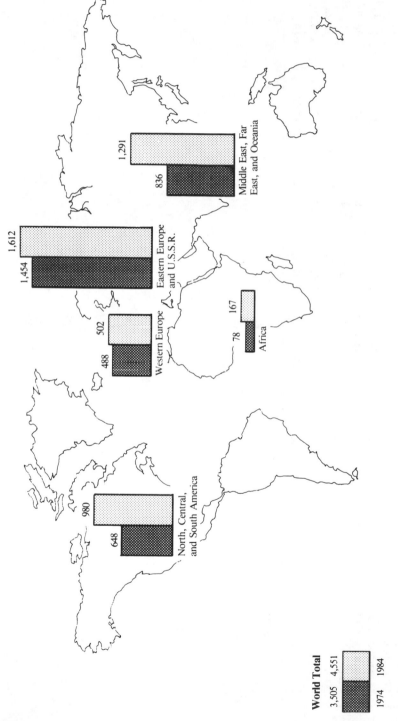

FIGURE 7–7 International coal production, 1975 and 1983, million tons. (Source: Energy Information Administration, *Annual Energy Review 1984*)

TABLE 7–5 World coal production, 1976–1986 (million tons).

Area and Country	1976	1981	1986
North, Central, and South America			
Canada	28	44	66
United States	685	824	890
Other	16	22	31
Total	729	890	988
Western Europe			
Spain	16	38	42
Turkey	11	19	48
United Kingdom	137	138	115
West Germany	247	241	222
Yugoslavia	41	58	72
Other	66	67	72
Total	518	561	571
Eastern Europe and U.S.S.R.			
Bulgaria	28	32	39
Czechoslovakia	130	137	141
East Germany	273	294	343
Poland	241	219	286
U.S.S.R.	784	776	825
Other	57	72	80
Total	1,513	1,529	1,714
Africa			
South Africa, Republic of	85	144	196
Other	6	5	6
Total	91	149	202
Middle East, Far East, and Oceania			
Australia	109	130	210
China	586	683	959
India	116	142	184
Other	101	114	130
Total	912	1,069	1,483
World Total	3,763	4,198	4,959

SOURCE: Energy Information Administration, *Annual Energy Review 1987*.

which the migrating petroliferous materials are blocked and in which they accumulate.

Different kinds of traps are recognized for oil and gas accumulation. The most common type is the well-known anticline (Figure 7–11), where oil or gas migrating upward in a permeable rock such as sandstone, and overlain by an impermeable rock like shale, is stopped at the crest and starts to accumulate. It

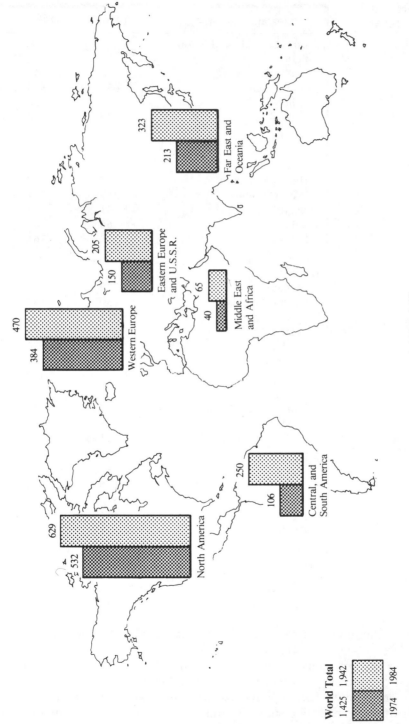

FIGURE 7–8   International hydroelectric power generation, 1973 and 1983, billion kilowatthours. (Source: Energy Information Administration, *Annual Energy Review 1984*)

North America
629
532

Western Europe
470
384

Eastern Europe
and U.S.S.R.
205
150

Far East and
Oceania
323
213

Middle East
and Africa
65
40

Central, and
South America
250
106

World Total
1,425   1,942
1974   1984

TABLE 7–6 World hydroelectric power generation, 1976–1986 (billion kilowatthours).

Area and Country	1976	1981	1986
North, Central, and South America			
Argentina	5	15	21
Brazil	82	130	180
Canada	213	263	308
Colombia	10	18	20
Mexico	17	25	27
United States	287	264	294
Venezuela	11	15	21
Other	21	29	40
Total	646	757	911
Western Europe			
Austria	20	31	32
Finland	9	13	12
France	49	73	64
Italy	41	45	44
Norway	81	92	96
Portugal	5	5	8
Spain	22	23	27
Sweden	54	60	60
Switzerland	27	36	34
West Germany	14	20	18
Yugoslavia	20	25	27
Other	19	26	29
Total	362	450	451
Eastern Europe and U.S.S.R.			
Romania	8	13	12
U.S.S.R.	134	185	211
Other	11	14	14
Total	154	212	237
Middle East and Africa			
Egypt	8	10	11
Zambia	7	10	10
Other	34	43	39
Total	49	63	60
Far East and Oceania			
Australia	15	q55	15
China	51	65	92
India	35	49	58
Japan	88	90	78
Korea, North	17	23	29
New Zealand	15	19	20
Other	21	28	44
Total	241	289	336
World Total	1,450	1,771	1,995

SOURCE: Energy Information Administration, *Annual Energy Review 1987*.

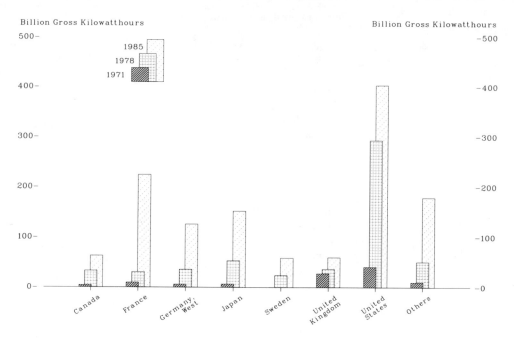

1985
1978
1971

Billion Gross Kilowatthours

Canada France Germany, West Japan Sweden United Kingdom United States Others

FIGURE 7–9   Nuclear electricity generation by non-communist countries. (Source: Energy Information Administration, *Annual Energy Review 1984*)

has been estimated that 85 percent of world petroleum reserves may be held in anticlinal traps. Other important traps include fault traps, limestone reefs (see Figure 7–11), stratigraphic traps, and many others.

In order for oil and gas to be found, there must have been favorable source rocks, rocks favorable for migration, and favorable reservoir rocks. If all three are not in communication, no oil or gas deposit can be found. Recognizing favorable rocks for source, migration, and reservoir are all part of modern petroleum exploration. Refined geophysical and geochemical techniques, combined with sophisticated geologic tools including satellite imagery and detailed aerial photographs, allow the explorationist to pick reasonable targets for detailed work, but that often is just the start. How seismic reflection and refraction work in exploration is shown in Figure 7–12. In some cases the source rocks were not connected to good rocks for migration, and the hydrocarbons are still diffuse in their sites of origin. In other places the hydrocarbons may have migrated inefficiently and become too disseminated. In still other sites the reservoir rocks may not be highly permeable so that the flow of hydrocarbons in them is too sluggish to allow good recovery. Further, in some environments, structures that are now favorable traps may not have had the same geometry when petroliferous hydrocarbons were likely to have migrated through them. Other structures that once contained oil and gas may have been breached or overturned and lost their hydrocarbons. In many instances, all the above factors are totally unknown, and only detailed pre-

TABLE 7–7   Nuclear electricity generation by non-communist countries, 1971–1987 (billion gross kilowatthours).

Country	1977	1982	1987
North America			
Canada	26.6	42.6	80.6
United States	264.2	298.6	477.9
Total	290.8	341.2	558.5
Central and South America			
Argentina	1.6	1.9	5.2
Brazil	0	0.1	1.0
Total	1.6	1.9	6.2
Western Europe			
Belgium	11.9	15.6	41.9
Finland	2.7	16.5	19.4
France	17.9	108.9	265.5
Italy	3.4	6.8	0.2
Netherlands	3.7	3.9	3.6
Spain	6.5	8.8	41.2
Sweden	19.9	38.8	67.2
Switzerland	8.1	15.0	23.0
United Kingdom	38.1	44.1	56.2
West Germany	36.0	63.4	130.2
Total	148.1	321.8	648.3
Far East and Africa			
India	2.8	2.2	5.5
Japan	28.2	104.5	182.8
Pakistan	0.3	0.1	0.3
South Africa	0	0	6.6
South Korea	0.1	3.8	37.8
Taiwan	0.1	13.1	33.1
Total	31.5	123.6	266.1
Total	472.0	788.5	1,479.1

SOURCE: Energy Information Administration, *Annual Energy Review 1987*.

liminary study followed by drilling can answer the elusive question of oil and gas occurrences. Figure 7–13 shows some critical steps for oil and gas formation. A typical oil well derrick and production system are shown in Figure 7–14.

Oil and gas are not only rare but erratically distributed in rocks. Some 80 percent of world petroleum production and reserves occur in less than 5 percent of known accumulations, and 65 percent in only 1 percent of known accumulations, namely 55 fields that are the supergiants—those fields with more than a billion barrels of petroleum or a trillion or more cubic feet of gas. Just two fields, the Ghawar field in Saudi Arabia and the Burgan field in Kuwait, contain 15 percent of known world reserves. Even in the United States, 60 percent of total production will come from less than 2 percent of the 23,400 known fields.

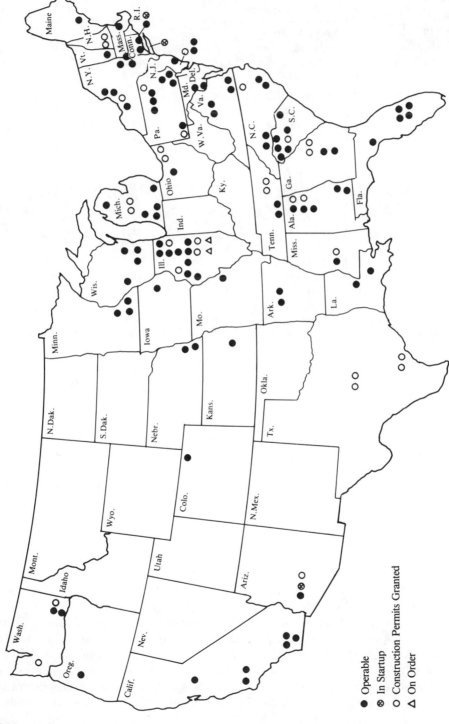

FIGURE 7-10 Status of nuclear reactor units, December 31, 1984. (Source: Energy Information Administration, *Annual Energy Review 1984*)

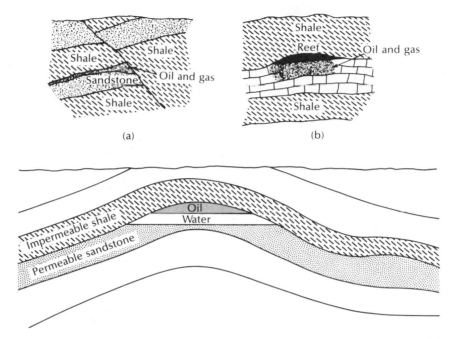

FIGURE 7–11   Structural traps for oil and gas accumulation. Almost 80 percent of the world's cumulative production and reserves are from simple structural *anticlines*. Because of their low density, oil or gas and water migrate through the permeable sandstone until further upward migration is stopped by an overlying layer of impermeable rock such as shale. Determining the structure of the subsurface by geophysical techniques maximizes the probability of drilling near the highest point of the anticline. Two other types of trap are important. In trap (a), a fault has allowed oil and gas to accumulate in sandstone on the footwall side of the fault where it is blocked by shale. In trap (b), a carbonate reef in a limestone layer has allowed oil and gas to accumulate. (Source: Brookins, Merrill, 1981)

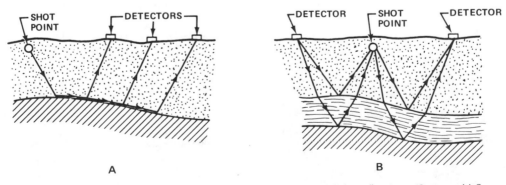

Figure 7–12   Seismic exploration methods: (a) refraction and (b) reflection. (Source: U.S. Department of Energy)

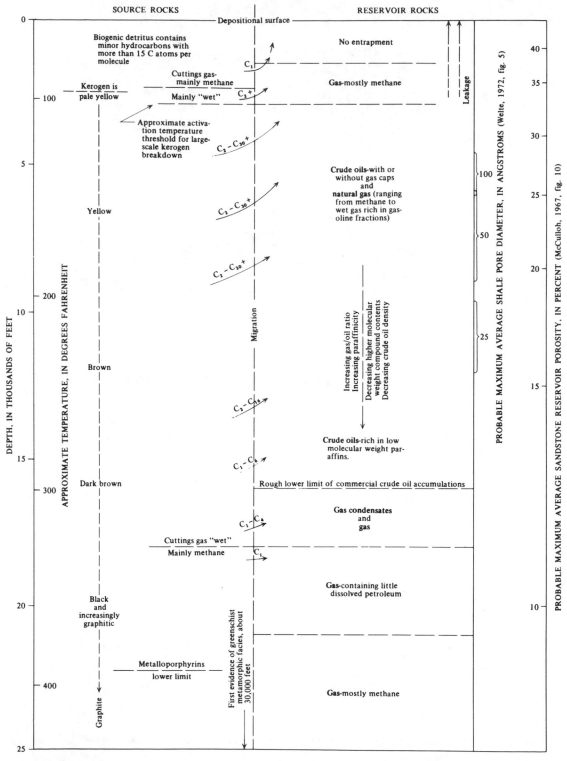

FIGURE 7–13 Selected diagnostic and critical steps in the genesis and migration of petroleum and natural gas from sedimentary organic matter. (Source: U.S. Geological Survey)

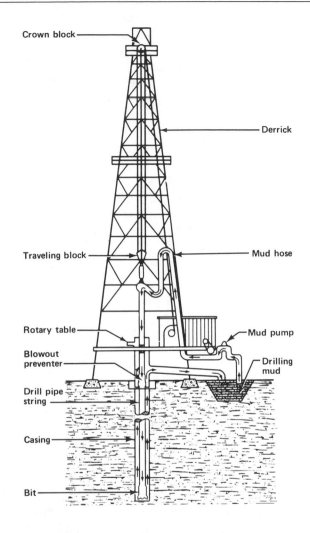

FIGURE 7–14 Petroleum drilling rig. (Source: U.S. Department of Energy)

Crown block

Derrick

Traveling block

Mud hose

Rotary table

Mud pump

Blowout preventer

Drilling mud

Drill pipe string

Casing

Bit

Not only are oil and gas unevenly distributed, but there are tendencies for oil and gas to occur more frequently and in greater abundance in geologically young rocks. Figure 7–15 shows the percent of world and U.S. reserves as a function of geologic age of host rocks. The world picture reflects the giant Middle Eastern fields of Mesozoic age. In the United States, most fields are in Cenozoic and Paleozoic rocks, and if the world picture holds here, then future U.S. production will probably show a tendency for Mesozoic finds to dominate over Cenozoic and Paleozoic ones.

Domestic petroleum production peaked in 1970, and only the very high prices from the early 1970s through the early 1980s allowed as much domestic production as has taken place. This is shown dramatically in Figure 7–16, an Energy Information Administration graph from 1984 showing that Alaskan oil will peak in 1995, outer continental shelf oil (OCS) will stay about constant, and

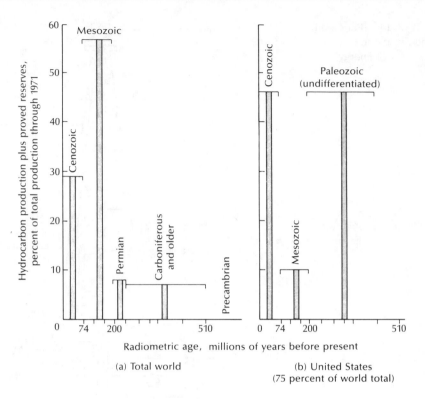

FIGURE 7–15   Age of rocks relative to oil and gas production. Most of the world's production and proved reserves comes from Mesozoic rocks, yet only 10 percent of U.S. production and reserves is from Mesozoic rocks. Thus maximum exploration in these rocks could improve U.S. oil and gas supplies (Source: Brookins, Merrill, 1981)

the lower 48 states onshore oil will either decline or, if incremental enhanced oil recovery (EOR) is feasible, will increase slightly. It is not at all certain now that the figure is accurate. A drastic lowering of the price of crude oil to near $11 per barrel by OPEC had a devastating effect on oil production in the United States, and it is likely that the production curves in Figure 7–16 for onshore 48 states may be lower than shown, thus the incremental EOR may not materialize at all.

In the United States, reserves of petroleum are inadequate to meet demand now and will be even more inadequate in the future. Despite the oil glut on the world market at this time, the outlook for the world is not optimistic either. At least one projection, that by M. K. Hubbard of the U. S. Geological Survey, calls for world oil to peak very early in the next century (Figure 7–17). Imports of petroleum, after a sharp peak in 1973, declined off and on through about 1981 and then have monotonously started to grow again. Based on projected demand and meager domestic reserves, imports can only be expected to increase. The

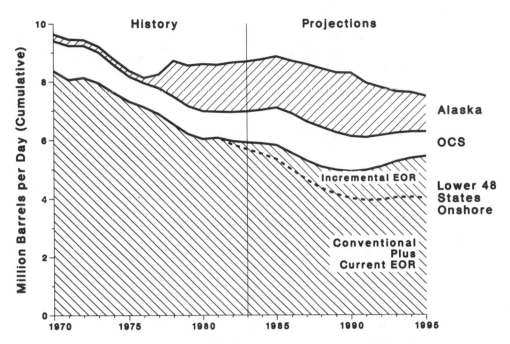

FIGURE 7–16   Oil production by source, 1970 – 1995. (Source: Energy Information Administration, *Annual Energy Outlook 1984)*

United States is very dependent on Middle Eastern oil, as shown in Figure 7–18: roughly 37 percent of our imports in 1984 were from Saudi Arabia and other OPEC countries, yet in 1973 Middle Eastern oil made up only 6 percent of our imports. Non-OPEC oil is imported by the United States from Canada, Mexico, the United Kingdom, and the Virgin Islands–Puerto Rico. In a hard-to-understand way, the United States has increased its exports of crude oil and petroleum products from 1975 to 1980 (Figure 7–19), although this may decline somewhat in the near future.

Petroleum Product Use

As shown in Figure 7–20, most petroleum products are consumed by the transportation sector. Breaking down the petroleum products by type (Figure 7–21), motor gasoline accounts for the largest fraction (44%), followed by distillate fuel oil (18%), and residual fuel oil and liquefied petroleum gases. The "other" category takes in jet fuel, petroleum feedstocks, and a variety of lesser items. Despite more fuel-efficient cars, the demand for gasoline will grow while the percent of total oil consumed will decline. This will be offset by an increased demand for distillate fuel oil.

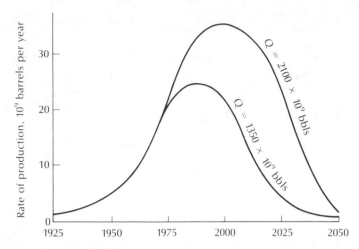

FIGURE 7–17   Cycle for world petroleum production. This graph shows two values for quantity (Q) of petroleum products. For $Q = 1350$ billion barrels, 80 percent of the total will be used by the year 2003 (to peak at 1990); while for $Q = 2100$ billion barrels, 80 percent of the total will be used by the year 2032 (to peak at 2000). These estimates have been made by experts from several federal and state agencies, the National Academy of Sciences, and acknowledged experts in industry. The world will have very limited supplies of petroleum products by the mid-twenty-first century. (After M. K. Hubbert, reproduced from *Resources and Man,* 1969, with the permission of the National Academy of Sciences, Washington, D.C., and M. K. Hubbert.)

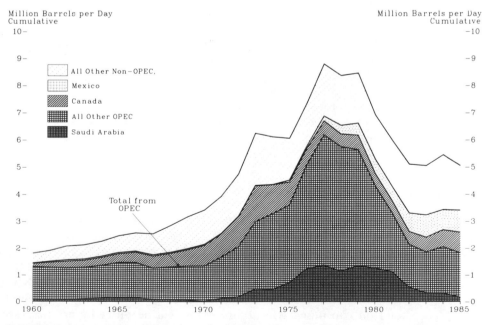

FIGURE 7–18   Imports of petroleum by country of origin, 1960–1984. (Source: Energy Information Administration, *Annual Energy Review 1984*)

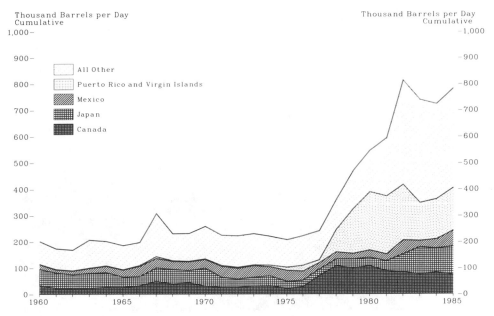

FIGURE 7–19   Exports of petroleum by major country of destination, 1960–1984.
(Source: Energy Information Administration, *Annual Energy Review 1984*)

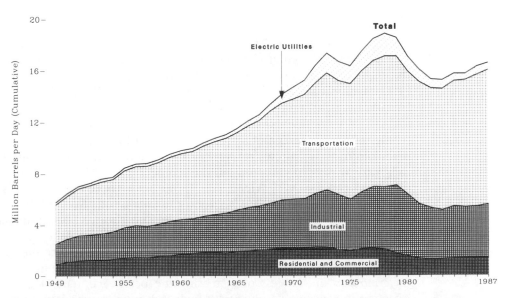

FIGURE 7–20   Petroleum products supplied to end-use sectors, 1949–1984. (Source:
Energy Information Administration, *Annual Energy Review 1984*)

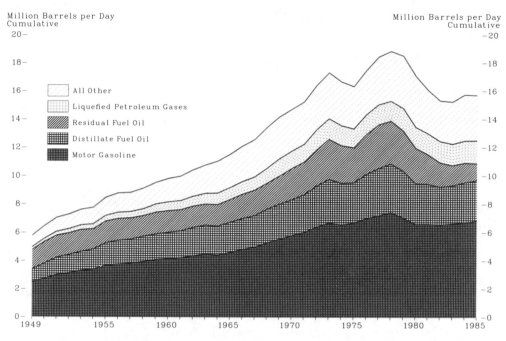

FIGURE 7–21   Petroleum products supplied by type, 1949–1984. (Source: Energy Information Administration, *Annual Energy Review 1984*)

Refining

Materials that can be refined from crude oil are divided into fuel and nonfuel components. The chief nonfuel components consist of solvents, lubricating oils, lubricating greases, waxes, and asphalt, among others. Conventional motor oil is a lubricating oil marked by its Society of Automobile Engineers (SAE) viscosity. A 10W (weight) oil is much less viscous than 30W oil.

Fuel components are separated by carefully controlled distillation, and include refinery gases, gasoline, kerosene, diesel fuel, and heavier fuel oils. Straight distillation allows these fractions to be separated, but since naturally occurring oils vary widely in composition, the temperatures at which the various fractions distill off can also vary widely (see Table 7–8). There are numerous processes to modify the various distillate fractions. *Cracking* results from adding chemicals that will crack the longer chain hydrocarbons into shorter ones, thus facilitating the distillation. Zeolites, some rare earths, and other materials are used. *Coking* is an extreme form of cracking used to break heavy distillates like fuel oil into lighter fractions and residual petroleum coke (used for electrolytic purposes in industry). *Reforming* is the process of increasing the octane content of gasoline, done by reacting low-octane gasoline with hydrogen over appropriate catalysts. *Alkylation* is the process of increasing octane of gasoline derived from paraffins. *Isomerization*

TABLE 7–8   Typical petroleum distillate fractions.

Fraction	Carbon Atoms per Molecule	Approximate Boiling Range (°C)	Major Uses
Gasoline (straight-run)	5–12	30–200	Gasoline engines
Kerosine	10–15	180–275	Diesel and jet engines
Fuel (or gas) oil	15–22	260–345	Heating
Lubricating oil	20–30	340–400	Lubrication
Residue	Large number	High	Heating, asphalt, etc.

SOURCE: U.S. Department of Energy.

is used to transform certain hydrocarbons into others without loss of carbons; that is, isomers of the original are produced. *Polymerization* is used to make more complex hydrocarbons from simple ones, that is, combining two low-carbon hydrocarbons over a catalyst of phosphoric acid to make a single more complex hydrocarbon. *Hydrotreating* is normally done by reacting hydrogen with different distillate fractions to remove sulfur by formation of H_2S gas. The H_2S is mainly trapped and sulfur is recovered from it (see Chapter 5). *Sweetening* refers to eliminating the often unpleasant odors of oil distillates, mainly due to the presence of sulfur-organics, by hydrotreating and by oxidizing the sulfur-organics.

Refining and Exploration Problems

The refining situation in the United States is very problematic. As shown in Figure 7–22 the refinery input and capacity have declined since 1979. Further, total input and output have also decreased over this same period (Figure 7–23). This results from many problems. First, there is a shift in the demand for different refining products, which mean expensive and time-consuming modifications to the refining processes. Second, the continued demand for unleaded gasoline creates difficulties because more and more domestic oils are heavier and contain more sulfur and other impurities than previously. Using lead helped make gasoline refined from such oils more efficient, but without it additional processes must be added to the refining. Third, there is a reluctance to build new refineries in view of uncertain market conditions. With the present glut of petroleum on the world scene and a severe drop in price, many domestic refineries must operate at reduced scales as is, and certainly no new refinery can be justified. Fourth, increased cost of materials, labor costs, and new environmental controls imposed on the refineries make their construction too expensive. The Department of En-

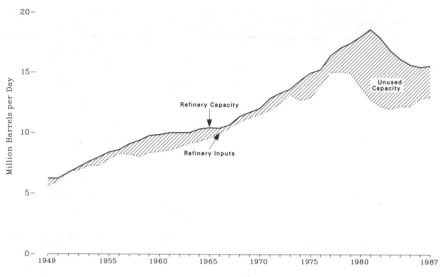

FIGURE 7–22   U.S. Refinery capacity and utilization. (Source: Energy Information Administration, *Annual Energy Review 1984*)

ergy estimates that the loss of refinery capacity will mean that the United States will become increasingly dependent on imports of refined products, not just crude oil, in the fairly near future. This is unfortunate as the United States' ability to quickly and efficiently produce refined oil products in time of emergency requires maximum refinery capacity.

Since the mid-1970s the United States has been stockpiling oil and gas for defense and defense-related purposes, much of this in salt domes in Texas and Louisiana. Some 450 million barrels of oil had been stored as of the end of 1984 (Table 7–9). All but about 33 million barrels of this reserve is from crude oil imports.

Exploration for oil and gas is also on hard times. There has been an increasing number of dry holes drilled for oil and gas, about 35 percent of the total. Additions of gas to the U.S. reserves have been fairly successful, but not so with oil wells. Further, the return of exploration, as indicated by barrels per foot of holes drilled, has been steadily decreasing for over the last 30 years. With low oil prices many operations, marginal at $20 per barrel, are totally subeconomic, and much geoexploration has been halted as well.

Enhanced Oil Recovery

The recovery of oil from rocks is often complex. If the pressure in the oil reservoir rocks is sufficient, then the oil is forced to the surface naturally when the well is completed. Often the top of an oil reservoir contains gas which facilitates the upward movement of the oil. After this initial surge (or flushing) occurs, the oil

225

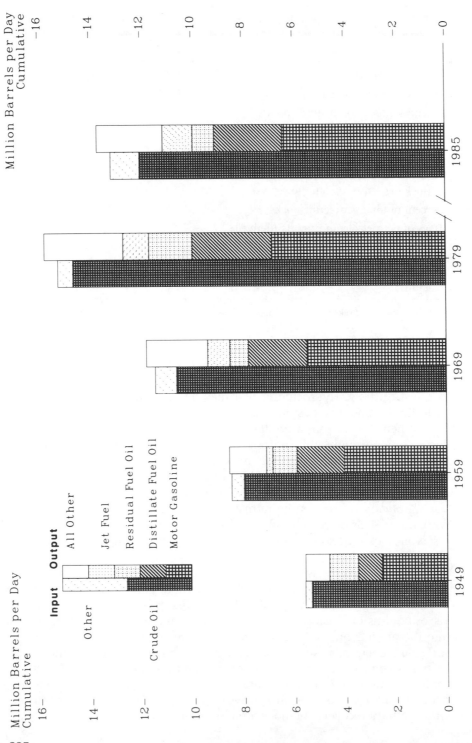

FIGURE 7-23 U.S. refinery input and output. (Source: Energy Information Administration, *Annual Energy Review 1984*)

TABLE 7–9   Strategic petroleum reserve, 1977–1987 (million barrels).

	1977	1982	1987
Crude oil imports	7.54	60.19	26.52
Domestic crude oil deliveries	0.37	3.79	2.40
Quantity (does not include imported quantities in transit to Strategic Petroleum Reserve terminals, pipeline fill, and above-ground storage)	7.46	293.83	540.65
Percent of crude oil stocks (includes lease condensate stocks)	2.1	45.7	60.8
Percent of total petroleum stocks	0.6	20.5	33.6
Days of net petroleum imports (derived by dividing end-of-year strategic petroleum reserve stocks by annual average daily net imports of all petroleum)	1.0	68.0	94.0

SOURCE: Energy Information Administration, *Annual Energy Review 1987*.

must be pumped from the reservoir rocks. Oil flow is often sluggish, and can be stimulated by several different processes and actions. The rocks may be fractured to speed up flow. Hydrochloric acid may be used in limestone or dolostone, or hydrofluoric acid in sandstone, to open up fissures and increase permeability.

When the oil from a producing well, and wells in a particular field, start to decrease significantly, this indicates that the readily recoverable petroleum in the rocks is diminishing. It does not mean that no more oil is left in the rocks, but rather that the rate of oil migration into the producing rocks is sluggish. There are several ways to obtain more oil under these conditions.

Secondary recovery is the term used when the oil-bearing rocks are flooded with water. This forces the oil ahead of the hydraulic head back into the initial reservoir rocks. In some cases, gas is injected back into the rocks to force the oil up.

Tertiary recovery is used when oil recovered by secondary recovery has started to taper off significantly. In this method, carbon dioxide is injected into the oil-bearing rocks. The carbon dioxide dissolves in the oil and decreases its viscosity, thus increasing its mobility. Additional flooding then forces this oil back into the production well areas. Other materials used for tertiary recovery include surfactant chemicals, such as detergents. When introduced to the rocks, these materials concentrate in the interface between the oil and rock, thus shutting off capillary action, and in turn facilitating the oil flow, especially when a polymer is added to the oil. Flooding then drives the polymerized oil upward in the rocks. Other tertiary recovery methods are also under study.

In 1984 the National Petroleum Council (NPC) reported that the United States could recover 14.5 billion barrels of otherwise unrecoverable oil by use of enhanced oil recovery methods. In addition to the secondary and tertiary recovery techniques mentioned above (accounting for 17 and 38 percent respectively), the NPC estimates an even larger amount by thermal recovery.

Natural Gas

The natural gas situation for the United States is better than for oil. The Department of Energy estimates reserves of 376 trillion cubic feet of natural gas for Canada, the United States, and Mexico, with an additional 1,211 trillion cubic feet of natural gas in the undiscovered resource category. The U.S. resources total 200 trillion cubic feet. Despite the magnitude of the last figure, the costs involved to get these resources into production are very large, especially since many gas deposits are deeply buried offshore, onshore, and in subarctic regions. The reserves are moderately large, roughly 34 trillion cubic feet, but no efficient way to transport the gas has been found. At present only about 277 billion cubic feet of Alaskan gas, all within Alaska, is consumed. Much of this gas is pumped back into the ground to accelerate oil recovery (this overproduction is one reason why Alaskan oil will peak about 1990), and the remainder is converted to liquid natural gas (LNG) and shipped to Japan. There are no west coast facilities to handle the LNG.

The disposition and supply of natural gas are shown in Figure 7–24. There has been a decrease in both since 1979, although some increases will no doubt occur in the late 1980s. Natural gas is both added to and removed from storage each year, as many formations do not hold gas well for prolonged periods of time. The United States stores roughly 6.5 trillion cubic feet of gas in this fashion, however. Use of natural gas by sector is shown in Figure 7–25 for the period 1949–1984; industrial uses account for the largest fraction.

COAL

Coal is the product of burial, compression, and alteration of residual plant matter that accumulates in freshwater to brackish swamps. The dead plant matter first forms peat, which consists of 90 percent water and 10 percent solids. Peat accumulations are common worldwide, with major amounts in the world's swamps. The Dismal Swamp of Virginia–North Carolina and the Okeefenokee Swamp in Florida contain large peat deposits. Peat is, in some countries, used as a low-quality fuel after first letting it dry thoroughly. In order for peat to accumulate, the climate must be warm and humid and the water table high so that plant material can accumulate under "rotting," or reducing, conditions. Swamplands on the edges of marine basins are good places for peat to accumulate as these lands commonly undergo gentle downwarping allowing peat to accumulate, then are periodically flooded, covering and thus protecting the layers with silt and

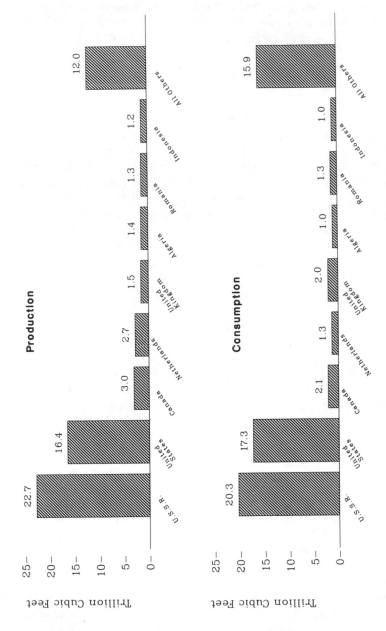

FIGURE 7-24 U.S. natural gas supply and disposition. (Source: Energy Information Administration, *Annual Energy Review 1984*)

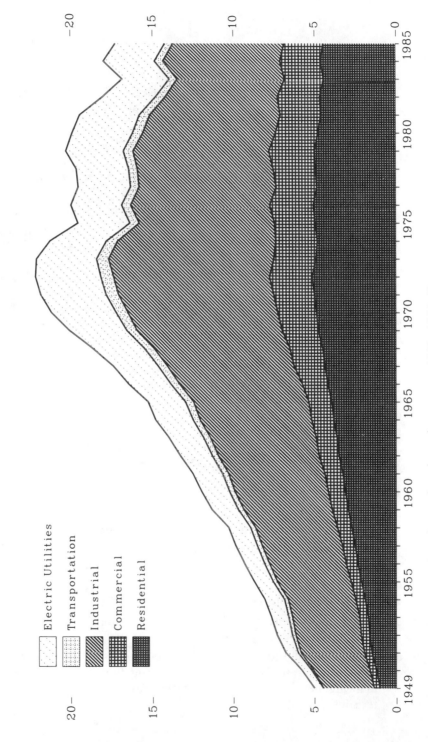

FIGURE 7–25 U.S. consumption of natural gas by end-use sector, 1949–1984. (Source: Energy Information Administration, *Annual Energy Review 1984*)

View of coal storage silos, 70 feet diameter by 220 feet high, each silo holds 14,000 tons of coal. Coal is gravity loaded into trains averaging 110 cars each. Each train can transport 11,000 tons of coal. Jacobs Ranch Mine, Wyoming. Courtesy of Kerr-McGee Corporation.

sand. For this reason the peat commonly occurs in layered sequences. Ultimately, the seas shift such that the entire sequence is covered. Later, more subsidence coupled with a marine regression may form another environment favorable for peat formation well over the earlier sequence.

Peat is transformed into coal as a result of burial by overlying sediment with accompanying chemical changes. This sequence is as follows:

1. Peat accumulates in fresh to brackish water.

2. The peat is compacted and dehydrated due to burial and chemical changes to form lignite (immediate precursor to combustible coal).

3. Continued burial of the lignite causes very low grade metamorphism to occur, and the lignite is converted to subbituminous coal.

4. Increased pressure and temperature, due to burial, convert subbituminous coal to bituminous coal.

5. Continued increased pressure and temperature due to additional burial convert bituminous coal to anthracite.

6. Further burial and accompanying pressure and temperature effects will alter the anthracite to first metaanthracite and then, ultimately, to graphite. Metaanthracite and subsequent materials rapidly lose their combustible characteristics.

Due to the very impure nature of swamps, the peat that accumulates is mixed with detrital material such as clay minerals, rock fragments, and silt, and the chemically reducing environment is favorable for formation of minerals such as pyrite or marcasite, clay minerals, and others.

The classification of coal, from lignite to anthracite, is referred to as *rank* of the coal. As rank of coal increases, the combustible matter content (or caloric value) increases, the moisture content decreases, and volatile matter first increases slightly from lignite to some bituminous coals, then rapidly decreases (Figure 7–26).

The *grade* of coal differs from rank in that grade depends on sulfur content, ash content, and other constituents. The detrital and authigenic (formed in situ) materials mixed with coaly material become the ash left upon combustion. The sulfur, mainly as pyrite or marcasite, but often as hydrous iron sulfate such as melanterite or as gypsum, is very problematic. In coal used for coke for the iron and steel industry, sulfur can lead to corrosion and other undesirable effects. In coal for combustion, it causes toxic emissions of sulfur dioxide and is responsible for acid precipitation. Coal that is washed to remove pyrite and other undesirable

FIGURE 7–26 Comparison on moist, mineral-matter-free basis of heat values and proximate analyses of coal of different ranks. (Source: U.S. Geological Survey)

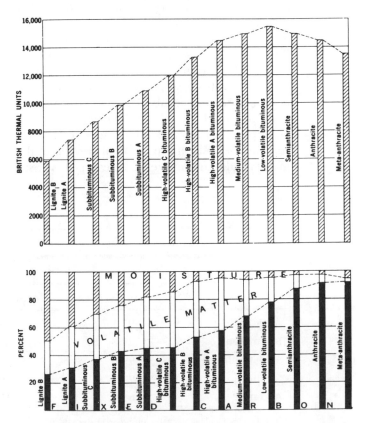

constituents results in large waste piles of these materials. Pyrite in these waste piles rapidly oxidizes to form sulfuric acid, causing acid soils, pollution of streams, and solubility of otherwise insoluble toxic elements. Sulfur contents in the various ranks of coal are listed in Table 7–10. Bituminous coal, the most widely used rank, also is highest in sulfur.

Bituminous coal represents the great majority of that consumed in the United States, accounting for 73 percent of the total of 890 million tons of coal produced in 1984. Subbituminous coal accounted for 19 percent, lignite 7 percent, and anthracite less than 1 percent. About two thirds of all coal produced is from states east of the Mississippi River; western coal has risen in recent decades from under 10 percent to over one third of U.S. production. Most coal is mined by surface mining techniques (60 percent), although underground mining still accounts for 40 percent. Figure 7–27 and Table 7–11 show a summary of ranks of coal mined, mining method, and geographic distribution for the period 1949–1986. Domestic consumption of coal in 1986 was 890 million tons. The United States exported some 95 million tons in 1986. Eighty-four percent of all coal consumed was for the electrical utilities, about 664 million tons. The industrial sector consumed 15 percent, over a third of it for making coke, and commercial and residential 1 percent. Coal flow for the United States for 1986 is illustrated in Figure 7–28.

Coal resources, as estimated by the U.S. Geological Survey, total 1,700 billion tons of coal at less than 3,000 feet depth. An additional 2,200 billion tons is estimated to occur at depths of 6,000 feet. However, coal recovery from any one area is only 40 to 80 percent or so, and many occurrences are in areas not favorable for recovery at all. The demonstrated reserve base, shown in Figure 7–29 and Table 7–12, is about 500 billion tons. Anthracite reserves are very small, amounting to only 7.3 billion tons, compared to 437 billion tons of bituminous coal, and 45 billion tons of lignite. Most of the bituminous coal will require mining by underground methods, especially in states east of the Mississippi River. In

TABLE 7–10   Distribution of sulfur in coals of different rank.

Rank	Sulfur Content (percent)		
	Low (0–1)	Medium (1.1–3)	High (above 3)
Anthracite	97.1	2.9	—
Bituminous	29.8	26.8	43.4
Subbituminous	99.6	0.4	—
Lignite	90.7	9.3	—
All ranks	65.0	15.0	20.0

SOURCE: U.S. Geological Survey (Brobst and Pabst, 1973).

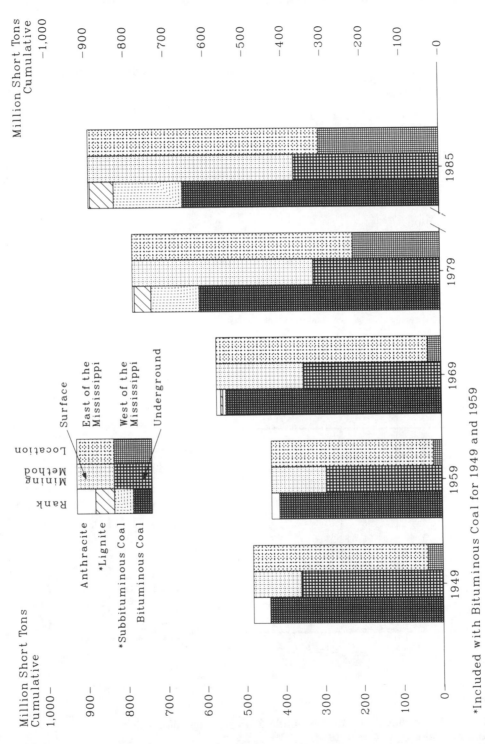

*Included with Bituminous Coal for 1949 and 1959

FIGURE 7-27 U.S. coal production. (Source: Energy Information Administration, *Annual Energy Review 1984*)

TABLE 7–11   Coal production, 1950–1987 (million tons).

	1950	1960	1970	1980	1987
Rank					
Bituminous	516.3	415.5	578.5	628.8	635.9
Subbituminous	*	*	16.4	147.7	198.9
Lignite	*	*	8.0	47.2	77.9
Anthracite	44.1	18.8	9.7	6.1	4.2
Mining Method					
Underground	421.0	292.6	340.5	337.5	373.0
Surface	139.4	141.7	272.1	492.2	543.8
Location					
West of Mississippi	36.0	21.3	44.9	251.0	336.6
East of Mississippi	524.4	413.0	567.8	578.7	580.3
Total	560.4	434.3	612.7	829.7	916.9

*Included in bituminous coal.

SOURCE: Energy Information Administration, *Annual Energy Review 1987*.

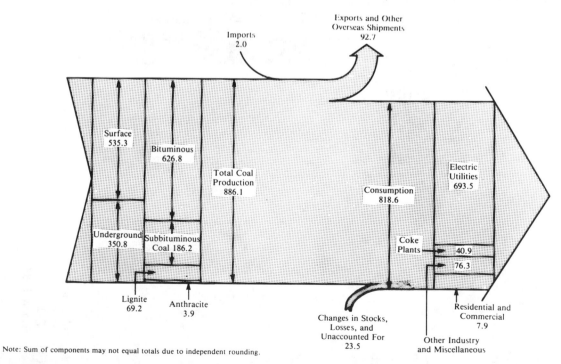

Note: Sum of components may not equal totals due to independent rounding.

FIGURE 7–28   U.S. coal flow, 1984 (million tons). (Source: Energy Information Administration, *Annual Energy Review 1984*)

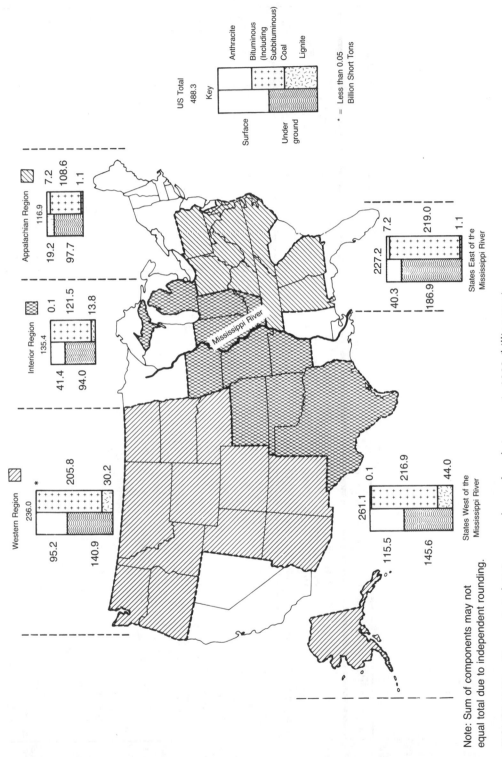

FIGURE 7-29 Demonstrated U.S. reserve base of coal, January 1, 1983 (billion tons).
(Source: Energy Information Administration, *Annual Energy Review 1984*)

Note: Sum of components may not equal total due to independent rounding.

Table 7-12 Demonstrated reserve base of coal,[1] January 1, 1983 (billion tons).

Region and State	Anthracite Underground and Surface	Bituminous Coal[2] Underground	Bituminous Coal[2] Surface	Lignite Surface[4]	Total Underground	Total Surface	Total Total
Appalachian							
Alabama	0	1.7	2.4	1.1	1.7	3.4	5.2
Kentucky, Eastern	0	17.2	2.1	0	17.2	2.1	19.3
Ohio	0	13.0	5.9	0	13.0	5.9	18.9
Pennsylvania	7.1	21.5	1.5	0	28.4	1.6	30.0
Virginia	0.1	2.3	0.8	0	2.4	0.8	3.3
West Virginia	0	34.0	5.1	0	34.0	5.1	39.1
Other[5]	0	1.3	0.4	0	1.3	0.4	1.8
Total	7.2	91.0	18.2	1.1	98.1	19.4	117.5
Interior							
Illinois	0	63.4	15.6	0	63.4	15.6	79.1
Indiana	0	8.9	1.6	0	8.9	1.6	10.5
Iowa	0	1.7	0.5	0	1.7	0.5	2.2
Kentucky, Western	0	16.9	4.0	0	16.9	4.0	20.9
Missouri	0	1.5	4.6	0	1.5	4.6	6.0
Oklahoma	0	1.2	0.4	0	1.2	0.4	1.6
Texas	0	0	0	13.8	0	13.8	13.8
Other[6]	0.1	0.3	1.1	(7)	0.4	1.1	1.5
Total	0.1	94.0	27.7	13.8	94.1	41.5	135.6

Western							
Alaska	0	5.4	0.7	(7)	5.4	0.7	6.2
Colorado	(7)	12.2	0.7	4.2	12.3	4.9	17.2
Montana	0	71.0	33.6	15.8	71.0	49.4	120.3
New Mexico	(7)	2.1	2.5	0	2.1	2.5	4.7
North Dakota	0	0	0	9.9	0	9.9	9.9
Utah	0	6.1	0.3	0	6.1	0.3	6.4
Washington	0	1.3	0.1	(7)	1.3	0.1	1.5
Wyoming	0	42.6	27.0	0	42.6	27.0	69.6
Other[8]	0	0.1	0.3	0.3	0.1	0.6	0.7
Total	(7)	140.9	65.3	30.2	140.9	95.5	236.4
U.S. Total	7.3	325.9	111.2	45.1	333.1	156.4	489.5
States East of the Mississippi River	7.2	180.4	39.4	1.1	187.5	40.6	228.1
States West of the Mississippi River	0.1	145.5	71.8	44.0	145.6	115.8	261.4

[1] Includes measured and indicated resource categories representing 100 percent of the coal in place. Recoverability varies from less than 40 percent to more than 90 percent for individual deposits. About one-half of the demonstrated reserve base of coal in the United States is estimated to be recoverable.

[2] Includes subbituminous coal.

[3] Includes 129.0 million tons of surface mine reserves, of which 113.5 million tons are in Pennsylvania and 15.5 million tons are in Arkansas.

[4] There are no underground demonstrated coal reserves of lignite.

[5] Includes Georgia, Maryland, North Carolina, and Tennessee.

[6] Includes Arkansas, Kansas, and Michigan.

[7] Less than 0.05 billion tons.

[8] Includes Arizona, Idaho, Oregon, and South Dakota.

Note: Sum of components may not equal total due to independent rounding.

SOURCE: Energy Information Administration, *Coal Production 1983*.

Electric shovel equipped with 36-cubic-yard bucket loading overburden into 170-ton haul trucks. Shovel can dig 3,050 bank cubic yards per hour. Jacobs Ranch Mine, Wyoming. Courtesy of Kerr-McGee Corporation.

1987 the United States consumed 917 million tons of coal, mostly bituminous coal, followed by subbituminous coal, lignite and anthracite (less than 1%).

Coal Uncertainties

Coal production in the United States is subject to several uncertainties, although the overall projections continue to call for ever greater amounts of coal to be used to meet our energy demands. Whether high or low economic growth is assumed, the difference is less than 10 percent of the total amount consumed. The transportation of coal is problematic, however, especially since the deregulation of railroad rates in 1980. Future rail rates are uncertain, and it is not clear if future regulations will be placed on coal transport by rail. Coal dust and particulates are readily lost during rail transport, and the environmental impact of these materials on the land is unknown. As more and more coal is mined in western states to meet demand in the eastern states, transportation costs will, of course, be greater. Coal mining labor productivity has also changed, in part for the better, with the use of mechanized mining processes and the phasing out of small, hand-dependent operations. Overall productivity in underground mines is down, however, due to (1) the Coal Mine Health and Safety Act (1969) which limits the amount of shift time per miner and requires more personnel; (2) wage agreements with the United Mine Workers union that called for additional personnel; (3) the surge in coal prices in the mid-1970s that made productivity a secondary issue; (4)

opening of new mines, which are more expensive to bring on line; and (5) strikes. The decrease in productivity for surface mines results from state regulations requiring extensive land reclamation, and the increase in coal prices made productivity a secondary issue here, too. The situation for coal exports and imports is also somewhat uncertain, although the United States should be healthy for coal exporting. Increased coal production abroad may decrease exports somewhat, but this should be minor. The increasing Japanese market, now mainly reliant on Canadian and South African coal, will also affect the United States.

The two major factors in addition to those mentioned above that may affect the coal industry are environmental effects and competing energy sources, notably nuclear power. These will be discussed shortly.

Coal Mining and Wastes

Surface mining of coal consists of removing surface vegetation by scraping, followed by removal of overburden, stripping of the coal, and land restoration. The main problem lies in the fact that trace elements enriched in subeconomic coal or in carbonaceous rocks just under the stripped area, as well as the redeposited overburden during land restoration, may become very mobile after mining. The new fill is not as compact as the rock originally removed, and the underlying rocks are cracked due to the blasting and stripping operations. Water will readily penetrate the restored site and may mobilize many trace elements, especially fairly soluble species of molybdenum, selenium, uranium, and arsenic. Any pyrite in these rules can oxidize to form sulfuric acid which, in turn, can cause even greater trace-element mobility.

In underground mining, a shaft is sunk which results in waste rock which must be deposited at the surface. The coal is removed as efficiently as possible, usually by following the seams and leaving natural rock pillars in place to provide roof support. Normally the seams are interlayered with clay-rich rocks, and this material is transported with the coal to the surface where it is separated. Mechanical mining becomes more expensive as narrower seams are encountered, a factor which has forced the closing of many coal mines in the eastern United States. Waste rock, much of which is pyritiferous and contains some coal, is a major problem. Many inactive sites contain huge areas of this waste rock, some of which is rich enough in combustible material to ignite. Burning coal waste piles occur in several areas of the United States. Two of the major problems of underground coal mining are subsidence and acid mine drainage. Subsidence occurs usually after mining operations have ceased as the column of rock over the mine settles. This can cause severe property damage on the surface, rupturing water and gas lines, causing foundations, walls, and other constructions to fail. It is possible to help prevent subsidence by pumping the waste rock back into the mine using a dense grid of drill holes. This technique has been successfully used in Wilkes-Barre, Pennsylvania, but most sites have gone untreated.

Coal mining is an extremely hazardous occupation due to fires, explosions, mine collapse, lung diseases, and other factors, all of which are discussed later in

Coal seams, Navajo Mine, New Mexico. Courtesy of Utah International Corporation.

this chapter. Most underground mines must be continuously pumped of the water which tends to accumulate in them, and when the mines are abandoned, they may flood. As water encroaches, the pyrite and marcasite in the exposed rocks oxidize with the formation of sulfuric acid. This acid mobilizes trace elements and is a major problem for surrounding waters. Acid mine drainage has been vigorously dealt with in many areas of the United States, but the National Academy of Sciences (NAS-NRC, 1980) estimates that some 5700 miles of waterways and tens of thousands of acres are still subjected to acid effects and pollution. Abandoned underground mines account for well over half of the acid mine drainage in the United States.

Coal treatment, no matter what the mining methods, is roughly the same. The coal is cleaned using water, taking advantage of the fact that the organic-rich coal is less dense than the minerals and rock fragments of the non-coal fraction. The trace-element content of the cleaning water becomes greatly enriched during

Undisturbed badlands prior to coal mining. Courtesy Utah International Corporation.

this process. The waste is sent as a slurry to settling ponds, while the coal is dried in separate pond areas or sometimes is transported as a water-rich slurry. The water associated with both the waste rocks and the coal is usually very contaminated and cannot be released to the environment. This poses a special problem for transportation of coal slurries as the water must be impounded to evaporate, being so impure that it can't be used for anything else.

Coal dust during storage, loading, and transportation activities presents a major problem. Trace elements are found to be enriched many times over background in some areas near these operations (NAS-NRC, 1980). Some coal dust, too, can be highly combustible.

Many coal-waste piles have been ignited, either intentionally or spontaneously. The NAS-NRC (1980) reported that 300 to 500 coal waste piles were on fire in the 1960s, possibly 50 percent of these within one mile of communities of 200 or more people, and 10 percent near communities with 10,000 to over 100,000 people. The emissions from these piles must be highly toxic, as they contain all the raw ingredients of coal without any treatment whatsoever (sulfur, nitrogen, and carbon oxides; mercury, arsenic, selenium, cadmium, and so forth). Yet the public health effects of burning waste coal piles have never been fully investigated.

There are approximately 3,000 to 5,000 abandoned coal-waste piles in the United States, and it has been estimated that 1.5 to 2 pounds of sulfuric acid and 0.5 to 0.7 pounds of soluble iron are produced per acre of refuse per day, and up to 300 pounds of sulfuric acid per day has been found at some highly mineralized sites. The effect of this acid mine drainage, oddly, is not known, but is presumed to be of major concern (NAS-NRC, 1980). The drainage has a pronounced ad-

Interburden drilling prior to blasting, Navajo Mine, New Mexico. Courtesy of Utah International Corporation.

verse effect on water supplies, including increased acidity; higher concentrations of dissolved species of iron, manganese, coal fines, and several other elements (U, As, Mg, Hg, Cd, Se, Mo): and high sulfates. Probably only a small amount of these reach humans by direct ingestion, but more is likely to enter the food chain as the trace elements are often selectively taken up by some plants (notably selenium and molybdenum).

Other health effects of coal mining and treatment are severe. Coal workers' pheumoconiosis (CWP) is an illness widely recognized in coal miners working underground, as are other chronic bronchitis and ventilatory impairments such as progressive massive fibrosis (PMF). A study of 1455 bituminous coal miners (reported in NAS-NRC, 1980) showed that 47 percent had CWP and 2.5 percent had PMF, while of 518 anthracite miners 60 percent had CWP and 14 percent had PMF. Both conditions can lead directly to pulmonary fibrosis. Some trace elements appear to correlate with high incidences of these respiratory diseases, including vanadium, magnesium, manganese, and germanium, but their effects on health are unknown.

The drying of coal prior to combustion in power plants also produces large amounts of dust that can be a health hazard to workers. Most of these plants are in West Virginia. The coal wash water or "blackwater" is produced in the amount of two tons per ton of coal washed. The residue contains heavy concentrations of trace elements, and the dried material is very susceptible to transport by wind action.

Coal Combustion

The combustion of coal in power plants results in several kinds of residues being generated. These include boiler bottom ash or boiler slag, fly ash (emitted to the atmosphere or trapped in the stacks), flue-gas desulfurization sludge (FGDS) due to SO_2 removal from the gases being emitted, gases emitted to the atmosphere, and waste waters. A flow chart showing where these wastes are generated is given in Figure 7–30. Many trace elements are released to the environment by the combustion of coal. Table 7–13 shows data for the amounts of several elements, including the amounts released to the atmosphere by coal combustion compared to the amounts released from all other sources in a typical year. The third column shows the ratio of the two numbers. For some elements, such as beryllium, lithium, arsenic, selenium, and vanadium, the amount released to the atmosphere by coal combustion is much greater than that from all other sources; significant quantities of barium, boron, and cobalt are produced as well. Further, although the amounts of mercury, lead, cadmium, chromium, molybdenum, and others are produced only in quantities much smaller than quantities consumed, it is nevertheless unknown what the actual consequences are of their release into the environment. Thus some elements may be fixed moderately close to coal-fired

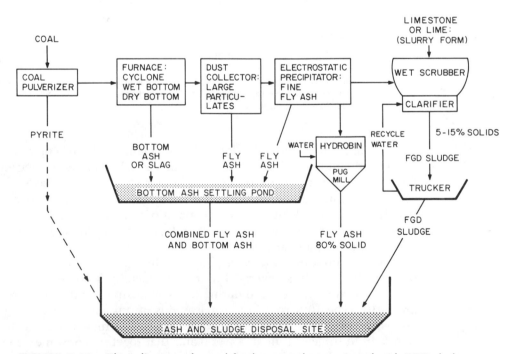

FIGURE 7–30 Flow diagram of a coal-fired power plant equipped with FGD sludge system. (Source: U.S. Environmental Protection Agency)

TABLE 7–13   Elements released from coal compared to releases from all other sources.

Element	Released from Coal (tons/year)	Other U.S. Releases (tons/year)	Ratio Coal/Other Use
Arsenic	17,000	12,600	1.35
Barium	265,355	757,000	0.35
Antimony	618	12,987	0.05
Beryllium	620	192	3.23
Cadmium	720	1,600	0.45
Cobalt	1,768	6,400	0.28
Chromium	6,200	256,000	0.02
Lead	2,800	1,300,000	0.01
Lithium	6,191	3,540	1.75
Mercury	68	1,932	0.04
Molybdenum	835	25,872	0.03
Selenium	1,300	530	2.45
Thorium	1,900	35	54.3
Uranium	1,300	—	—
Vanadium	18,000	8,550	2.11

SOURCES: U.S. Bureau of Mines; NAS-NRC (1980). The figures for the coal use column are subject to large error, perhaps 25 percent. Coal consumption of 890 million tons per year (data for 1984) is assumed.

power plants and be available for uptake by plants and thus enter the food chain. Even small amounts of some elements, such as selenium, may concentrate in this fashion. The health effects of these and other elements are discussed in Chapter 8. A perplexing situation arises in the case of uranium and radioactive uranium daughter elements. While uranium content of most coals is low, it is nevertheless above the crustal average, and the overall amounts of these elements released to the environment are large (Table 7–13).

The amount of wastewater produced during coal combustion is small compared to the amounts generated from coal cleaning, transport, and slurries. The bottom and fly ash produced during coal combustion is 90 percent retained (Figure 7–30). The 10 percent that is released is of unknown environmental impact. The NAS-NRC (1980) reported data for several elements released to the atmosphere from three coal-fired power plants that together consumed 25,000 tons of coal per day. These data are shown in Table 7–14. It must be remembered that, first, some 900 million tons of coal is consumed in the United States each year, but that, second, the trace-element contents of different coals vary widely. Thus

TABLE 7–14 Combined daily atmospheric discharge of elements from three coal-fired steam plants.

Element	Atmospheric Discharge (kg/day)	
Al	510	
As	3.4	
Ba	5.1	
Br	96	(gaseous)
Ca	170	
Cd	0.2	
Co	0.3	
Cr	5.1	
Cs	0.2	
Fe	3,400	
Hg	1.7	(gaseous)
K	153	
Mg	850	
Mn	3.3	
Na	68	
Pb	3.3	
Rb	1.2	
S	150,000	(gaseous SO_2)
Sb	3.3	
Se	6.9	(90% gaseous)
Th	0.2	
Ti	68	
U	0.3	
V	6.8	
Zn	34	

SOURCE: U.S. Environmental Protection Agency.

the data of Table 7–14 cannot be directly applied to overall trace elements added by coal combustion fly ash. It is noteworthy, though, that for the three plants in question, some 150,000 kg of sulfur dioxide is added to the atmosphere every day.

Bottom and fly ash possess low density, thus they are well suited for fill, for lightweight aggregates, and similar purposes. Cement plants in some areas now use fly ash instead of clays or shale in the cement manufacturing process. In part the availability of certain elements to concentrate in the fly ash depends on their relative volatility compared to coal. Coal is normally combusted at about 1550°C, which vaporizes arsenic, barium, bismuth, cadmium, chromium salts, lead, selen-

TABLE 7–15   Percentage of elements entering with subbituminous coal discharged in various coal residues.

Element	Sluice Ash[1] (22.2%)	Precipitator Ash[2] (77.1%)	Flue Gas[3] (0.7%)
Aluminum (Al)	20.5	78.8	0.7
Antimony (Sb)	2.7	93.4	3.9
Arsenic (As)	0.8	99.1	0.05
Barium (Ba)	16.0	83.9	< 0.09
Beryllium (Be)	16.9	81.0	< 2.0
Boron (B)	12.1	83.2	4.7
Cadmium (Cd)	<15.7	80.5	< 3.8
Calcium (Ca)	18.5	80.7	0.8
Chlorine (Cl)	16.0	3.8	80.2
Chromium (Cr)	13.9	73.7	12.4
Cobalt (Co)	15.6	82.9	1.5
Copper (Cu)	12.7	86.5	0.8
Fluorine (F)	1.1	91.3	7.6
Iron (Fe)	27.9	71.3	0.8
Lead (Pb)	10.3	82.2	7.5
Magnesium (Mg)	17.2	82.0	0.8
Manganese (Mn)	17.3	81.5	1.2
Mercury (Hg)	2.1	0	97.9
Molybdenum (Mo)	12.8	77.8	9.4
Nickel (Ni)	13.6	68.2	18.2
Selenium (Se)	1.4	60.9	27.7
Silver (Ag)	3.2	95.5	1.3
Sulfur (S)	3.4	8.8	87.8
Titanium (TI)	21.1	78.3	0.6
Uranium (U)	18.0	80.5	1.5
Vanadium (V)	15.3	82.3	2.4
Zinc (Zn)	29.4	68.0	2.6

[1]Bottom ash sluiced to a settling pond wire.

[2]Ash obtained by electrostatic precipitator.

[3]Fly ash in the flue gas.

SOURCES: Schwitzgebel and others (1975) and U.S. Environmental Protection Agency (1975).

ium, antimony, tin, strontium, thallium, rubidium, and potassium; hence these elements are easily emitted into the atmosphere. More refractory elements and compounds such as uranium, thorium, titanium, silica, alumina, copper, beryllium, cobalt, nickel, and manganese are concentrated in the ash. Table 7–15 shows how elements are enriched in sluice ash, precipitator ash, and flue gas in a typical coal operation.

The flue-gas desulfurization sludge (FGDS) is produced in the attempts to prevent sulfur dioxide (SO_2) emissions to the atmosphere. This sludge, like the bottom and fly ash, is enriched in many trace elements (Table 7–16). FGDS is

TABLE 7–16   Chemical composition of bottom ash and FGD sludge (ppm) and element enrichment ratios in a 330-MW subbituminous coal-fired power plant.

Element	Bottom Ash	FGD Sludge (Solids)	Enrichment Ratio Bottom Ash / Coal	FGD Sludge / Coal	FGD Sludge / Bottom Ash
Mn	3300	2200	41	28	0.7
P	1800	1700	2.8	2.7	0.9
Sr	970	860	7	6	0.9
Sb	0.49	0.82	1.1	1.9	1.7
As	5.4	17	4.2	13.1	3.2
Ba	320	110	2.5	0.9	0.3
Be	11	7.5	5.9	3.9	0.7
B	800	530	7.3	4.8	0.7
Cd	0.67	0.89	1.1	1.4	1.3
Cr	110	180	3.8	6.2	1.6
Co	26	17	4.4	2.9	0.7
Cu	120	340	4.8	13.6	2.8
Pb	8	27	2.8	9.3	3.4
Mo	35	23	2.4	1.6	0.9
Ni	28	19	2.2	1.5	0.7
Se	0.87	5.8	0.5	3.6	6.7
Th	13	18	5.0	6.9	1.4
W	3.7	5.0	1.1	1.5	1.4
U	23	7.2	9.2	2.9	0.3
V	180	280	2.5	3.8	1.6
Zn	10	66	11.5	76	6.6

SOURCE: U.S. Environmental Protection Agency (1975).

produced in stacks where the emitted gases are sprayed with a chemical absorbent such as lime or limestone; this process referred to as scrubbing, captures 60 to 90 percent of the SO_2. Because of the increased concerns over SO_2, and the expected increase in coal combustion, the FGD sludge is expected to reach amounts greater than all municipal sewage sludge by the end of the century (OTA, 1981).

In predicting how trace elements reach the environment, their solubility in water is critical. When first wetted, fly ash has a very high pH, about 12, and under these conditions many elements are not very soluble, although some may be enhanced. As soil waters encroach on the ash, the pH is lowered to 7 or 6, and there is a pronounced increase in solubility, especially for sodium, barium, strontium, molybdenum, uranium, and selenium. The leachate from fly ash and from FGDS typically contains levels of some trace elements (including fluorine, barium, boron, molybdenum, and selenium) above the recommended maximums for both public drinking supplies and for irrigation.

Atmospheric fallout near coal-fired power plants is measurable, but the effects of this fallout on the food chain are not known. The amounts of contaminant elements in fallout are often less than background (soils), yet they may enter the food chain more readily from the fallout material.

Fly ash has been added to soils as a source of trace nutrients. While this is successful for several nutrients such as zinc, the amounts of selenium, molybdenum, cobalt, strontium, barium, and cesium selectively taken up by plants are dangerous to animals (NAS-NRC, 1980). Fly ash has also been used in pilot studies to neutralize acid mine drainage and to remove phosphorus from lakes, with some success.

Radioactive species in coal may warrant more attention than they have had. It has been shown, for example (see NAS-NRC, 1980), that stored or spread fly ash from just one 1000-megawatt coal-fired power plant over its 30-year assumed operational lifetime could result in radon releases to the air from 3.4 to 15 times greater than background. This will be discussed in more detail in Chapter 8.

Synfuels and Coke

Synthetic fuels, or synfuels, are produced by the decomposition of coal upon heating in the absence of oxygen. They represent the volatile fraction, while the residual char, or coke, is used as a fuel and reductant for iron and steel manufacturing. Several steps for synfuel production are shown in Figure 7–31. In most coke-making operations the volatiles are not of concern, and many leak from furnaces and during discharge and quenching operations. The fate of this material, and any trace elements that may be present, is unknown, and data are lacking.

Coking operations are known to be correlated with lung cancer, presumably due to organics and major pollutants (NAS-NRC, 1980), although the role of various trace elements in this is not known. Synfuel production is done using leak-resistant systems, such that occupational exposure is small. Even a small leak, however, could have serious consequences.

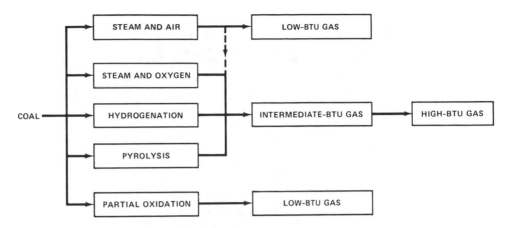

FIGURE 7–31   The coal gasification process. (Source: U.S. Department of Energy)

Acid Deposition and Air Pollution

These two topics, acid deposition and air pollution, are treated together because both are closely related to contaminants produced by burning of fossil fuels, mainly coal, and from emissions from vehicles. *Primary air pollutants* are sulfur dioxide (SO_2), nitrogen oxides (NO_x) (the "x" because several oxides of nitrogen may form), and hydrocarbons. *Secondary pollutants* are sulfuric acid, nitric acid, and ozone; SO_2 is the source for the sulfuric acid, NO_x the source for the nitric acid, and NO_x and hydrocarbons the source for ozone.

Sulfuric and other acids produced in the atmosphere are transported by prevailing winds and removed by precipitation (Figure 7–32), and there exists a strong correlation between acidity of precipitation and dissolved sulfate and nitrate in the rain. In addition, there is a certain amount of acid fallout under dry conditions. it is difficult to pinpoint exact sources of particular pollution and resultant deposition. Figure 7–33, a map of eastern North America divided into four regions, shows a breakdown of sulfur dioxide emissions and total sulfur deposition. The conclusion is reached by scientists (OTA, 1981) that all regions contribute to deposition in other regions, that each region contributes to its own deposition more significantly than do other regions, that substantial amounts of emission have originated from more than 500 km away, and that more pollutants are transported from west to east and south to north than from east to west or north to south. This is important as it is commonly assumed that coal-burning power plants in the upper Mississippi Valley (region IV) has exported significant acid deposition to region I to the northeast.

The effects of acid deposition are most apparent on small lakes and streams with little capability to neutralize acids. When the pH reaches about 5 this condition is toxic for fish, and major changes in the lake ecosystem occur. Metals mobilized under lower pH, including aluminum, lead, cadmium, and mercury, may also have a pronounced effect on aquatic life.

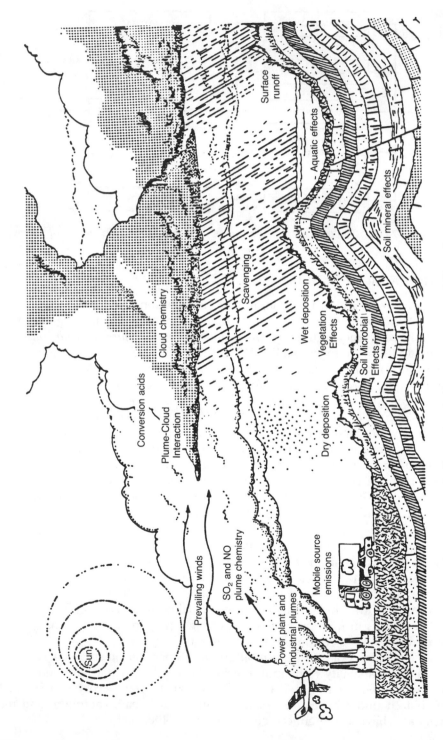

FIGURE 7-32   Processes and environmental effects of acid deposition. (Source: Office of Technology Assessment)

Sun

Prevailing winds

Conversion acids

Plume-Cloud Interaction

Cloud chemistry

Scavenging

Surface runoff

SO_2 and NO plume chemistry

Power plant and industrial plumes

Mobile source emissions

Wet deposition

Aquatic effects

Dry deposition

Vegetation Effects

Soil Microbial Effects

Soil mineral effects

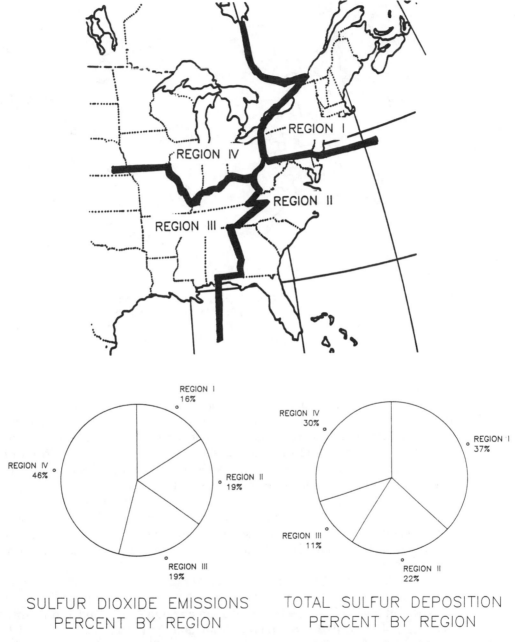

REGION I
16%

REGION II
19%

REGION III
19%

REGION IV
46%

SULFUR DIOXIDE EMISSIONS
PERCENT BY REGION

REGION IV
30%

REGION I
37%

REGION II
22%

REGION III
11%

TOTAL SULFUR DEPOSITION
PERCENT BY REGION

FIGURE 7–33   Emissions and deposition of sulfur in eastern North America. (After U.S. Office of Technology Assessment)

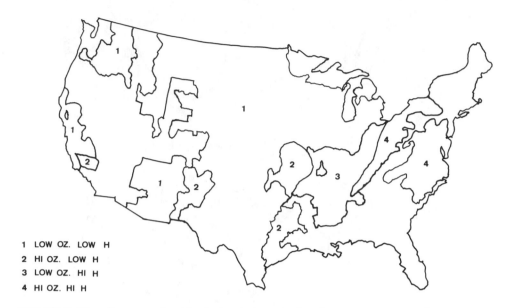

1 LOW OZ. LOW H
2 HI OZ. LOW H
3 LOW OZ. HI H
4 HI OZ. HI H

FIGURE 7–34   Ozone and acid rain levels in agricultural areas (counties with greater than 30 percent agricultural land) for 1978. "High ozone" is greater than 60 parts per billion June–September maximum 7-hour daily values; "high acid" is greater than 0.50 kg per hectare annual deposition of H^+ ion. (Source: U.S. Office of Technology Assessment)

     Figure 7–34 shows information about areas identified as being sensitive to acid deposition and ozone. The lakes total more than 17,000 and streams total more than 117,000 miles of first and second order streams. The estimate by OTA (1981) is that for a 10 percent increase in acid deposition by 2000, some 5 to 15 percent of these lakes and streams will worsen—either become acidified or prone to acid inputs. This same study emphasizes that cessation of acid deposition can allow most of these waters to recover within a few decades. The long-range effect of the acidity on lake and stream bottom soils, however, is not known.

     Acid deposition effects on land vegetation are more difficult to assess. The role of ozone in this situation is also not well understood. In some cases the acid deposition may actually prevent damage by ozone; in other cases the opposite may be true.

     Burning of fossil fuels in the United States adds roughly 30 million tons of SO_2, of which in the eastern 31 states some 74 percent is from fossil fuel burning electrical utilities (Figure 7–35). In this same region NO_x is added primarily from vehicles (44%), with large amounts from utilities (34%) and industrial boilers (17%) as shown in Figure 7–36. The focus in the area of acid deposition has traditionally been on these 31 eastern states, although some recent research suggests that the western states may be experiencing the adverse effects of acid deposition.

     Motorized vehicles produce significant amounts of both NO_x and hydrocarbons. These two primary pollutants in part go through chemical reactions to form

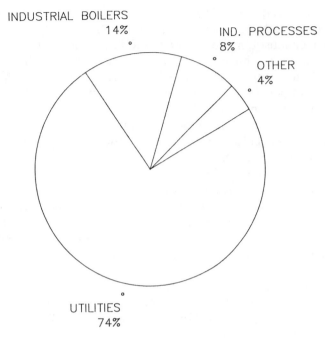

INDUSTRIAL BOILERS
14%

IND. PROCESSES
8%

OTHER
4%

UTILITIES
74%

FIGURE 7–35   Sources of sulfur dioxide emissions in the 31 eastern states, 1980. (After U.S. Office of Technology Assessment)

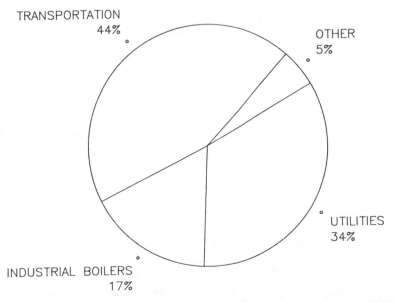

TRANSPORTATION
44%

OTHER
5%

UTILITIES
34%

INDUSTRIAL BOILERS
17%

FIGURE 7–36   Sources of nitrogen oxides emissions in the 31 eastern states, 1980. (After U.S. Office of Technology Assessment)

ozone, and some NO_x goes to form nitric acid, a component of acid deposition. The contribution of vehicle pollution to acid deposition is not clear, since most of the exhaust fumes are close to the surface and not readily dispersed to allow them to reach the mixing layer of the atmosphere. In the Los Angeles Basin, though, acid deposition has been linked to vehicle pollution. Motorized vehicles account for 40 percent of the NO_x emission in the United States, 38 percent of hydrocarbon emissions, and just 3 percent of SO_2 emissions.

The effects of ozone alone are difficult to assess, especially since they may not be recognized for some years. The combined effect of ozone and acid deposition on vegetation is known to some degree, however. Figure 7–37 shows a summary of some of the positive and negative effects of combined ozone and acid deposition on forest growth.

The U.S. Environmental Protection Agency (EPA) has clearly demonstrated that use of emission controls on motor vehicles can drastically reduce NO_x and hydrocarbon pollution. Better emission devices on new vehicles and inspection and maintenance, with corrections where needed, could result in a reduction of these pollutants by 20 percent by 1995. Yet in many parts of the country these procedures are not followed.

The terrestrial effects of ozone and acid deposition are apparent, but difficult to quantify. The OTA (1981) reported that 90 percent of pollution effects on crops is caused by ozone and sulfur dioxide. Yields for corn, soybeans, wheat, and alfalfa would have been significantly higher were it not for ozone and related pollution, a loss of about $3 billion worth of crops. Acid deposition on crops, with

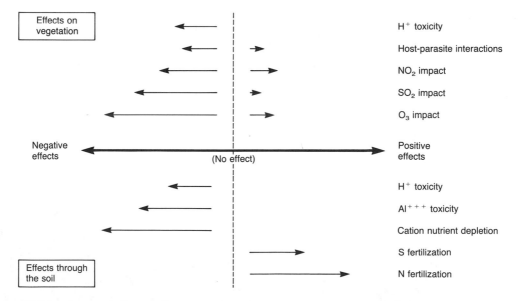

FIGURE 7–37   Relative effects of the mechanisms that affect forest growth due to deposition of air pollutants. (After U.S. Environmental Protection Agency)

and without increased ozone, is shown in Figure 7–34. In general, the lower the pH the more damage to crops, with a wide range of variability. The effect of acid deposition on forests is sometimes more pronounced, in part due to the fact that, unlike agricultural areas, the soil in a forest area is not managed. In the northeastern United States, a noted decrease in growth of red spruce, oak, and several species of pine has been attributed to acid deposition, although proof has not been firmly established. It is known, however, that the growth decline correlates with increased soil and rainfall acidity. The mechanisms perturbing the forest ecosystem are partially known. Some nutrients such as calcium and magnesium are lost under acidic conditions, and some toxic metals are mobilized. The added sulfuric and especially nitric acids in the forest soils could provide these nutrients to the soils, but no such correlation of positive effects with acid deposition is known.

The effect of acid deposition on aquatic life has received disproportionate attention in the United States. Negative effects of acid deposition on fish are well known, but not so well known are the effects on algae, water plants, and certain amphibians. Indirect effects are also known, such as decreasing the food supply to aquatic and bird life as well as mammals.

Fish do not reproduce below a pH of 4.5, and usually do best when pH is 6.5 or slightly higher. Elements such as aluminum mobilized by acid deposition also can affect fish longevity and reproductive capability (OTA, 1981). Studies of the lakes of the Adirondack Mountains in New York state and several lakes in Canada clearly show that pH has dropped over the last 40 years or so, and that when pH reaches about 5, the fish die off and no new fish live in the waters. The effect of acidity increase during spring snowmelt times is well known and somewhat predictable; but life returns after the acidic snowmelt is neutralized. In the case of the acid deposition, however, the acid supply is too large and the neutralizing agents insufficient to counter the acidity. Salamander and frog populations are especially sensitive to pH lowering, and deaths and deformity result (OTA, 1981). Crayfish and mollusks do not survive in the lower pH waters. Algae pose a special problem. Under acidic conditions, more acid-resistant algae are produced, but these species are not edible by the zooplankton which would normally feed on them, and in turn are the source of food for smaller fish. The effects of decreasing pH on aquatic organisms are shown in Table 7–17.

Health Effects of Sulfur Dioxide and Nitrogen Oxides

Sulfur dioxide is added to the atmosphere in quantities of about 30 million tons per year in the USA, most from the burning of fossil fuels. By itself, SO_2 is only very harmful in concentrations about 2 ppm; the environmental standard is 0.14 ppm over a 24-hour period. The main environmental concern comes from the secondary sulfates that SO_2 forms, including sulfuric acid and ammonium sulfate. These sulfates form extremely small particles, commonly under 0.001 mm diameter, and they can be carried great distances in the atmosphere. Further, due to their small size, they can reach the very deep passages of the lungs when inhaled.

TABLE 7–17   Effects of
decreasing pH on aquatic
organisms.

pH	Effect
8.0–6.0	In the long run, decreases of less than one-half pH unit in the range of 8.0 to 6.0 are likely to alter the biotic composition of lakes and streams to some degree. However, the significance of these slight changes is not great.
	Decreases of *one-half to one pH unit* (a three- to tenfold increase in acidity) may detectably alter community composition. Productivity of competing organisms will vary. Some species will be eliminated.
6.0–5.5	Decreasing pH from 6.0 to 5.5 will reduce the number of species in lakes and streams. Among remaining species, significant alterations in the ability to withstand stress may occur. Reproduction of some salamander species is impaired.
5.5–5.0	Below pH 5.5, numbers and diversity of species will be reduced. Reproduction is impaired and many species will be eliminated. Crustacean zooplankton, phytoplankton, mollusks, amphipods, most mayfly species, and many stonefly species will begin to be eliminated. In contrast, several invertebrate species tolerant to low pH will become abundant. Overall, invertebrate biomass will be greatly reduced. Certain higher aquatic plants will be eliminated.
5.0–4.5	Below pH 5.0, decomposition of organic detritus will be impaired severely. Most fish species will be eliminated.
4.5 and below	In addition to exacerbation of the above changes, many forms of algae will not survive at a pH of less than 4.5.

SOURCE: International Joint Commission, Great Lakes Advisory Board (1979), after Hendrey (1979).

Prolonged exposure to sulfates can cause chronic lung disease, and at least block the lungs and cause difficulty in breathing. The Office of Technology Assessment (1981) reported that there is a direct correlation between ambient sulfate concentrations and mortality. It also reported the findings of Brookhaven National Laboratory scientists who present strong arguments and data that 2 percent of fatalities in the United States and Canada might be due to atmospheric sulfates, in turn caused by increase in SO_2 in the atmosphere. It is known, for example, that in cases of severe air pollution with high sulfate concentrations, there is a pronounced increase in deaths among people with heart and lung diseases. The elderly and children are also prone to sulfate-induced respiratory problems. Laboratory studies verify that the sulfate particles penetrate deep into the lung's passages, and also that sulfates render lung tissue more susceptible to carcinogenic effects of certain organic compounds. The EPA has also documented that high sulfate levels can aggravate heart conditions, asthma, and chronic bronchitis. In all these studies, and especially so for studies of pollution in cities, the one common denominator of all possible pollutants is that of the sulfate particles, especially sulfuric acid. The sophisticated study by the Brookhaven workers strongly suggests that there are roughly 50,000 premature deaths in the United States each year from SO_2 emissions to the atmosphere, assuming 1980 census figures and 1978 coal emission data. Since the amount of coal burned now is much greater than in 1978, and our population has also increased, then this 50,000 death estimate is a minimum figure. Further, the nature of the scientific model is that this figure could be in error by 50–75 percent. These same workers projected an increase in the mortality figure to 57,000 per year by 2000 without increased SO_2 removal, and even when SO_2 is reduced by 30 percent below 1978 levels the projected deaths per year are still a staggering 40,000.

Of more possible impact, however, is the realization that for every possible death from SO_2-induced sulfates, there may be as many as five times as many people afflicted with lung disease, aggravated heart disease due to lung dysfunction, and sulfate-induced respiratory infections. No estimates of life shortening for this category can yet be made. The same report also notes that both the United States and Canada export about 2000 to 3000 fatalities to each other each year; most of the U.S. exports are due to fossil fuel burning, while the exports from Canada are due to metal sulfide ore roasting.

The role of the nitrogen oxides is less clear. It is known that both NO_2 and NO can cause acute respiratory infections in children and the elderly, but the Brookhaven workers declined to attempt any mortality figure for nitrogen oxides. Chronic bronchitis may be caused by nitrogen oxides, in part from nitric acid formed from the oxides. Again, though, no fatality figures for this cause can safely be derived.

HYDROPOWER

Hydropower is the electric power that is generated from water flowing from a high to a low level. Roughly 13 percent of all U.S. electrical energy is generated from hydropower plants. The principles are simple. Water that accumulates and flows from a higher to a lower level contains a large amount of potential energy which can be converted to mechanical energy. Water is delivered by way of a windpipe (Figure 7–38) to a hydraulic turbine at a low level where the flowing water energy is converted to rotational mechanical energy. The turbine runs an electrical generator.

FIGURE 7–38   Hydroelectric
power system. (Source: U.S.
Department of Energy)

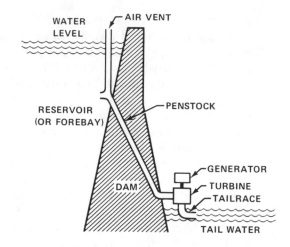

FIGURE 7–38   Hydroelectric power system. (Source: U.S. Department of Energy)

There are two main types of hydropower plants, storage types and run-of-the-river types. The run-of-the-river plants are located where there is a sharp topographic gradient and year-round water supply. The power plant at Niagara Falls is a classic example of such a plant. These plants do not usually have facilities for water storage, thus they can be somewhat prone to the amount of water flow. In the storage hydropower plants, the water is stored in a lake or reservoir, usually created by building a dam across the river. Water level in the impounded area is controlled by a spillway which allows excess water from the reservoir to be removed. During periods of drought when water levels fall, the hydropower plants must be shut down or run at reduced loads. The cost of building dams is very high, but power-plant fuel (water) is abundant and operation and maintenance costs are low.

There are severe limits to hydropower in the United States. It is estimated that in addition to the current generating capacity of 65,000 megawatts, an untapped 135,000 megawatts is available. Yet at least 30 percent of this is in Alaska and not readily available, and much of the remainder is in areas already over-dammed or in wilderness areas. There is a public reluctance to open wilderness tracks for energy exploitation, and in other areas there is objection to the flooding of large areas of land. In addition, the damming of streams has a pronounced effect on the stream ecosystem, and there is environmental concern about the consequences. Untapped sources of hydropower are the numerous dams built for storage purposes in many parts of the United States. While it is estimated that the resource is potentially large (perhaps 54,000 megawatts), not all of these reservoirs are located where there is a need for power.

NUCLEAR ENERGY AND THE NUCLEAR FUEL CYCLE

Nuclear energy is the result of controlled fission of ^{235}U (the isotope of natural uranium of atomic mass 235) in a nuclear reactor. Naturally occurring uranium contains two important isotopes, with ^{238}U making up 99.3 percent and ^{235}U the remainder. Both are radioactive with half-lives of 4.5 billion years for ^{238}U and 0.7 billion years for ^{235}U. Under the nuclear reactor technology employed in the United States, the ^{235}U content must be increased to 3 to 4 percent by weight, thus the naturally occurring uranium must be *enriched* in its ^{235}U content. The enriched uranium is fabricated into uranium oxide (UO_2) fuel rods. The nuclear fuel cycle is shown in Figure 7–39. After use in a nuclear reactor, there are several alternate paths that nuclear fuel might follow. The original plan was to reprocess spent uranium fuel rods, with separation of plutonium and uranium for recycle and the remaining wastes to be disposed of. At present, however, the policy in the United States is to bury the spent fuel rods directly without any reprocessing.

Uranium deposits in the earth's crust range in age from Archean to recent. Prior to about 1.8 Ga (billion years ago), the time when banded iron formations stopped and red bed iron deposits started (see Chapter 4), the oxygen content of the atmosphere was not sufficient to oxidize weathering uranium minerals. Consequently, accumulations of heavy minerals commonly contained minerals such as pitchblende (uranium oxide). Other uranium in the Archean and Early Proterozoic was deposited in veins, especially at unconformities. The world's richest

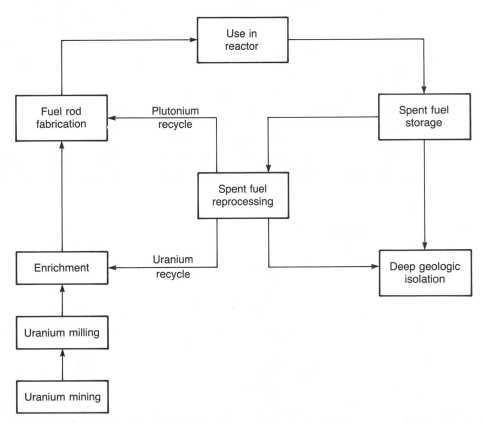

FIGURE 7–39   The nuclear fuel cycle. Reprocessing and recycling are not being done in the United States.

uranium deposits are found in areas of Precambrian rocks, mostly as unconformity-vein types or in fossil conglomerates. In the United States, though, most of the uranium deposits are found in sandstone, and these collectively may contain 20 percent of the world's reserves. A word on how these form is in order here.

Uranium is a fairly rare but widespread element in the crust; it occurs in most rocks and minerals and is present as U^{4+} (uranium ion with a charge of +4). It either forms accessory uranium-rich minerals such as uraninite, or else the U^{4+} substitutes for other ions of similar size or charge. Thus uranium is found in zircon substituting for Zr^{4+}, and also in minerals such as apatite and sphene, substituting for Ca^{2+}. Uranium is also commonly found in the same sites in minerals as thorium, as Th^{4+} is identical in charge and almost identical in size to U^{4+}. Table 7–18 shows the abundance data for uranium and thorium in several common rocks; notably, coaly rocks like lignite can contain high concentrations of uranium, up to 1 percent by weight. Formation of uranium deposits in sandstone requires several things to happen. First, there must be favorable source rocks, even if uranium is only a part per million or two above average, and this uranium must be readily leachible from the rocks. In iron-bearing rocks, when Fe^{2+} is oxidized to Fe^{3+}, yellow to red Fe^{3+}-oxyhydroxides are formed in the rocks. It takes less energy to oxidize U^{4+}, which is highly insoluble, to U^{6+}, which is highly soluble. Hence if rocks contain limonite or goethite or hematite due to weathering, the conditions that formed these minerals were also favorable for release of uranium. This uranium is released as U^{6+} in oxyions such as UO_2^{2+} (uranyl ion) or $UO_2(CO_3)_2^{2-}$ (uranyl dicarbonate ion). At the same time other metals such as vanadium, molybdenum, and selenium form oxyions which have roughly the same solubilities. These metal oxyions are commonly transported together as a group. If solutions containing these metal oxyions percolate into a permeable rock such as sandstone, normally under surface oxidizing conditions, they migrate downward until they encounter chemically reducing conditions. Here, where there is insufficient oxygen to maintain a high oxidation potential,

TABLE 7–18 Uranium and thorium concentrations in some common rocks.

	U (ppm)	Th (ppm)
Igneous Rocks		
Granitic rocks	3.5	18
Intermediate rocks	1.8	7
Basic rocks	0.8	3
Sedimentary Rocks		
Sandstone	0.5–1	2–5
Shale	3.7	12
Bauxite	9.3	53.1
Limestone	1.3	1.1
Lignite	to 10,000	—

SOURCE: Modified from Gera (1975).

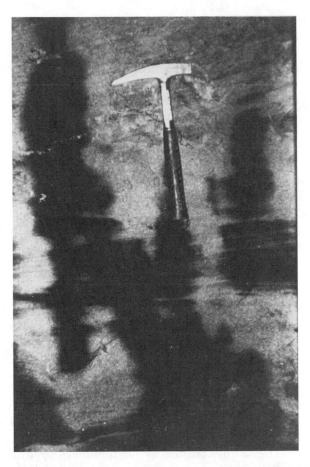

Underground uranium ore from a mine in the Laguna District, New Mexico. This ore, which consists of dark uranium minerals mixed with organic matter, is called "rod ore" because it has seeped downward across bedding planes of the host sandstone. The scale is indicated by the rock hammer.

many metal oxyions undergo reduction from high- to low-charge species. Thus U^{6+} is reduced to U^{4+}. The low-charge species are insoluble, hence minerals containing U, V, Mo and Se form readily at the "oxidation front" which marks the break between oxidized and reduced rocks. This break is easy to see in sandstones, as in the oxidized rocks hematite or other brightly colored iron minerals are found, along with bleached clay minerals and detrital quartz, feldspar and other rock grains; whereas in the reduced rocks they are typically rich in organic carbon, pyrite (which contains the iron), and other sulfides. Thus the exploration geologist can, with due caution, use color of rocks as a guide to looking for sites favorable for uranium ore accumulation.

Once the uranium has been found, it can be mined by underground, open pit, or solution mining techniques. Deep high-grade deposits can be mined by underground methods and much lower grade, shallow deposits by open pit methods. Solution mining is used where the host rocks to the uranium are permeable, and where aquifers will not be degraded. Current law requires that the rocks be restored to groundwater quality conditions as good as those present when the mining was undertaken. Interestingly, at some sites the quality of the groundwater has improved after the removal of ore. Solution mining produces little waste, quite different from conventional mining operations. Typical sandstone ores contain 0.1 to 0.5 percent uranium oxide (1000 to 5000 ppm).

Uranium is recovered from sandstone ores by a simple milling process. Rocks are crushed to a fine size, usually blended based on the radioactivity of the ore so as to homogenize the feed, then leached with sulfuric acid. Rocks rich in carbonate are leached with sodium hydroxide. The leaching solution, which contains sodium perchlorate, a strong oxidizing agent, oxidizes the insoluble U^{4+} to highly soluble U^{6+}, which complexes with the sulfate ion present. This U-sulfate complex is moved through large vats containing resin, with the uranium progressively enriched by stripping and loading. When sufficiently rich, it is separated, dried, washed, and converted to a yellow uranium oxide powder known as yellowcake. All the rest of the starting material, plus, the chemicals and other materials added during the milling process, is waste. After the original crushing, the rock is divided into sand-rich and clay-silt rich (slime) fraction. The slime fraction contains most of the uranium. The sand is moved to an impounded area and built into a tailings pile. The slimes are pumped as a slurry to the top of the tailings piles and allowed to evaporate, leaving a residue that contains clay minerals, pyrite, sulfate, carbonate, chloride,* and many other elements such as vanadium, selenium, molybdenum, and other chalcophile elements. Prior to mining, many of these elements were locked tightly in the rocks; after mining, however, they are prone to attack from the surrounding acid-rich medium.

The tailings created by the mining of uranium ore are not large compared with tailings from other mining operations, but they must be carefully handled. They are weakly radioactive, for example. If a rock containing 4000 ppm uranium is milled, and the milling successfully removes 90 percent of the uranium (3600 ppm), the waste rock still contains 400 ppm. This amount of uranium, while not large, is nevertheless many times over background. Further, one of the intermediate decay products of 238-uranium is 226-radium, and the radium from all the rock is waste. While some radium is removed by barium chloride treatment, much of it winds up in the piles. The daughter product of radium is 222-radon, a gas whose radiation health hazard is discussed later in this chapter. The radon flux over most tailings piles ranges from a few picocuries per liter of air to several tens or even hundreds of picocuries per liter (background values are near 1 pCi/L). For most uranium mill tailings in the United States, background conditions are met within 1 to 3 or 4 kilometers from the tailings pile. The radon tends to

---

* Barium chloride is added to remove radium salts from reaching the tailings piles, and thus adds chloride to the slurry.

be transported in more concentrated fashion if the topography is rugged, and is quickly disseminated in flat-lying areas.

Conversion and Enrichment

Yellowcake is cleaned by removing sodium and ammonia to yield a fairly pure mix of U_3O_8 and UO_3. The uranium oxides are then purified to remove neutron poisons—elements (such as Cd, B, Hf, and Fe) that absorb uranium's energetic neutrons and reduce its fuel value. These oxides of uranium, however, contain uranium of normal isotopic composition, about 0.7 percent ^{235}U and 99.3 percent ^{238}U. Only 235-uranium fissions in nuclear reactors, and it must be enriched to 3–4 percent. First the uranium oxide mixture is converted to a gas, uranium hexafluoride, in a three-step chemical process:

$$(U_3O_8 + UO_3) + H_2 \rightarrow UO_2 + H_2O \qquad \text{step 1: reduction of } U^{6+} \text{ to } U^{4+}$$

$$UO_2 + 4HF \rightarrow UF_4 + 2H_2O \qquad \text{step 2: formation of } UF_4$$

$$UF_4 + F_2 \rightarrow UF_6 \qquad \text{step 3: formation of } UF_6$$

The conversion step is not considered problematic since the precautions taken to handle the conventional chemicals are more than enough to prevent release of radioactive chemicals. Historically, the number of fatalities from the conversion step of the nuclear fuel cycle is under ten (for all operations at all plants), much less than the fatalities for similar or related operations in the chemical and other industries.

To enrich it in 235-uranium, the uranium hexafluoride (UF_6) gas is passed through porous membranes called diffusion cells. The lighter molecule, $^{235}UF_6$, after many thousands of passes will tend to be separated from the heavier $^{238}UF_6$ as shown in Figure 7–40. When the concentration of $^{235}UF_6$ reaches 3–4 percent, this material is separated for fuel rod fabrication. The remaining UF_6, now more enriched in ^{238}U, is sent to special disposal areas or sold for industrial use. This ^{238}U is a potentially valuable resource as it is the starting material for the breeder reactor. (It should be pointed out that there is a major difference between nuclear reactor enriched uranium and nuclear weapons uranium. Reactors use fuel enriched in 235-uranium to 3 to 4 percent whereas nuclear weapons, which depend on instantaneous total fission with release of all energy, use material enriched to over 90 percent 235-uranium. The uranium fuel of nuclear reactors cannot sustain instantaneous total fission, and it will *not* cause a nuclear explosion.) Enrichment can also be achieved by centrifuge and laser methods (see references at end of chapter).

To fabricate fuel rods for nuclear reactors, the enriched UF_6 is reconverted to uranium dioxide, UO_2, using a complex reaction involving water and then ammonia treatment, followed by heating and reduction under hydrogen. This material is powdered, then cold pressed into pellets about 0.9 cm diameter and 1.5 cm long. These pellets are then sintered at 1700°C to increase their density,

FIGURE 7–40 Enrichment
of ^{235}U from ^{238}U by use of
diffusion cells. The gaseous
species $^{235}UF_6$ and $^{238}UF_6$
pass through the cells at
different rates, allowing their
separation.

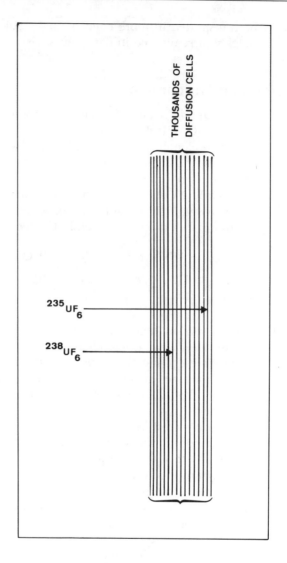

THOUSANDS OF DIFFUSION CELLS

$^{235}UF_6$

$^{238}UF_6$

then they are ground to size and packed end-to-end in zircaloy (a zirconium
alloy) tubing to form fuel rods. The zircaloy is known as cladding.

Nuclear Reactors

Although there are several types of nuclear reactors, only the light-water reactor
(LWR) will be discussed here as it is the most common type in use in the United
States. The LWR contains a central core, as shown in FIgure 7–41, holding the
fuel rods. In the presence of a moderator, the fission reactions that take place
inside the fuel rods produce fission products (new atoms of lower mass than
uranium), neutrons, and other subatomic particles. The energy of the reactions is
sufficient to heat the moderating medium. The neutrons promote additional fis-

FIGURE 7–41   Principle of nuclear reactor. (Source: U.S. Department of Energy)

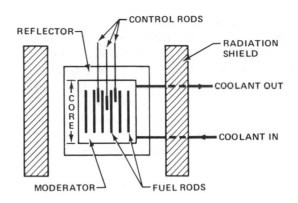

sion reactions until, when the neutron density is sufficient, a carefully regulated, nonexplosive chain reaction takes place. The point where net energy is produced is called *criticality*. The core also contains the *moderator*, the medium which slows down the fission neutrons emanating from the fuel rods. Ordinary or light water is the moderator in LWRs. (Heavy water contains an enriched amount of deuterium, the isotope 2-hydrogen.) Surrounding the core is a *reflector*, whose purpose is to minimize loss of neutrons. A *coolant* is used to remove the heat generated by the fission reactions, and it is circulated through the reactor core. In LWRs, water is the coolant as well as the moderator and reflector, thus making the design and operation simple. The heated water bathing the reactor core is pumped through a heat exchanger hooked into a turbine for power generation.

To control a nuclear reactor requires simply adjusting the neutron density in the core. This is achieved by use of a *neutron poison:* some material that efficiently absorbs neutrons without becoming radioactive. Boron and cadmium are two such elements. These are fabricated into control rods (Figure 7–41), which are fully inserted before reactor operation and at shutdown. By slowly withdrawing the control rods, the neutron density of the core and thus the power output can be controlled. The rods are removed to allow criticality to be achieved. The reactor core, with its control and fuel rods and water, is totally surrounded by a radiation shield consisting of 2 to 3 meters of concrete, which is sufficient to absorb the core-generated neutrons and gamma rays (and beta particles).

Reactors in the United States do not follow the complete nuclear fuel cycle shown in Figure 7–39. In theory, spent fuel rods are reprocessed to reclaim unfissioned uranium (to be made into new fuel rods) and prepare fission products and other radioactive species for disposal. Even though the United States does not presently use reprocessing, other countries such as France and the USSR do this on a routine basis.

The Breeder Reactor

In nuclear fission, two to three neutrons are released on the average for each fission. Only one is required to strike another atom and induce yet more fission, and the others are usually lost by capture or escape. If, on the other hand, these neutrons are prevented from being captured or escaping, they may form a new *fissile* (fissionable) isotope through neutron capture by a fertile isotope. Thus if neutrons bombard 238-uranium, the most abundant uranium isotope, the transuranic product 239-plutonium forms, which is fissile. Further, as the 239-plutonium fissions, it produces more neutrons which interact with even more 238-uranium. Because more fuel is created than consumed, this process is called *breeding*. A nuclear reactor that uses this process is called a breeder reactor. To make this work well, special coolant and moderator fluids must be used that prevent excess neutrons from escaping, such as liquid sodium. The acronym LMFBR stands for liquid metal fast breeder reactor. The core of a breeder reactor contains separate fuel and blanket components. The fuel rods are a mix of 15–20 percent fissile material and the remainder fertile material. The blanket surrounding the fuel rods is fertile material only. As the reactor operates, fissile material is generated from fertile material in both the fuel rods and the surrounding blankets. From time to time both fuel rods and blankets are removed and reprocessed: fertile and fissile materials can then be reused.

Breeder technology has very attractive features. For example, the reactor uses 238-uranium as fuel, which is 99.3 percent of total uranium in rocks. Thus even though the amount of uranium mined to power our light-water reactors is small relative to other energy options, roughly one hundredth the amount of uranium needed for a LWR is needed for an equivalent LMFBR. Very simply, this means that uranium already on hand (either as enriched uranium, stockpiled uranium, or depleted uranium tails) is adequate to meet LMFBR needs for well over a century. In foreign countries the breeder reactor is seen as a necessity. France has been using the breeder since the early 1970s, and the French now are the world leaders in breeder technology. The USSR, UK, Germany (FRG), and Japan are all developing breeder technology. In the United States there has been concern over the fact that plutonium (which is poisonous, carcinogenic, and can be used to make nuclear weapons) is produced in the breeder reactor, and the domestic program has been stopped indefinitely.

Radioactive Wastes

Several types of radioactive wastes are generated from commercial nuclear power plants, from the production of nuclear weapons, and from the mining and milling of uranium ore. These include spent fuel rods, cladding, high-level radioactive waste, transuranic waste, low-level radioactive wastes, and uranium mill tailings.

When nuclear fuel rods can no longer sustain efficient nuclear fission reactions, and when the fission products have built up to a high level, the fuel rod is pulled from the reactor. If the fuel rod is not reprocessed, it is referred to as SURF for spent unreprocessed fuel. These fuel rods are very radioactive and heat generating, and they are stored temporarily in pools of water adjacent to the reactors.

High-level radioactive waste (HLW) comes from the reprocessing of the fuel rods. These processes can extract the fission products, the plutonium and other transuranic elements, from the remaining uranium. In the United States reprocessing is used only for defense-generated wastes, although several other nations, such as France and the USSR, now reprocess power plant wastes.

Material contaminated with long half-life, alpha-particle emitting radioactive isotopes is called transuranic waste (TRU) if its total concentration is greater than 100 nanocuries per gram (100 nCi/g). The defense industries generate most of the TRU.

Low-level radioactive wastes (LLW) are those not classified as SURF, HLW, or TRU. The LLW is very weakly radioactive, or else has come in contact with radioactive materials (the box which contains LLW sealed in a can, for instance, is itself classified as LLW). Basically, most LLW is just trash. Often the most hazardous part of the LLW is the nonradioactive material present. Much of the LLW is generated by medical and educational-research institutions, as well as from commercial and defense facilities.

Uranium mill tailings contain small amounts of uranium (although usually above background) and, commonly, moderate amounts of other nonradioactive elements such as selenium, molybdenum, chlorine, sulfate, and others.

Table 7–19 compares the amount of nuclear wastes generated by the commercial and defense sectors. Through 1985, nearly all of the highly radioactive wastes (SURF and HLW), all of the TRU, and much of the LLW are generated by the defense industries. Commercial facilities are responsible for all of the uranium mill tailings.

The disposal of highly radioactive wastes in the United States is covered by the Nuclear Waste Policy Act of 1982. Both HLW and SURF will be disposed of in deep, mined geological repositories, to be licensed by the U.S. Nuclear Regulatory Commission. These wastes must conform to the following requirements:

1. HLW or SURF waste packages must keep their integrity for 300 to 1000 years after closure of the repository. This time will allow many of the highly radioactive species to decay to innocuous levels.

TABLE 7–19 Quantities of nuclear wastes.

Type	Thousand cubic meters	
	1985	2000
Commercial waste		
Spent fuel rods	5	16
High-level waste	2	8
Low-level waste	1,160	2,441
Mill tailings	100,000	146,500
Defense waste		
High-level waste	355	346
Transuranic waste	286	376
Low-level wastes	2,181	4,043

SOURCE: U.S. Department of Energy DOE/RW-0140 (1987).

2. Following the containment period, the release of radioactive isotopes from the waste system shall not exceed 0.001 percent of the 1000-year inventory (the amount remaining at 1000 years after the repository closure).

3. The release of radioactive isotopes to the accessible environment must adhere to the limits set by law for 10,000 years after disposal. These limits are different for different elements (depending on the element's toxicity) and are set to be very conservative.

Disposal of Radioactive Wastes in Rocks

In the 1950s, the Atomic Energy Commission (AEC) requested the prestigious National Academy of Science–National Research Council to investigate the matter of disposing of radioactive wastes in rocks. The NAS-NRC, after a long and detailed study, recommended that salt beds (see Chapter 5) were the most favorable. Their reasons were many and included:

Favorable thermal conductivity (salt can absorb heat readily)

Low porosity (water can't flow)

Low permeability (low water flow)

Virtual absence of included water

Plasticity (if a fracture, such as a fault, were to cut the salts, salt would flow into and seal the fracture)

After a false start in Kansas, bedded salts of southeastern New Mexico were selected for further study. In the mid-1970s, the Waste Isolation Pilot Plant (WIPP) project was officially endorsed. This site has many favorable features, both geologic and socioeconomic. The geology and hydrology are very well known. The rocks contain water only in the form of microscopic fluid inclusions, water of crystallization of some minerals, and isolated brine pockets; and the fluid inclusions and brine pockets are saturated with salt (i.e., once a solution is saturated with respect to a solute, no more of that solute can dissolve in that solution; thus the brines and fluid inclusions are incapable of dissolving their host bedded halite). There has been active potash mining in the area for 50 years, and technology to mine and maintain openings in bedded salt at depth is very well developed. Geochronologic studies show that the bedded salts have been closed chemical and isotopic systems for 200 million years. The surface value of the land is minimal. The WIPP site requires about one-eighth of one square mile to be permanently withdrawn from possible other use. Since the Environmental Protection Agency (EPA) allows seven cattle to graze per square mile in this arid terrain, the WIPP site, on paper, will require the relocation of one cow. The land is not suitable for growing crops, and the WIPP site occurs in an area of very low population density. Access to the site by both rail and road is excellent. While potash is present in the WIPP site, it occurs in low-grade polyhalite ores which are totally subeconomic at present and not likely to ever be economic. Even so, they could be mined up to within a kilometer of the WIPP shaft. Finally, the southeastern New Mexico area sorely needs the employment and other economic benefits that WIPP would bring to the area.

The WIPP site, however, is a Department of Energy (DOE) facility. It was decided in the late 1970s tht the site would be used only for the disposal of government (mainly defense) generated transuranic wastes (TRU), which are low-level wastes containing one or more transuranic elements (plutonium, neptunium, americium, curium) and uranium. The material to be stored is essentially garbage: old gloves, bottles, other containers, packing, and so on that in one way or another contains small amounts of TRU. Some high-level waste (HLW) and spent fuel rods (SURF) will be temporarily emplaced at the WIPP site to study their effects on salt. By an act of Congress, the site is only licensed to receive TRU for permanent disposal. After the experiments are completed, the HLW will be removed, but it is likely that this site meets all criteria for disposal of HLW and SURF as well.

Over the last few years the responsibility for the disposal of radioactive wastes has been divided between DOE, the U.S. Nuclear Regulatory Commission (NRC), and the EPA. The EPA is responsible for standards, the NRC for licensing and adherence to the standards, and DOE for the selection and building of the repository. The regulations involved are extremely rigid, and great pains are being taken to ensure that a high-level repository will be environmentally acceptable.

Nine sites were selected by DOE for detailed investigation in 1982, including salt domes, bedded salt, basalt, and tuff. By 1988 these nine sites had been narrowed to the tuffaceous rocks at Nevada Test Site (NTS). While we know most about bedded salts, the NTS site was selected in part because of the presence of government facilities; nuclear weapons testing has been ongoing at NTS since the 1940s.

The bedded tuffs of Yucca Mountain at NTS present a unique disposal site in that the waste would be disposed of above the water table. As such, flow of water in the rocks is both rare and erratic. It is also, unfortunately, unpredictable. While unsaturated rocks may offer little chance for water to remain in them, the path that any waters might take is not well known. Contaminant flow in unsaturated media is thus receiving a great deal of attention. The tuffs appear to be easy to mine and, in the absence of water, very stable.

This site is, however, in the western United States. There is a concern that since most of the nuclear power plants are in the eastern United States (see Figure 7–10), there should be a disposal site there as well. The DOE has thus initiated new work on crystalline rocks (here used to designate granitic rocks and metamorphic rocks) through its Office of Crystalline Rock Depository (OCRD) program, and is looking into sedimentary rocks other than bedded salt through its Sedimentary Rock Program (SERP). These studies are not very far along in this country, although in other countries both options are being vigorously explored. Belgium, France, and Switzerland are studying shales, and Sweden, Finland, Switzerland, France, Japan, and the UK are investigating granitic rocks for radioactive waste disposal.

There is much more to the disposal of radioative waste than just finding a good site, however. The U.S. approach is to use the multibarrier system, in which the waste is surrounded by a series of barriers, all or each of which can prevent water from reaching the waste or prevent escape of waste. A generic design is shown as Figure 7–42. The SURF will be encapsulated into some kind of waste form. Borosilicate glass, phosphate glass, ceramics, metal-ceramic blends, and synthetic minerals have all been proposed as waste forms, and are all receiving study at this time. The encapsulated waste will be very resistant to chemical attack and, based on laboratory studies, will be resistant to loss of radionuclides for thousands of years (or longer). The waste form is surrounded by a container made of steel or some other steel-based alloy, itself resistant to most chemical attack for an additional long period of time, perhaps in the many thousands of years range. The canister is surrounded in most schemes by a metal sleeve, then surrounded by an engineered barrier: a mixture of rock and mineral and other materials designed to further isolate the waste. This engineered barrier is typically designed to be rich in clay minerals and to contain some pyrite and carbon (or activated charcoal). The pyrite and activated charcoal ensure a low oxidation potential, and the clay minerals effectively prevent water flow and sorb any radionuclides that might somehow escape. Laboratory studies of some engineered barrier materials indicate retention times of 250,000 years or longer. When the entire system of

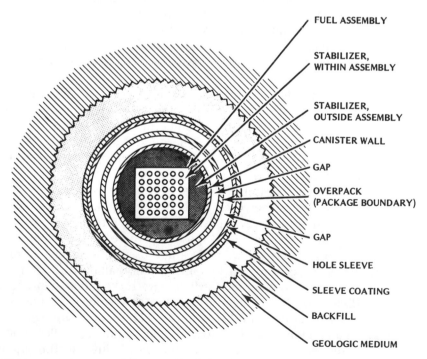

FIGURE 7–42   Conceptual design of high-level nuclear waste repository. (Source: U.S. Department of Energy)

barriers is put together, and when one remembers that the ultimate barrier is the rock surrounding the respository, the isolation of the radioactive wastes is truly impressive. The requirements currently call for isolation, or at worst releases of radionuclides to the environment not to exceed background; and the multibarrier system appears to be adequate to ensure isolation for much more than the 10,000 years required.

Transportation of SURF (and HLW) will be by specially designed carrier containers, designed to resist the most severe possible accidents. Testing of these carriers has included impacting the truck into a cement wall backed by an earthen wall at 75 miles per hour, impacting a train head-on with each going 75 mph, and even dousing the truck and its payload in rocket fuel and burning it. In all cases the canister of simulated high-level wastes did not leak.

The transportation of TRU is carried out in special trucks designed to each carry forty-two of the stainless steel 55-gallon drums containing the TRU. The truck is designed to resist impact and fire as well.

Transportation of LLW is normally by rail or by specially designated commercial carriers, and while standard precautions are taken, the ultrasafe measures for SURF (and HLW) and TRU are not deemed necessary.

Low-level radioactive wastes (LLW) are, by volume, after uranium mill tailings the largest of the different radioactive wastes. This category includes all used materials and supplies contaminated or in some cases merely exposed to low levels of radioactive materials. It differs from TRU in not containing any long-lived transuranic isotopes. LLW is generated everywhere, including educational and medical facilities, the military, the other parts of federal and state and local governments, and industry. Where to put this mountain of garbage is problematic. At present only three facilities are licensed to receive LLW: Beaty, Nevada; Barnwell, South Carolina; and Hanford, Washington. Older sites were at West Valley, New York; Maxey Flats, Kentucky; and Sheffield, Illinois. These three older sites are in areas that receive considerable precipitation, and they were not as carefully selected as desirable. There has been leakage of tritium for example, at Maxey Flats. While the leakage is minimal, better site selection would have prevented it. Now these sites must be carefully monitored, and plans made for remedial action should further leakage occur. Sites for LLW are being actively sought in all 50 states, and the utmost care is now taken to ensure environmental safety. Many states have formed compacts, with each state taking its turn to dispose of LLW for all compact members. Because of its low radioactivity levels, LLW may be disposed of in steel drums emplaced in 30-foot-deep trenches and then covered by earth fill.

An interesting idea for LLW disposal involves the poisoned land/inland sea concept. This concept argues that there exist many areas in the United States that have been naturally or otherwise poisoned, and that many of these areas are in closed hydrologic basins in areas of no resources, isolated from population centers. Many playas in Nevada and elsewhere are good examples: elements such as arsenic have accumulated to the point where it is damaging, if not fatal, to plant and animal life. If these playas are closed hydrologic basins, then adding LLW to these sites allows use of otherwise unproductive land and at the same time eliminates part of the LLW problem. Interestingly, while LLW may not be disposed of in such sites, these sites are also just as favorable for chemical waste disposal.

Alternate Methods of Disposal

Numerous methods for disposal of radioactive wastes have been proposed over the last 30 years or so.

1. Rocketry. The idea is to encapsulate the waste and fire it into orbit. The main pro for this option is removing the waste from the earth; the cons are expense, transportation costs, the risk of rocket failure, and lack of retrievability.

2. Antarctica. Use of the Antarctic for radioactive waste disposal would require a United Nations action on an old resolution barring such disposal. Further, there are severe logistical and other drawbacks to this option. Safety on the sea lanes is only one factor. Dual port facilities to load and unload waste and transportation over the Antarctic land and ice are expensive and cumbersome. Retrievability would not, in most scenarios, be possible.

3. Deep wells. This has been used for some chemical wastes in the Rocky Mountain area. The intent is to isolate the wastes by injecting them into very deep wells drilled from essentially impermeable rocks. This method poses problems of possible contamination of aquifers and other geologic-hydrologic risks, however.

4. Rock melt concept. Here the radioactive wastes would be concentrated and encapsulated and allowed to melt their way into the earth. This is theoretically possible, but not a good option due to unknown geology, possible interactions with water, and many other factors. Again, there is no retrievability.

5. Subduction zones. Putting radioactive waste on the lip of land about to be subducted looks good on paper to nongeologists. Unfortunately, the rate of subduction is so slow, and the opportunity of the wastes to react with ocean waters so great, that this method is not likely to ever be considered seriously.

6. Ocean floor disposal. This method has the main advantage that it probably would successfully isolate the wastes from the surface environment. Putting the wastes into the abyssal plain muds, in areas of extreme tectonic stability, meets many of the criteria for disposal. Technically, the problems are few. However, the expense, risks of sea transports, and lack of retrievability are major factors to consider. Also, the United Nations must agree on the legal status of the seafloor.

Natural Analogs for Radioactive Waste Disposal

It is often stated that in the disposal of radioactive waste, we are conducting an experiment for which there is no precedent. Yet this is not true. There are natural analogs for all the elements (and isotopes of the elements) involved. Let us review for a minute the elements involved. Uranium is the fuel, and small amounts of the transuranic elements neptunium, plutonium, americium, and curium are formed by nuclear reactions in the fuel. The uranium and the transuranics decay ultimately to isotopes of stable lead (206, 207, 208) and bismuth (209), and all the usual actinide daughters are generated during this overall decay. Fission products are produced in some quantity; the isotopes of concern involve those of krypton, xenon, strontium, zirconium, tin, technetium, antimony, iodine, cesium, barium, and the rare-earth elements. Knowing the mix of uranium isotopes, we can predict the mix of the product isotopes at any subsequent time; likewise from the products we can infer the source composition.

The branch of geology that deals with the dating of rocks and minerals according to their mix of isotopes is called geochronology. Rocks and minerals are dated by any number of methods, but the ones of interest here are uranium-lead, thorium-lead, rubidium-strontium, potassium-argon, and samarium-neodymium. Specifically, 238-uranium decays to 206-lead, 235-uranium decays to 207-lead, 232-thorium decays to 208-lead, 87-rubidium decays to 87-strontium, 40-potassium decays to 40-argon and 40-calcium, and 147-samarium decays to 143-neodymium. For example, if a rock has been dated by all these methods with concordant dates, we know that this particular rock has remained a closed system— it has not lost or gained parent or daughter isotopes in the time since it formed to the present. Geochronologists have dated materials as old as the earth itself (4.55 Ga, which is obtained on meteorites) and the most ancient crustal materials (3.8 Ga in Greenland and northern Canada).

From these studies we know that rocks do behave as closed systems in many cases to uranium and thorium, which as actinides are analogs for the transuranics as well due to their similar chemistries. They have also remained closed systems to actinide daughters (protactinium, radium, radon, polonium, bismuth, and lead) and the eventual stable lead isotopes; for, if this were not true, the ages would be too young and wildly discordant. They have also remained closed systems to rubidium and strontium, and thus to cesium as well since rubidium and cesium behave nearly identically in rocks and in solution. Further, barium is chemically very similar to strontium, so if a rock is a closed system to strontium it is also probably closed to barium. Potassium-argon dates tell us that the noble gas argon is also retained in many rocks. Since noble gases behave alike in nature, we can safely say that krypton and xenon would also be retained. The date from the samarium-neodymium method involves two rare-earth elements, hence all the rare earths with their nearly identical chemistries must also have behaved the same.

Collectively, radiometric age determinations tell us that rocks and minerals can and have retained their chemical and isotopic integrity for very long periods of geologic time, much longer than the isolation time for buried radioactive waste now required, and that these systems are only disturbed when exposed to chaotic conditions of thermal metamorphism or plutonism, at very high temperatures and pressures and for tens of thousands of years. Buried radioactive waste will have initial temperatures near or less than 150°C, and this will drop rapidly as the highly radioactive species decay away. These studies have shown no migration for zirconium, molybdenum, niobium, cadmium, strontium, rubidium, uranium, thorium, lead, antimony, tin, barium, rare earths, and others.

But how about other elements such as technetium, antimony, and iodine? One approach is to examine the zones of contact metamorphism surrounding igneous dikes and plutons, and see what effect the emplacement of molten rock at extremely high temperatures (800–1200°C) and high pressure—and always without benefit of barriers—has on elemental distribution in these affected zones. In these cases, for the most part, the redistribution of elements is by no more than millimeters, even when the thermal regime is maintained for 10,000 years and in the presence of fluids. This argues that at the much lower temperatures of our own radioactive waste, with the barriers described earlier, with the absence of available fluids for reaction with the waste, and with the decay of the radioactive species, that there will be essentially no migration of elements from the emplaced canister to pose a threat to public health.

The best natural analog is the Oklo natural reactor. This is the name given to a very ancient uranium deposit found in the Republic of Gabon near the small town of Oklo. Nuclear fission reactions in a body of uranium ore there sustained criticality lasting up to 500,000 years, during which time hundreds of tons of fission products formed. All of these "wastes" can be accounted for in the local rocks. The uranium in the rocks was deposited in concentrated form at about 2 billion years ago, when uranium deposited in rocks contained 3.25 percent 235-uranium. (Because 235-uranium has a shorter half-life than 238-uranium, the percentage of 235-uranium is progressively greater the older the rock.) Thus at Oklo two billion years ago the ore was actually rich enough to serve as a fuel. But other criteria had to be met, too. Evidently the original low-grade ore was deposited in sandstone, but the sandstone fractured at depth and the surrounding shale wedged into the fractures between the slabs. During this time some of the low-grade uranium was dissolved and reprecipitated in these shale-filled fractures, and in very rich amounts, sometimes to 70 percent uranium. Thus another important criterion, high uranium content, was realized. Yet the story doesn't end there. For a reactor to work there must be a proper moderator, and apparently just enough water was present—about 15 percent—to serve as moderator. Further, there must have been a lack of neutron poisons. All of the conditions were favorable at Oklo, and consequently a long-time nuclear reaction was started, soon after the uranium was deposited and remobilized.

But how do we know the rocks at Oklo went critical? The evidence first surfaced at the French gaseous diffusion plant at Pierrelatte, where a sample of African ore was found to be significantly depleted in 235-uranium. The ore was traced to Oklo, and a high-grade sample was tested at 0.4 to 0.5 percent 235-uranium rather than the usual 0.71–0.72 percent; in other words, the particular samples were strongly depleted in 235-uranium. The only natural process by which this can happen is nuclear fission. The French Atomic Energy Commission, under the investigatory leadership of Robert Naudet, immediately argued that the Oklo ores had fissioned, and that, if so, the evidence for fission products in the rocks would have to be found. This can be done quite easily in many instances, because the isotopes of many fission elements are produced in different amounts than are found in the natural element. Rubidium, for example, contains isotopes of mass 85 and 87, and the natural ratio of 85/87 is 2.59. But nuclear fission produces rubidium with an 85/87 ratio of about 0.35. Today's mass spectrometers can measure such isotopic ratios to better than ± 0.00001, so even very small anomalies can be detected. Naudet and his coworkers found that they could account for the presence and retention of most of the fission products, as well as for the actinides. These studies were expanded to include investigators from outside France, including the United States, and major conferences on the Oklo Natural Reactor or Phenomenon were held in Libreville, Gabon in 1975, and in Paris in 1977.

The results of these studies, which are still ongonig, show that the Oklo body acted not only as a nuclear reactor but also as a nuclcar waste repository, and, most important, *all* the radionuclides of concern were kept in the rocks without any appreciable migration. The Oklo data convincingly show that rocks can and do retain radioactive wastes, and it should be noted that the Oklo rocks are not especially good candidates for a repository as they are fractured, impure, and somewhat porous and permeable. Yet they have retained their waste for two billion years. All in all, Oklo is a fascinating scientific gold mine, and it is receiving continued study. Yet Oklo did not act as a high-powered reactor. It sustained fission for at least 500,000 years, produced some 16,000 megawatt-years of energy, and produced hundreds of kilograms of fission products, all of which stayed put in the rocks or migrated no more than a few meters (and at a depth of 500 to 100 meters or so). Oklo is the strongest testimony known for rocks' ability to retain radioactive wastes.

Uranium Mill Tailings

Uranium mill tailings result from the waste rock generated by processing of uranium ores. In the abundant sandstone-type uranium deposits of western United States, only 0.5 to 0.2 percent of the rock is uranium oxide, and the rest consists of quartz, feldspars, and rock fragments. Many metals are also fixed with the uranium in and with organic carbonaceous matter and pyrite.

The tailings generated in the United States through 1984 total some 200 million metric tons, and cover over 1300 hectares at 51 sites, 26 abandoned and 25 active. Of these sites many are on standby at the present time, but they have not been officially closed.

The abandoned sites are covered by the Uranium Mill Tailings Remedial Action (UMTRA) program. All of these sites are in the western United States (Colorado, New Mexico, Wyoming, Idaho, Arizona, North Dakota, Oregon, Utah, and Washngton) except the site at Cannonsburg, Pennsylvania; also, there are two active mills in Florida associated with uranium by-product recovery from phosphate mining.

As mentioned earlier, sands constitute some 65 percent of the tailings by volume and slimes 35 percent, but the slimes—because they contain residual uranium and most of the radium—contain 85 percent of the total radioactivity. The slimes also contain most of the molybdenum, vanadium, selenium, arsenic, and other trace metals. Furthermore, the solutions accompanying them are high in chloride, sulfate, sodium, some organics and have pH values of 1–2.

Some of the problems or potential problems from the uranium mill tailings are:

1. Generation of 222-radon from the piles. Both 230-thorium and 226-radium, the grandparent and parent to 222-radon, have long half-lives and are present in the piles, hence there is a potentially large radon flux. The concern is that the 222-radon flux from the piles constitutes a health hazard (see ''Indoor Radon,'' this chapter).

2. Removal of radioactive particles from the piles by wind. The concern is that these particles would be inhaled by people living near the piles and constitute a health hazard.

3. Contamination of water by dissolved heavy metals, radionuclides, or other toxic materials (such as organics) seeping into the ground beneath the piles. The very acidic tailings waters contain large amounts of dissolved toxic materials, and how far some of these materials may be transported needs to be established.

4. When tailings are used for construction or other industrial indoor purposes, there is a potential for creating a new indoor radon source that may be of potential threat to public health.

As with any other part of the nuclear fuel cycle, there is very wide disagreement on the magnitude of the threat to public health. On one extreme, tailings are considered to be the most serious type of radioactive waste because of their large volume and the presence of long-lived radionuclides. On the other extreme, a negligible radiological hazard to public health is advocated because of the very low radioactive content of the piles (at or just barely above background) and, in most cases, their remote locations.

In 1984 the National Academy of Science's National Research Council was asked to look into the question of uranium mill tailings in terms of the science behind the EPA standards, and to make an overall assessment of environmental mill tailings problems, including public health. Their findings over a two-year period may be summarized as follows:

1. Health risks from radon to the general population in the United States are trivial if not negligible, and these risks are minimal for most people living close to piles, although, in some special instances, individuals living very close to certain piles might be subject to a greater risk.

2. Each site is distinctive in terms of geology, topography, hydrology, climate (direction of winds, precipitation) and other factors such that risk-management strategies must be tailored to each site.

3. A low-level surveillance of the piles should be started and maintained over a long term, and must have provisions for corrective measures should they become necessary.

In addition, the panel concluded that risk from particulates was small, though a covering designed to prevent active wind attack should be investigated for future use, but that the measure advocated by the EPA and NRC—putting 6 meters of soil over the tailings—is unwarranted. The panel noted that given the potential for groundwater and surface water contamination near the piles, a monitoring system should be put in place, with provisions for remedial action if necessary. To date, such contamination is well in check and is being monitored. Physical moving of piles was deemed not necessary for most tailings piles; the overall toxicity of the tailings is very low, but a large-scale moving operation would pose much more risk in terms of disturbance of stored radioactive and other hazardous materials, loading risks, transportation risks, and so on. Only the Vitrio Pile in Salt Lake City, in an industrial park area, has been moved, although the Cannonsburg, Pennsylvania pile is now being moved.

To leave uranium mill tailings where they are may come as a surprise to many, but the NAS-NRC panel made this recommendation primarily because most piles are in isolated, arid locations and are, as such, not a threat to public health or the environment. However, they did recommend an active low-level monitoring program with accompanying recordkeeping to ensure that there is no human intervention. They also recommended that tailings not be used for building materials, and if used for other types of construction, that the sites be monitored.

One concern is that of flooding, which is of greater significance at some tailings piles than others. Short of heroic measures, it is probable that future 100-year floods will have a pronounced effect on at least some of the piles. In the late 1950s part of the Green River, Utah pile was washed away by a flood. Studies in the area find no trace of the tailings removed, showing that nature's action in this case effectively distributed the tailings and their constituent chemicals over such a wide area that all were infinitely diluted.

In summary, uranium mill tailings are large by volume and should be monitored, but the threat they pose to public health and to the environment is very small.

Background Radiation

Everything is radioactive, and therefore, by definition, everything emits some radiation, admittedly most of which is very small. Yet we should not lose sight of the fact that radiation and radioactive materials surround us and are in us, and our lives are involved with both in many, many ways. Let us first review some ways of discussing radiation and radioactivity. The most commonly used unit for radioactivity is the curie, which is defined as 3.7×10^{10} disintegrations per second. A more convenient unit is the picocurie, or one-trillionth (10^{-12}) of a curie.

Thus, for indoor radon, discussed later, a typical range of radon radioactivity in air is about 0.8 to 1.5 picocuries (pCi) per liter of air.

Background radiation, on the other hand, is most often measured in terms of rems or millirems, where 1 rem is 100 ergs of absorbed radiation per gram of matter multiplied by some quality factor for different types of radiation. For alpha radiation, the quality factor is 20; for gamma and beta radiation this factor is near unity. Now let us put this into some convenient framework. In the United States, background radiation from all sources is about 100 millirem (mrem) per year. Radiation received from what we eat and breathe and drink (mainly from 40-potassium in food and drink and 14-carbon from the air) is about 22–28 mrem. Cosmic rays give an average individual in the United States about 35 mrem per year, and of course it is higher in states such as Colorado with a higher mean altitude. The exposure we receive from trace amounts of radioactive materials in our building materials and from soil and rock is also about 35 mrem per year. Finally, the average radiation value per individual in the United States from X-ray diagnostic work is about 70 mrem per year; thus a typical resident in the United States will receive about 90 to 160 mrem per year from natural sources, plus additional amounts from medical X-rays.

Our next question, of course, is what constitutes a safe level of radiation? The question is a scientific controversy, but much of the discussion of it in the mass media owes little to the actual arguments. It is a known fact that there is a definite correlation between adverse health effects and radiation dose at high dose levels. What is not known is whether there is a linear relationship between dose and adverse health effects at low levels (Figure 7–43). For example, if someone takes 100 strong painkillers, and an adverse health effect results, then the linear hypothesis tells us that the same adverse health effect for one person would probably occur if 100 people took just one of the strong painkillers each. Applied to radiation, the linear hypothesis is that any dose, no matter how small, will cause some measurable effect on public health. Others argue that the linear hypothesis holds only at moderately to very high radiation doses, and that there is some kind of threshold below which radiation doses are unimportant.

Some data exist for high-dosage, acute radiation effects. At a dose of 25 to 50 rem the first detectable signs of physiological effect on humans appear. Doses of 400 to 500 rem will be fatal to 50 percent of humans so exposed. Yet single X-ray treatment to specific organs for medical treatment may be as much as 5000 rem. What these data suggest is that there is a dose-to-health effect relationship at doses equal to or greater than 25 rem or so. The federal agencies thus, in a very conservative approach, set 5 rem as an allowable limit of radiation dose for so-called low-level radiation. That is, anything under 5 rem is by definition low level.

There are no good nor comprehensive studies on background radiation and health effects in the United States, at least in terms of a statistically large sample for study. In fact, only one such comprehensive study has yet been undertaken anywhere in the world. The People's Republic of China conducted a truly monumental study in the early 1970s. The aim of the study, in rural Guangdong

FIGURE 7–43   The top graph shows the ideal linear relationship between dose and adverse health effects. The bottom graph shows the linear relationship between dose and health effect for a case involving a threshold.

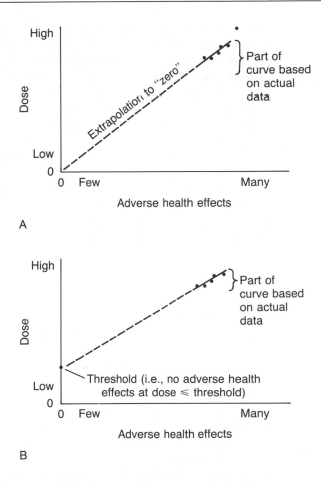

province, was to see if there were appreciable health differences between a large population group under average background radiation relative to an equally large group exposed to much higher background radiation. Two control groups of 70,000 people were chosen, consisting of people whose families had lived in their respective areas for two or more generations. One group lived in a river valley with normal background radiation of about 100 mrem per year. The other group lived on a highland plateau where the cosmic ray exposure is high and where the rocks have a high thorium and uranium content; the resultant background radiation is about 250 mrem per year. The linear hypothesis suggests that the group with higher background radiation should show an increased incidence of birth defects, leukemia, and other cancers, should have generally shorter life expectancy, and should be less healthy on the whole than the group with low background radiation. The people in both groups wore dosimeters over a five-year period (1970–1975), removing them only to bathe or sleep. The dosimeters were collected after this five-year period, measured, and the data studied for three years. Statistically, there was no difference in any aspect of health between the two groups; if anything, the group at the slightly higher background radiation was healthier. This single study argues against, though by itself does not disprove, the linear hypothesis as it pertains to low dose and health effects. Studies of this type are notoriously imprecise, yet the results do suggest that a few millirems exposure above background is probably of no consequence at all. The Chinese study should be much more widely known, for in the absence of any new investigations it remains the one comprehensive study of background radiation ever undertaken.

In the United States, the states with the highest background radiation are Colorado (170 mrem/yr), New Mexico and Wyoming (154), Utah (142), Idaho (134), Montana (133), Nevada (128), Arizona (125), South Dakota (123), and North Dakota (121). There is no correlation of increased cancers or other health effects with background radiation in these states; in fact, North Dakotans have the nation's longest life expectancy.

Indoor Radon

There is, however, one aspect of background radiation that is of extremely serious concern. All earth materials contain at least trace amounts of uranium, and, with time, some of the uranium radioactively decays. As 238-uranium decays to 206-lead, with a half-life of 4.5 billion years, it proceeds through a series of daughter isotopes as shown in Figure 7–44. One of these, 222-radon, is a gas and can easily escape from the soil and rock. In fact, there is a flux of radon everywhere on the surface of the earth, most of which escapes into the atmosphere where it is essentially infinitely diluted. 222-radon by itself is not very hazardous, but with a half-life of four days it decays five separate times through radioactive isotopes of lead, polonium, and bismuth to 210-lead. These radioactive species, here called radon daughters, are extremely carcinogenic and, if formed in the lungs, can cause lung cancer. We all breathe in radon gas, and most of it is exhaled as well. A small fraction of this radon decays to radon daughters in the lungs, but the danger is that the chemically active daughters quickly lodge in the lung where they release a strong dose of radiation in a few minutes of time. We also inhale some of the radon daughters attached to particulates in small quantities. Yet if the radon content is high in the air, so too is the amount of radon daughters.

The indoor radon problem results from the fact that many houses act as traps for radon gas that otherwise would escape into the atmosphere. If a house is built upon high-uranium rock, for example, the radon flux will be very high. Yet even in areas of normal radon flux, a house that is heavily insulated and tightly sealed, usually for energy efficiency, also becomes a trap. The EPA has initiated major programs to address the indoor radon problem: they report that 5,000 to possibly 30,000 lung cancer deaths per year are due to radon daughters formed indoors from 222-radon. This makes indoor radon one of the top killers in the nation, and means that much more research is needed and widespread monitoring in the United States to combat the problem.

FIGURE 7–44 The decay chain for the ^{238}U series. The vertical lines denote alpha decay, in which the nucleus emits a helium nucleus and changes both its weight and its atomic number. The diagonal lines denote beta decay, in which a neutron in the nucleus emits an electron and becomes a proton with no change in weight. (Source: Brookins, *Geochemical Aspects of Radioactive Waste Disposal*, Springer-Verlag, 1984)

^{238}U Series decay chain

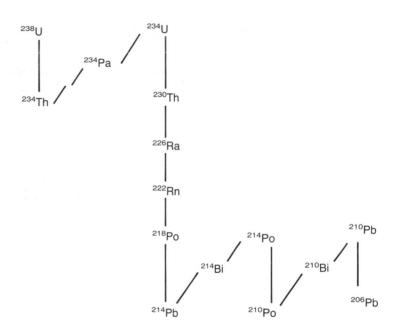

One study in Albuquerque, New Mexico found the soil radon to be about twice the average for the rest of the country. This isn't surprising as the soil in the area is largely granitic detritus, very immature, and geologically young. Its inherent uranium has not been removed as it has in more mature soils. Solar homes are also popular in the area; these dwellings are typically well insulated and sealed. While this practice is great for energy efficiency, it also makes the homes greater potential traps for pollutant gases of all kinds, including radon. Of 200 homes, 53 had at least one reading above 4 picocuries per liter of air (4 pCi/L) which is the standard the EPA uses. This is roughly equivalent in lung-cancer risk to smoking four cigarettes a day. Outside air is usually in the range from 0.8 to 1.5 pCi/L, and the average indoor air in the United States is somewhere between 1 and 2 pCi/L. The figure of 4 pCi/L is considered a reasonable limit for indoor radon for new structures by the EPA, yet it is probably not being enforced nationwide. How radon may get into a home is shown in Figure 7–45.

While health workers have long been concerned over the risk of indoor radon, only in the last few years has the public become increasingly aware of the problem. Interestingly, part of the problem was discovered when a worker at a nuclear power plant set off radiation detectors, and it was found that the source of the radiation was on him when he came to work. Subsequent sleuthing showed that the water he washed with was high in radon, and that measurable amounts of radon daughters were fixed on him. A radon survey of his home

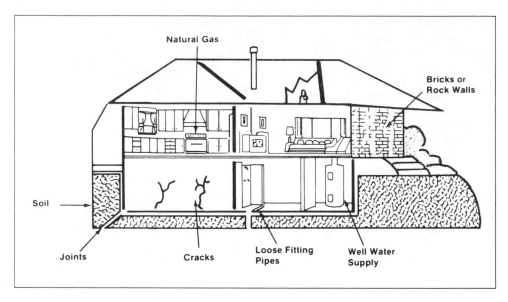

FIGURE 7–45   Ways that radon enters dwellings. (Source: U.S. Environmental Protection Agency)

found very high values. Still later, other houses locally, then regionally, were found to contain high radon values. The Reading Prong, a Precambrian gneissic basement rock complex in the northeastern United States, is fairly uraniferous, and many homes are built on bedrock or on soil barely covering bedrock. Because of the cold winters most homes are heavily insulated, adding to the problem. Several northeastern states soon began full-time studies of the radon problem. Other regions should not consider themselves any safer; given the general lack of data, radon is a nationwide problem of major importance. At present, most of the states have ongoing radon studies, usually coordinated between the EPA and state health agencies. There is a possibility that areas more prone to the radon problem may be identified by analyzing data from the old Project National Uranium Resource Evaluation (PNURE), in which an attempt was made to determine uranium levels for the whole country, using at least one sample per ten square kilometers. Although the intent was to find areas of high uranium background for use in uranium exploration, the data can also be used to zero in on areas of possible high radon flux.

Once high levels of indoor radon have been found, the solution to the problem is often quite simple. In some instances, a simple window fan can flush the air adequately; in others, sealing basement or foundation pipes and cracks will effectively keep radon out. In others, an air-exchange system may be necessary. In cases where large amounts of radon are coming from building materials, such as the case in Grand Junction, Colorado, where uranium mill tailings were used to make construction bricks, the buildings may have to be abandoned or torn down. Again, it is likely that only a few percent of homes in the United States may be affected by potentially hazardous amounts of indoor radon; but without monitoring them it will be impossible to tell. And 6,000 to 30,000 fatalities per year, most of which can be avoided, is of major concern.

Risk from Nuclear Power

In the early 1970s two doctors, Arthur Tamplin and John Gorfman, proposed that there would be an additional 32,000 cancer deaths per year due to radiation from nuclear power plants. Their prediction assumed an excess 170 mrem per person per year from nuclear power alone over a 30-year period for the entire population of the United States. Their assumption was totally wrong: The annual per capita dose of radiation emitted from all nuclear power plants amounts to only 0.28 mrem, and even the most conservative estimate is that nuclear power plants may cause about 0.002 percent of all deaths from cancer. This is shown in Table 7–20, where it is noted that of some 400,000 cancer deaths per year in the United

TABLE 7–20 Annual radiation exposure to the U.S. population and statistical projections of cancer fatalities.

Source	Average Individual (mr em)	Total to U.S. Population (person-rem)	Estimated Cancer Fatalities[1]
Natural background	100	20,000,000	2,670
Technologically enhanced	5	1,000,000	130
Medical diagnostics	90	18,000,000	2,400
Nuclear power (general public)[2]	0.28	56,000	8
Nuclear power (workers)	500	50,000	7
Nuclear weapons (development and fallout)	7	1,400,000	190
Consumer products	0.03	6,000	1
Total from all radiation sources		40,512,000	5,406
Total cancer fatalities from all causes			400,000
Percent from all radiation sources			1.4
Percent from nuclear power			0.002

[1]Estimates are that 1 to 2 cancer fatalities occur per 10,000 person-rem. The estimates above are based on an average of 1 cancer fatality per 7500 person-rem.

[2]Includes all operations of the nuclear fuel cycle.

SOURCE: Based on data from U.S. Department of Health, Education, and Welfare, Interagency Task Force on the Health Effects of Ionizing Radiation, *Report of the Work Group on Exposure Reduction* (1979); National Academy of Sciences—National Research Council, Division of Medical Sciences, Advisory Committee on the Biological Effects of Radiation, *The Effects on Populations of Exposure to Low Levels of Ionizing Radiation* (1972); Leonard Sagan, "Radiation and Human Health," Electric Power Research Institute *EPRI Journal* 4, no. 7 (1979): 6. Estimates by the Interagency Task Force differ slightly from estimates of the National Academy of Sciences BEIR Committee.

States, 5,400 are due to radiation-induced cancers, and almost all of these are due to natural background radiation and medical diagnostics. Televisions emit enough radiation to cause 130 cancer deaths. Only 15 excess cancers are predicted from nuclear power. Compare this to excess deaths per year of at least 50,000 for coal, 2,500 for petroleum, 1,000 for gas, 50 for hydroelectric, and (if solar were a major energy option) a potential 2,000 from solar.

The nuclear industry has a black eye in the mind of the public, but the fact is that no energy option has a better track record. Nuclear power presents the least risk to public health and the least overall environmental damage of all the major energy options. Nothing else even comes close to its enviable safety record.

So why the scare? And why are the media so intent on focusing in on nuclear while ignoring the other energy options? This is difficult to answer. Certainly part of it is lack of education. Many science writers for papers have little (if any) science training, and even fewer have background in atomic physics (even at the beginning high school level).

Let me give two examples of media hype on the nuclear issue. A. C. Upton (see reference at end of chapter), in a 1980 article on low-level ionizing radiation, gave a list of topics to several groups with the instruction to enumerate them from greatest risk, in actual fatalities per year, to least risk. The topics included well-known killers such as smoking (150,000) fatalities per year, alcohol (100,000), motor vehicles (50,000), handguns (18,000–34,000), electric power (14,000), and others ranging from 3,000 (for swimming and motorcycles) down to 100 fatalities and under. The groups to whom this list was given included the League of Women Voters (LWV), various college students, and Business and Professional Club members (BPC). Both the LWV and college students listed nuclear power as the topic with the most actual fatalities per year, and the BPC listed nuclear as eighth out of the 25 topics. In actuality, nuclear with 100 deaths per year (from mining, construction, and transportation—all non-radiation related) ranked twenty-fourth of the list. The LWV and college students cited as their sources of information the media, which allowed them to reach this totally wrong conclusion.

Nevertheless, the risks of nuclear power have been given media attention far out of proportion to their numbers. When Donald Winston inspected the prestigious *New York Times* for 1980, for example, he noted that there were 50,000 people killed on the highways in 1980, yet only 120 stories about fatal car accidents, or one report for every 417 deaths. There were some 12,000 fatal industrial accidents that year with only 50 stories, or one story per 240 deaths. And for 4,500 deaths by suffocation there were 25 news stories or one for every 800 fatalities reported. Yet for nuclear power, with zero fatalities, the *Times* ran 200 stories about death from nuclear power radiation.

Whereas a number of scientists do not support nuclear power, they are a very small minority. One research team (Rothman and Lichter, 1982) selected at random 1000 names from *American Men and Women of Science*, of whom 741 responded to their survey. The researchers broke the group down into subcategories of scientists in energy-related fields, scientists in fields very close to nuclear energy, and those who did not fit either category. On the subject of growth of nuclear energy, 53 percent of the total advocated proceeding rapidly with nuclear energy, 36 percent would proceed slowly, and only 10 percent would halt development or dismantle the existing plants. Seventy percent and 92 percent of the scientists in energy and nuclear-related fields, respectively, would proceed with rapid nuclear power development. This same breakdown also held for questions such as: Are risks of nuclear power acceptable? Would you be willing to have nuclear power located in your community? Is there enough knowledge to solve nuclear problems? The energy scientists also indicated that nuclear should be pursued as an active energy option, behind coal and oil, and over biomass and all solar options.

One item of importance is that most pro-nuclear scientists publish in quality, refereed journals, while most anti-nuclear scientists publish in popular, non-refereed magazines. The anti-nuclear scientists and spokesmen also more frequently address non-technical audiences, and the media is a willing participant. The questions, Should scientists restrict public statements on science policy matters to areas of expertise? and Should research findings be accepted by professional journals before being reported in the popular press?, invoked endorsement by the pro-nuclear scientists and strong disagreement from the anti-nuclear group. Clearly the lines are very distinct. Seventy percent of the scientists writing for popular journals believe that there are serious problems with nuclear power plants, while only 15 percent of those publishing in refereed journals hold this view.

Rothman and Lichter also found, not surprisingly, that science journalists, and especially television reporters, are much more skeptical of nuclear energy than scientists. The entire sample of scientists is very positive on support of nuclear power. Science journalists for many prestigious presses support nuclear power, but less so than the scientists. Science journalists at the *Washington Post* and the *New York Times* are not supportive of nuclear power, nor are television reporters and producers.

Nuclear waste disposal is often brought up in antinuclear arguments. Recently, an in-depth study was undertaken of workers in the field of radioactive waste management by Princeton University sociologists. The 153 people canvassed were from government agencies, national laboratories, private firms, universities, and other groups, all with a high degree of visibility in virtually all aspects of radioactive waste management. When asked to respond to the statement, "Scientific/technical knowledge at present is sufficient for radioactive waste disposal," 80 percent agreed with it. This should not be taken lightly. These results should indicate to one and all that the disposal of radioactive waste is feasible both from a scientific and technical viewpoint.

Nuclear and Nuclear-Related Accidents

Nuclear explosions from conventional light-water reactors or from nuclear waste are not possible. It should be remembered that reactors use 235-uranium that is enriched from 0.7 percent to 3–4 percent, and that nuclear waste contains even less 235-uranium due to its fission. Weapons-grade uranium must be enriched to 90 percent, an order of magnitude higher.

Although nuclear power plants will not become atom bombs, like any complex piece of high-energy machinery they can malfunction. This section reviews two of the most notorious accidents at nuclear power plants, Three Mile Island and Chernobyl.

Three Mile Island

While the nuclear industry has the best safety record of any energy-producing industry, it is not without risk. The most serious nuclear power plant accident in the United States occurred at Metropolitan Edison's Three Mile Island–2 Reactor in Harrisburg, Pennsylvania, on March 28, 1979. When coolant was lost to the core of the reactor and a backup system failed, there was serious damage to the core, a large amount of radioactivity was released to the containment building, and some radioactivity entered the atmosphere. Although media coverage of the accident was intensive and often sensational, it is true that the public was not well informed. More importantly, the accident demonstrated that human error coupled with some valve breakdowns and other minor instrument failures caused the coolant loss. With routine inspection on a regular basis as required by law, the accident could have been prevented. People living near Three Mile Island were not adequately informed of what had happened, what measures were being taken to correct the situation, or whether they should evacuate the area.

The total amount of radiation and radioactivity emitted to the environment was small. Levels of 131-iodine were found to be well below the limit set by the Environmental Protection Agency (EPA). Other radioactive species such as 137-cesium were not detected above background levels. Two million people live within 80 kilometers of the reactor; and the average individual dose in this area was 1.5 mrem, well under the 100-mrem level set by the Nuclear Regulatory Commission (NRC) and the EPA. From a 100-mrem dose there is a 1-in-50,000 chance that a person will develop cancer—compare this to the 1 in 7 normal lifetime incidence of cancer in the population. Health-effect calculations indicate that one excess cancer fatality is likely in the area, compared with 325,000 fatalities to be expected over the lifespan of 2 million people.*

The Kemeny Commission appointed by the president investigated the incident and offered the following recommendations:

1. The NRC must be restructured to ensure smoother functioning.
2. The nuclear industry must improve its attitude toward safety and regulation of nuclear power plants.
3. Greater emphasis must be placed on operator training.

---

* R. A. Knief, *Nuclear Energy Technology*. New York: Hemisphere Publishing Co., 1982.

4. Person-to-machine control must be improved with more emphasis on risk assessment, instrument corrections, and backup systems.

5. More research is needed on low-level radiation and its effects.

6. Emergency planning and response by the utilities and national, state, and local governing boards must be upgraded and coupled with an all-out effort to educate the public on radioactivity and nuclear power.

7. The public has a right to ''up-to-the-minute'' information if an emergency occurs, and steps to prevent erroneous or misleading information must be taken.

Much has been done to correct these shortcomings, but it should be emphasized that these recommended measures are for the most part only enforcing regulations in effect at the time of the accident.

Obviously there is no bright side to a nuclear accident. However, backup systems did work sufficiently well to prevent serious core damage. Thus this ''billion-dollar experiment'' demonstrated that certain catastrophic events (core meltdown, release of excessive amounts of radiation with loss of life and permanent damage to the environment, and evacuation of hundreds of thousands of people) did not happen as some opponents of nuclear power had predicted.

The effects of the Three Mile Island accident can be divided into two types: tangible and intangible. The tangible effects are those described earlier, plus an ultimate cost of about $1 billion for investigation time, reactor downtime, cleanup procedures, repair, inspections, return to licensing, and lost revenue. Hence the accident is called a ''billion-dollar experiment'' based on the approximate expense to make the reactor operative again. The intangible effects are more difficult to assess. It is well known that many people fear radiation and that many individuals are reluctant to continue living near the reactor. Decreased property values and other effects on area residents cannot adequately be put into dollars and cents, nor can the costs of nuclear power development that is delayed, or designed to overcautious safety standards, because of public fears. Nuclear power advocates have reinforced their points that catastrophe did not happen at Three Mile Island and that nuclear power accidents can be prevented by the NRC, EPA, and other groups. If other accidents occur, they will probably be similar to the one at Three Mile Island, where the ultimate damage is to the economy rather than to public safety or the environment.

Chernobyl

In the early morning hours of April 26, 1986, there was a sudden power increase at the Chernobyl Reactor No. 4, some 100 km north of Kiev in the USSR. This power surge resulted in emission of a large quantity of hydrogen gas which exploded, and in turn damaged the reactor, thus starting a very serious fire and ultimately releasing an unknown but large amount of radioactivity. The reactor was nearly totally shut down when the accident occurred because a series of experimental tests were being conducted that violated the most basic safety practices.

The Chernobyl No. 4 reactor is a cumbersome type with huge graphite rods as moderators, quite different from the designs used today in the United States. Further, although there were concrete containment walls around the individual reactor units, the ceilings and coolant towers were not designed to withstand any kind of pressurized condition. The explosion blew through the poor quality roof, ignited the non-concrete building materials, and allowed radioactivity to escape.

Graphite reactors of the Chernobyl type (heterogeneous water-graphite channel-type reactor) in the USSR are, for the most part, old and have never been updated in terms of reactor technology and especially safety. Faulty construction materials were used in the concrete containment structures. One consequence of the explosion at Chernobyl was that, in part due to the high temperatures at which these reactors operate (700°C, as opposed to typical U.S. reactor temperatures of 300° to 400°C), many of the graphite moderator rods were ignited by the explosion, making it even more difficult to contain and put out.

Although graphite is an efficient moderator, it is sensitive to sudden temperature increases. When such a temperature increase occurs, more of the water between the moderator and fuel boils and its neutron-absorbing ability decreases; this is a case of positive feedback which creates a greater potential for a Chernobyl-type accident. In American reactors, temperature increase results in power decrease in a negative feedback that minimizes potentially dangerous situations. Further, the core of the Chernobyl-type reactor is very large, and difficult to monitor and control.

The total amount of radioactivity released is unknown. In the days following, an investigatory team from the International Atomic Energy Agency (IAEA) visited the site; during a flight of 800 meters over the reactor, the IAEA team found peak radiation was measured at 350 mrem per hour. Radiation levels at the time of the accident were reported to the IAEA as being between 100 and 750 rem, and some workers in the immediate site could have received 1000 rem. By late May, over 100 people had been hospitalized and 30 had died from radiation and burns.

The accident should never have happened, and human ignorance and stubbornness played key roles, as did the lack of built-in safety devices in the reactor. Briefly, the reactor operators took advantage of a routine reactor shutdown to see if auxiliary electrical generators could be run from the inertia of the dying turbogenerator of the reactor. The reactor operators reduced the operational reactivity below a permissible level, they ran their experiment at a power level below that specified in the test program, all circulating pumps were on and some exceeded authorized discharge, the shutdown signal from both generators was overridden, as were the water level and steam pressure trips, and the emergency core cooling system was switched off. Any one of these items was foolhardy; collectively they destroyed the reactor. The unauthorized experiment placed the core in an extremely complex configuration with which the reactor operators were unfamiliar or inexperienced. A sudden rise in temperature drove off the coolant, and the core began to generate the hydrogen gas that exploded, starting the chaotic chain of events.

The radiation released that caused the most concern in western Europe was not of significant levels to pose an acute public health hazard, but it was much higher than background. On May 6, experts from the World Health Organization stated, "There is no, and has not been any, immediate danger outside the immediate accident area." Still, several countries issued warnings about drinking fresh milk and washing vegetables carefully, and in Italy fresh produce was ordered destroyed. The European Economic Community banned imports of fresh produce and food from several countries within 1000 km of Chernobyl, including Bulgaria, Poland, Czechoslovakia, Hungary, Romania, Yugoslavia, and the USSR.

The Soviet response to the Chernobyl accident is hard to fully evaluate due to lack of precise information. On April 27, some 36 hours after the accident, 25,000 people living within 10 km of Chernobyl were evacuated. A 38-km radius around the plant was ordered evacuated on May 2, and was completed by May 4. Some 90,000 people in all were evacuated. The IAEA estimated that the peak radiation levels in the 38-km zone were 10–15 mrem/hr, which had dropped to 2–3 mrem/hr by May 5, and to 0.15 mrem/hr at the perimeter of this zone by May 8. Normal background is 0.01 mrem/hr.

It is still perhaps too early to tell either the physical or socioeconomic aspects of Chernobyl. Like Three Mile Island, there was a great deal of fear on the part of the public associated with Chernobyl, and in this case the concern crossed international borders. However, unlike Three Mile Island where the risk to public health was near zero, the Chernobyl accident resulted in loss of life with a strong probability of more lives to be lost in the not too distant future. The Dutch, Austrian, and Yugoslavian governments have placed restraints on nuclear power development in view of Chernobyl, although France, Sweden, the UK, West Germany, Italy, Spain, and others have noted that, due in large part to their different reactor designs and better safety practices, there would be no decrease in their nuclear power effort.

The Chernobyl situation, as viewed one year after the incident, has been discussed in detail by Moore and Dietrich (1987; see references at end of chapter), who note that there will be no demonstrable health effects of the nuclear accident outside the USSR, and even in the USSR these will be confined to the 30 or so kilometers around the site of the explosion. The only positive aspect of Chernobyl is that finally the USSR will cooperate with the IAEA in its program of reactor monitoring and notification of, and information reports about, nuclear accidents.

There are over 500 nuclear power plants either in full or partial operation or nearing completion, and these are found in 38 countries. Most of these plants are of current design and have good safety standards. Most countries cooperate with the IAEA in reactor safety assessment. Yet Chernobyl is proof that some older reactors can be dangerous. In all fairness, however, it must be reemphasized that in the overall nuclear power picture Chernobyl is a small and isolated factor, and nuclear power still has the most enviable safety record of energy options. To reinforce this, all one has to do is compare the fatalities per year just in the United States from major energy options, as is done at the end of this chapter.

Nuclear accidents do happen. Most are so trivial they are hardly worth mentioning, although they are rigorously and thoroughly reported and investigated in the United States by the NRC. The only serious nuclear power plant accident in the last 20 years is that at Chernobyl, and that could have, and should have, been prevented.

ALTERNATIVE ENERGY OPTIONS

Solar Energy

Solar energy is the energy received in the form of radiation that can be converted into other usable forms of energy, such as heat or electricity. Use of solar energy can be divided into three general categories:

1. Direct thermal applications for space heating of residences, for hot water, heat for agricultural operations, and so on.

2. Solar electric applications are those in which the solar energy is converted directly into electrical energy; this includes photovoltaics and solar thermal methods to drive turbines. Also included here are wind power and ocean thermal power, which are discussed separately in this chapter.

3. Fuels from biomass (see "Biomass Energy," this chapter).

All solar methods rely on collectors. A typical flat collector is shown in Figure 7–46, although curved, spiral, and many other geometries are known. In the flat-plate collector, sunlight is absorbed by a flat plate under which are tubes containing a heat transport fluid. Transparent covers allow sunlight in but trap the heat from escaping, and insulation lies below the absorber and heat transport tubes. The absorber is commonly of metal painted black. Sunlight is absorbed on the black surface, converted to heat, and transferred to the heat-transport fluid.

FIGURE 7–46 Cross section of a typical flat-plate solar collector. (Source: U.S. Department of Energy)

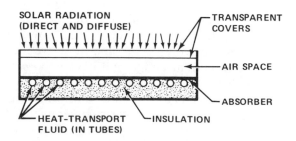

Water is the most common heat-transport fluid, but air and other fluids can be used. In the winter problems may arise from water freezing in the tubes, which can be prevented by adding antifreeze to it, although this tends to foul up the heating system. Corrosion of the tubes is also a problem.

Direct thermal applications of solar energy are widespread in areas of the United States where there is abundant sunshine. In passive solar energy systems, no pumps, blowers, or other mechanical devices are used; air is circulated past solar heated surfaces and through the building by convection. In active systems, fans and pumps regulate the air circulation. A view of a typical active solar energy system is shown as Figure 7–47. A heat storage system is necessary to provide heat (or cooling) during days when the sun isn't shining or at night. Water is used for the heat storage when water is the heat transport agent, and pieces of rock when air is the heat transport agent. Direct thermal applications of solar energy in the United States are used for space heating (and sometimes cooling) of residences and other buildings, for agricultural and industrial process heat, and for other purposes. In areas of the world where there is abundant sunshine and warm temperatures, both active and passive solar energy could, and perhaps should, be an integral part of most new buildings being constructed.

One problem, though not unique to solar energy, can affect both passive and active solar energy systems. Most systems include extensive insulation of the building to ensure that it is as energy-efficient as possible. While this makes the solar energy system more effective, it also increases the likelihood that the building will act as a trap for undesirable gaseous pollutants such as radon.

In order to make solar energy important to the nation's fuel picture, efficient and economic ways to convert solar energy into electricity must be realized. Solar photovoltaic cells can convert the solar radiation directly into electrical energy. Modern photovoltaic cells make use of semiconductor materials, usually silicon. When doped with elements such as arsenic, phosphorus, or antimony, the silicon becomes a so-called n-type material. If doped with boron, gallium, aluminum, or indium it becomes a so-called p-type material. When n-type and p-type materials are joined, they form an electrical barrier (Figure 7–48) across which electricity will flow. This same join can be created using two semiconductors of different materials, such as cadmium sulfide and cuprous sulfide. When exposed to sunlight, some of the radiation energy will cause electrons to be removed from the

FIGURE 7–47 Schematic diagram of an active solar-heating system. (Source: U.S. Department of Energy)

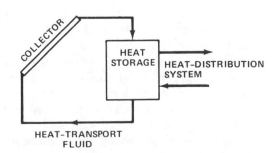

FIGURE 7–48 *Generation of electric current by a solar cell. (Source: U.S. Department of Energy)*

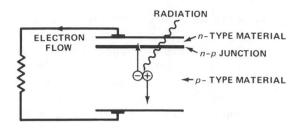

semiconductor, producing free electrons and holes or electron deficits which can be treated as positive particles. The electrical barrier at the *n-p* junction then results in free electrons moving into the *n*-type material and holes into the *p*-type material; and if electrical contacts are connected to each semiconductor part, the electron flow can be regulated into electricity.

Solar photovoltaic cells have some obvious shortcomings. Of the available sunlight, only about 45 percent is capable of producing electrons and holes in the semiconductor materials, and after internal losses the efficiency of converting the solar radiation into electricity in the photovoltaic cell is only about 10–14 percent. Thus the power output of each cell is low. Materials for the photovoltaic cells are problematic. Single-crystal silicon is expensive, and so are the doping techniques used, although amorphous silicon promises great cost savings to offset its lower efficiency. Some of the chemicals used in photovoltaic cells are known or suspected carcinogens, and their handling and processing could introduce major health risk. Arsenic and cadmium compounds are of prime concern; two of the best semiconductor materials, for example, are gallium arsenide and cadmium sulfide. B. L. Cohen calculated in 1982 that the 300 tons of cadmium imported for use in just one large-scale solar electric generating plant would result in 12,000 cadmium-caused cancer deaths over a 30-year period. The overall health effects of large-scale solar electrical energy are discussed at the end of this chapter. Large-scale solar electric generators would require very large amounts of materials, perhaps even straining existing reserves. Finally, the costs of the components, both individually and in total, are still high. Solar electrical energy is not cost competitive with other energy sources for large facilities, nor is it likely to be in the foreseeable future. Photovoltaics are finding uses in specialized applications such as spacecraft, in automatic monitoring instruments such as remote weather stations, and in some consumer products.

Because solar systems only work during daylight, they must be hooked into large storage networks, and they must have a backup power supply. In most scenarios for large-scale solar electric systems, coal-fired power is considered as backup. This, of course, greatly increases the adverse environmental aspects of solar energy. When properly constructed, both active and passive solar direct thermal applications have been and will continue to prove successful on a moderate scale in some areas of the United States and of limited success for the overall U.S. energy picture. Large-scale solar electrical energy generation does not appear to be practical, economic, or environmentally acceptable. However, innovative uses of solar energy for water heating and space heating and for some electricity generation are very important in developing countries, especially in Africa and Asia.

Geothermal Energy

A potentially useful reservoir of essentially inexhaustible dimensions occurs in the upper 10 kilometers or so of the earth's crust. This is the heat that is stored in this zone. It is due, in turn, to higher heat at depth, reflected in the presence of molten rock. In some parts of the earth, the molten rock is close to the surface and in places reaches the surface as volcanoes. If groundwater reaches high-temperature rocks, a geothermal reservoir of steam or hot water can result, where there is a concentration of extractable heat. If these reservoirs can be tapped, their energy can be used for generation of electricity, for space heating of buildings, and for industrial processes.

Worldwide areas of active or recent volcanism contain most of the known geothermal resource areas. For the United States, most of the geothermal resource areas are in the West. (Figure 7–49). Often areas are identified by the presence of steam or hot springs at the surface. Development of geothermal resources is expensive, as the waters in the rocks are not only very hot but highly corrosive. Typical temperatures can reach 350°C, which is much higher than in oil and gas drilling. Drilling muds may react and break down under these conditions.

There are five categories of geothermal resources; hydrothermal convective systems, geopressurized systems, hot dry rocks, magmas, and normal rock areas.

FIGURE 7–49   Geothermal areas in the western United States. (Source: U.S. Department of Energy)

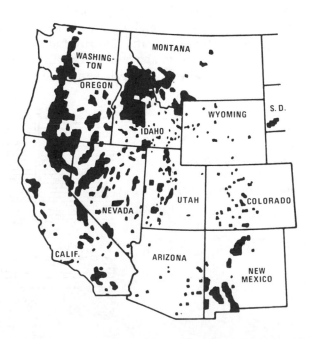

The *hydrothermal convective systems* include reservoirs at moderate depths which contain steam or hot water under pressure at high temperatures. If temperature is high enough, these can be used to generate electricity as is done at The Geysers in northern California, Lardarello in Italy, and Wairakei, New Zealand. If the temperature is lower, they can still be used for space heating, as in Iceland. Hydrothermal convective resources are subdivided into steam (or vapor) dominated and liquid dominated, depending on whether they contain dominant steam or liquid water. The vapor-dominated fields (Geysers and Lardarello) are the easiest to use for the generation of electricity (Figure 7–50). The electricity is generated by use of a turbine as shown in Figure 7–51, and the exhaust steam is condensed, cooled, and pumped back into the ground as shown. Environmental effects of these operations are not fully known. The steam is quite corrosive, containing high contents of hydrogen sulfide (4–5 percent), carbon monoxide, methane (up to 1 percent each), and significant amounts of mercury, arsenic, other metals, and radon. Much of this material is recycled back into the rocks from the cooling towers.

Liquid-dominated hydrothermal convective resources are known at Wairakei, in Iceland, and in the Salton Sea area of southern California. Electricity is generated at Wairakei.

Geopressurized areas consist of salt waters held in rocks at high temperatures and very high pressures, mainly in a belt in Texas and Louisiana parallel to the coast of the Gulf of Mexico. These hot water (brine) reservoirs formed by accumulation of heat in trapped seawater in porous sedimentary rocks over long periods of time. Geopressurized reservoirs also contain abundant methane gas. The economic future of these resources is uncertain.

The *hot dry rock* (HDR) method was developed at the Los Alamos National Laboratory in the mid-1970s. It takes advantage of the fact that when magma is closer to the surface that in surrounding areas, the rocks above the magma chamber are heated to above-average temperatures. Hot dry rocks exist because they are essentially impermeable to water, or else water doesn't reach them. The

FIGURE 7–50
Hydrothermal convective region. The porous rock may be filled with vapor (steam) or liquid (hot water). (Source: U.S. Department of Energy)

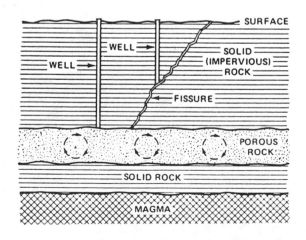

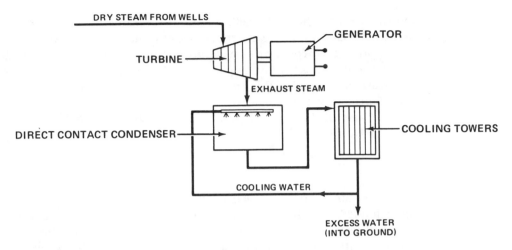

FIGURE 7–51   Hydrothermal power plant. (Source: U.S. Department of Energy)

method, shown in Figure 7–52, calls for drilling a well into the hot dry rocks, fracturing them, then introducing water to be returned to the surface as steam or very hot water. The potential of this method is very high, although there are numerous problems such as clogging of the fractured rocks by secondary products, control of the fractures, effective dissolution of the rock, and other factors.

Magmas are an attractive geothermal resource. How to extract the heat without destroying the extracting apparatus, however, is a major problem. Ways to do this are under study in Hawaii and other areas of active volcanism.

Normal geothermal areas are, at first, not likely candidates for geothermal energy, yet rocks everywhere increase in temperature at depth. A 3-kilometer drill hole in the western United States in an area of high heat flow, for example, might

FIGURE 7–52   Heat extraction from hot dry rocks. (Source: U.S. Department of Energy)

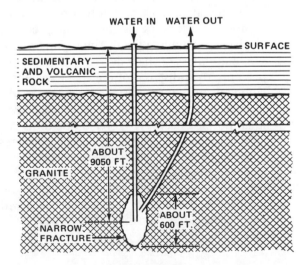

return steam by the HDR method, but at the same depth in the eastern United States no steam would be generated because the rocks at that depth do not contain enough heat. These rocks will, however, generate hot water suitable for space heating and for agriculture applications. The DuPont Company of Wilmington, Delaware is now doing this, and it is hoped that other major industries will follow. Since 40 percent of total energy consumption by industry goes for space heating, use of this technology for space heating would save large amounts of fossil fuels or electricity.

Oil Shale

Under certain conditions, such as in stagnant, swampy lakes that are quiescent for extended periods of geologic time, dead organic life forms accumulate in a stratified fashion. Upon burial, compaction and alteration occur to transform the organic matter into kerogen, a solid of complex and heterogeneous composition consisting of complexly bonded carbon (80 percent), hydrogen (10 percent), nitrogen, sulfur, oxygen, and others (combined, about 10 percent). The resulting rock is oil shale. Kerogen heated in the absence of air undergoes a process known as pyrolysis (destruction distillation) where combustible gas is produced along with a liquid hydrocarbon mixture and spent shale. Oil shales classified as high grade yield 25 gallons or more per ton of shale, or one barrel of oil per 1.7 tons of shale. The combustible nature of oil shales has been known for well over a century, and oil shales serve as a local fuel in Scotland, Manchuria, and elsewhere. In the United States there are tremendous reserves of oil shale, and the U.S. Geological Survey estimates that they may contain 2 trillion barrels of oil equivalent (at 15 gallons per ton). The oil shales of prime interest are found in various nonmarine basins in Utah, Wyoming, and Colorado. The highest grade oil shales are found in the Piceance Creek basin, Colorado, although the largest oil shales are in the Green River Formation of Wyoming (Figure 7–53).

The shale must be retorted to recover the petroliferous materials. This can be done in retort vessels after mining the shales, or in situ by fracturing the rocks in place and igniting them. The mining of these shales is very problematic because while the mass of shale decreases after retorting, its volume increases by some 12 percent. Thus disposal of the spent shale is a major problem. Further, many carcinogenic and toxic substances have been identified in oil shales. The requirement of large volumes of water (3–4 barrels of water per barrel of oil), especially when it cannot be effectively recycled due to its contamination, requires a readily available water supply. But in the areas of interest, water is a scarce commodity, and it is not likely that any substantial amount could be diverted to this purpose. In situ retorting has the advantage that there is no spent shale. The rocks are fractured in place and ignited, and the vapors which collect at the top are cooled and pumped to the surface. A major problem with this method is that most shales are so impermeable that the collection of the retorted vapors is highly erratic and inefficient.

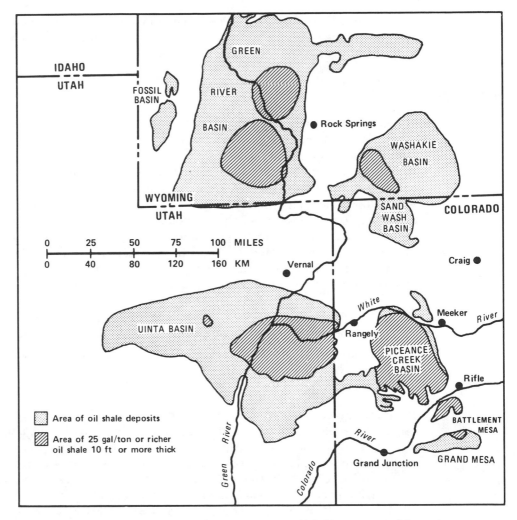

FIGURE 7–53   High-grade oil shale areas. (Source: U.S. Department of Energy)

The main ingredient recovered from retorting of oil shale is a high-sulfur, high-nitrogen crude oil. This can be improved by hydrotreating and cracking, but it still needs extensive refining before it can be used. Thus the oil from these rocks is very expensive.

Oil shales pose a very frustrating problem for the United States. On the one hand, the potential crude oil in these rocks is very large, in theory enough to meet demands in this country for over a century. Yet their requirements of water, waste disposal, decontamination of water used, and other factors make them virtually impossible to use. The pilot plants initiated to explore ways to use this resource have been terminated. It is likely, though, as the world petroleum peaks early in the next century, that some oil shale will prove to be economic, even with the environmental and other obstacles.

Wind Power

Use of the wind's kinetic energy to do work dates back to the first windmills and other devices. Wind is a form of solar energy, originating from temperature differences of the surface of the earth.

Devices to utilize the wind's energy are known as wind turbines. These include sails, vanes, or blades, all radiating from some central axis or shaft. As the wind blows, the shaft is rotated. This allows ready conversion of the wind's energy into electrical energy simply by hooking up a turbine to an electrical generator. This system is commonly referred to as an aerogenerator. Use of wind power to generate electricity dates back to the nineteenth century in Denmark, and many countries have used this power source since. To make a site effective and efficient for wind-generated electricity, the following criteria must be met:

1. Strong winds should be prevalent.
2. No tall buildings or other obstructions upwind from the turbine.
3. The site should preferably be on an open plain, along a shore line, on a smooth hilltop above a plain, on an island, or in a narrow valley in which the wind is channeled.

Batteries must be hooked to the electrical generating system to store energy not used, in case production of electricity in strong winds exceeds demand.

At present, the DOE is conducting a series of pilot tests of wind/electrical-generating power plants. These are part of the wind energy conversion (WEC) plan, funded jointly with NASA. The purpose of this work is to identify areas where wind energy is workable, and to develop technology to make it even more efficient. As a local source of energy, wind power will work in many areas, but it is doubtful that it will produce significant quantities of electricity to make a dent in national needs. Wind energy is harnessed in Hawaii and in the San Francisco Bay area to generate electricity.

Ocean Thermal Energy

The solar energy stored in the surface waters over colder deeper waters is a potential source of ocean thermal energy. To use the energy stored in the surface waters, machines could utilize the temperature gradient between the warm heat source and the larger, colder heat sink to convert mechanical energy into electrical energy. One method showing how this works is illustrated in Figure 7–54. In some parts of the ocean, temperature differences between surface and deep waters reach 20°C. The site for such an operation must, of course, be within a few kilo-

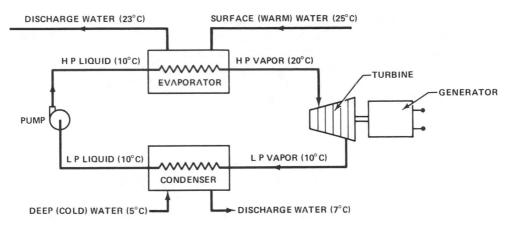

FIGURE 7–54   Schematic diagram of ocean thermal energy conversion. LP, low pressure; HP, high pressure. (Source: U.S. Department of Energy)

meters of shore so that the electrical energy generated can be transferred to the shore via underground cables. There are several problems with the ocean thermal energy conversion (OTEC) method. The heat exchangers currently in use are expensive and tend to corrode in seawater, and their efficiency is less than desirable. Microorganisms grow readily (biofouling) on the cooler side of the heat exchanger, thus decreasing their efficiency. The facilities to harness the ocean thermal energy will have to be immense. For a 100-megawatt operation, the facility must be able to handle a flow of 500 million gallons of water, and the surface area of the heat exchangers would have to be on the order of 10 million square feet (roughly 1 square kilometer). And even then the output is only one tenth that of a large coal-fired or nuclear power plant. At the time of this writing (1986), OTEC does not appear practical, based, of course, on present-day technology.

Tidal Power

Tidal power is the electric power that can be generated from the rise and fall of ocean water due to tides, using a generator hooked up to a hydraulic turbine. Tides vary daily, seasonally, and with geographic location. The largest maximum tidal range is 15 meters at the Bay of Fundy in eastern Canada. The world's only major tidal power plant is located in the Rance estuary in France, where the tidal range is 12 meters.

Electricity can be generated by the drop or rise of water flowing through a barrier built across the estuary, much like a conventional hydroelectric power plant. The energy that can be generated is dependent on the volume of water that flows to the turbine, the range of the tide, and the area of the enclosed basin. In the United States conditions are probably suitable only at Passamaquoddy Bay in Maine (range 8 meters). The DOE estimates, however, that the maximum power output would be 800 megawatts, and even then the facility could only operate a few hours per day. The total energy output would be about one half that of a conventional coal-fired or nuclear power plant of the same capacity. At present, although there is ongoing research and interest in a joint Canada–United States venture to harness Bay of Fundy tidal power (which includes Passamaquoddy Bay), tidal power is not economic.

Tar Sands

Tar sands are sandstones that have been impregnated with an organic, tarry substance called bitumen. Bitumen is a very complex hydrocarbon which upon processing and refining can yield synthetic crude oil (syncrude). These deposits presumably originate by biodegradation of oil that has migrated into the sandstones, then been trapped and altered. Reserves of tar sands are very extensive in Canada, in the USSR, and Venezuela; U.S. reserves are small. The largest of these deposits is in Utah, with smaller occurrences in California, New Mexico, and Texas.

Shallow deposits can be mined by open pit methods. The Athabasca tar sands in Alberta, Canada are the most heavily exploited deposits in North America. Hot water and small amounts of sodium hydroxide are added to the mined material, which releases the bitumen that is then carried as a froth to the water surface. After separation from the water, the bitumen-rich fraction is pyrolyzed and the vapors collected to form a synthetic crude oil, which then has its sulfur removed by hydrotreatment. This method, which works well in the Alberta deposits, may not work as satisfactorily in the U.S. deposits because the bitumen is more strongly fixed on mineral grain surfaces in the latter than in the former. Solvent extraction is currently being tested for Utah and Texas tar sands.

For deep deposits, in situ methods must be employed. Pilot studies with steam have been partially successful, and so has combustion of the underground bitumen, creating a front of liquid bitumen which flows toward a production well. Use of different solvents to extract the bitumen is also under study.

The attempts at getting large amounts of syncrude from tar sands in the United States have not been successful. Although these deposits contain some 10 billion barrels of resources, they are as yet uneconomic. Further, there are several environmental problems with the deep recovery processes.

Biomass

The fuels that can be obtained from living matter (mainly plants and plant products and residues) are referred to as biomass fuels. These materials contain energy that is derived from photosynthesis. Chlorophyll-containing plants take up energy from solar radiation, and then synthesize compounds of carbon, hydrogen, and oxygen from atmospheric carbon dioxide and water. Many compounds are produced during photosynthesis, but glucose (a simple carbohydrate) is the most common. The reaction by which glucose forms, in simplified form, is

$$6 \; CO_2 \; + \; 6 \; H_2O \; \xrightarrow[\text{energy}]{\text{Solar}} \; C_6H_{12}O_6 \; + \;\;\; 6 \; O_2$$

Carbon   Water   Glucose   Oxygen
dioxide

Glucose is the building block for most of the complex carbohydrates found in all plant matter. When these materials are burned in air, a reverse reaction takes place, with water and carbon dioxide forming and energy as heat given off. This is not an efficient process, but nevertheless it is important as biomass is a renewable resource. Biomass sources for energy include high-yield crops, rapidly growing trees, forestry, and agricultural residues. For example, crops of sugar beet and sugar cane can be grown to produce ethanol as a fuel or fuel additive. Using short-rotation forestry techniques is also possible, assuming fast-growing trees could be planted and harvested at five- to seven-year intervals. The Department of Energy estimates an annual harvest whose energy yield is equivalent to 180 million metric tons of coal or 850 million barrels of oil. The residues of forestry and agriculture contain abundant biomass. On farms, in particular, these could be used to provide local energy sources with some efficiency.

The DOE also is conducting research on ocean-grown biomass in the Ocean Farm Project off the California coast. Here giant kelp is grown, tied to nylon lines, some 10–35 meters below the ocean surface. This kelp can be a source of methane.

Getting the energy from the biomass can be done by thermochemical or biochemical conversion. In thermochemical conversion, the biomass can be combusted directly, or pyrolyzed in the absence of air to produce gases that can yield methane, methanol, or ethanol, for example, or hydrogenated to form liquid hydrocarbon fuels.

Biochemical conversions include anaerobic bacterial action on biomass which produces methane (the same process that produces marsh gas); plant residue and animal wastes both yield substantial amounts of methane by this process. Fermentation of carbohydrates yields ethanol, which can be used as fuel directly or as an additive to gasoline.

Municipal Wastes

In the United States some 154 million tons of solid municipal waste is generated each year, about half of which is combustible materials like cardboard, paper, and textiles. If 50 percent of this could be burned as an energy source, it would be equivalent to 130 million barrels of crude oil. This waste has long been used as landfill throughout the United States. Energy can be extracted from municipal wastes by incineration, by pyrolysis, and by biochemical conversion. With incineration, if the combustible fraction is separated prior to burning, the noncombustible fraction can be processed to remove recyclable materials as well as ash from the burning. All of this, in theory and as demonstrated on a pilot scale by the U.S. Bureau of Mines, can be utilized. When municipal waste is referred to as urban ore, the description is quite accurate!

In pyrolysis the waste is heated to 500–1000°C in the absence of air, and gaseous and liquid fuels can be obtained. Methane, methane-rich, and combustible hydrocarbon-rich gases and liquids are obtained by any of several methods.

Biochemical conversion takes place after the municipal waste is mixed into a slurry, usually with domestic sewage, and bacterial action generates methane. Some of the waste could be treated to separate the cellulose-rich fraction of the waste, from which ethanol can be obtained.

Municipal wastes present serious environmental problems (see Chapter 8), and utilizing large amounts of this waste to provide energy would eliminate part of the disposal problem, provide badly needed energy, and create more easily disposable refuse.

Nuclear Fusion

Perhaps no alternative energy option has captured the imagination of the public, and perhaps much of the scientific community as well, as nuclear fusion. In this process, two (or more) atoms of one or more elements are fused together to form a heavier atom. This process does not work at the low temperatures on earth because the repulsive forces of the nucleus of the atom prevent fusion. Fusion is an important process inside the sun, however. Hydrogen isotopes are considered ideal as fusion fuel because they are the lightest isotopes known in nature, they are abundant, and they have a single charge when ionized. The three isotopes are hydrogen (H) with mass 1, deuterium (D) with mass 2, and tritium (T) with mass 3. Tritium is radioactive and has a half-life of 12.3 years. An especially favorable reaction for fusion is

$$D \quad + \quad T \quad \longrightarrow \quad {}^4He \quad + \quad n \quad + \; 17.6 \; MeV$$

$$\text{Deuterium} \quad \text{Tritium} \quad \text{Helium} \quad \text{Neutron} \quad \text{Energy}$$

To put this reaction into perspective, if 0.4 kilogram of deuterium were mixed with 0.6 kilogram of tritium and fusion were to occur, some 94 million kilowatt-hours of energy would be generated. This energy is four times as great as that from fission of a kilogram of uranium.

The above reaction requires only naturally occurring water, which contains small but easily separable quantities of deuterium, and lithium, which by neutron bombardment yields tritium.

The advantages of creating fusion energy by D-T fusion are:

1. There is an essentially limitless supply of deuterium.
2. There is a very large reserve of lithium.
3. The actual fuel in a fusion reactor at any time is, almost by definition, very

small so that even in the case of an accident the release of energy of matter to the environment is small.

4. It is a safe procedure with no chance of overheating.

5. Radioactive species are produced, but they are minor.

6. There is a high thermal efficiency of the reaction to convert thermal energy into electrical energy because of the high temperatures involved.

This last point is one of the problems, however. For D-T fusion to successfully work, temperatures on the order of 100 million degrees Celsius are required. In the hydrogen bomb, for example, the temperature to instantaneously cause hydrogen fusion is provided by a fission nuclear bomb. For energy, the intent is to control the fusion so that the energy is both predictable and manageable. To do this, use is made of the fact that at very high temperatures, electrons become separated from the nuclei of atoms and form a plasma. Once the components of the plasma have been ignited, the reaction is self-sustaining in part because of the tremendous energy generated. In order for fusion to work, the plasma must be kept from touching the sides of the containment vessel. This can be done, for example, by use of magnetic fields or by use of lasers on small targets. How to heat the fuel to 100 million degrees under controlled conditions is, of course, one of the major problems of fusion energy research. So, too, is development of alloys for the magnetic confinement system which can withstand the conditions. Niobium, a scarce element often found in pegmatites, is one such metal, and it is being stockpiled accordingly (see Chapter 4).

The best estimates are that a workable fusion reactor will not be available until well into the next century.

ENVIRONMENTAL AND HEALTH RISKS FROM MAJOR ENERGY OPTIONS

No energy option is without environmental risk in a broad sense, and without risk to public health. This section will examine the health risk and other environmental risks for the major energy options, including oil, gas, hydropower, nuclear, and coal. Solar energy is included here for comparative purposes, though this source accounts for less than one half of 1 percent of total energy consumption in the United States, because interest in it is so great.

Hans Inhaber in 1979 reported on his intensive study of risks for oil, gas, coal, hydropower, nuclear, and solar, all based on comparison of equivalent 1000-megawatt power plants. The available data allow a good comparison of oil, gas, hydropower, coal, and nuclear, but the solar information must be estimated. Inhaber argued that all risks must be considered, including those associated with mining, processing, handling and transportation of materials; construction, maintenance, wastes, land involved, chemicals used; and disposal of wastes. Some of these components are the same for all the energy options discussed. All power plants require, for example, concrete, steel, and fabricated parts; and there is a calculable risk in terms of deaths, accidents, and other adverse health effects for these materials. For equivalent size power plants, solar energy requires the greatest amount of materials, some 150 percent higher than the others, and the longest construction time. Consequently solar has the highest construction risk.

In addition, solar energy requires a backup to handle power requirement when there is a lack of sunshine, and a large energy storage system is required as well. The lowest risk backup system for solar is nuclear, but most plans for solar backup call for coal, hence the coal risk (which is high) is added to an already high risk for solar.

Inhaber calculated comparative risks for oil, gas, coal, hydropower, nuclear, and solar in terms of person-days lost per megawatt-year net output over the energy option lifetime. The days lost include deaths, nonfatal accidents, and diseases, and may be subject to a 10 percent error. His findings are summarized in the following list:

Gas: approximately 1.3 to 6.5

Nuclear: approximately 1.4 to 8.5

Hydropower: approximately 30 to 50

Solar (photovoltaic): approximately 110 to 600

Oil: approximately 10 to 1400

Coal: approximately 110 to 3000

At the time of Inhaber's study, the health effect of air pollution due to burning of fossil fuels was not well known, hence the high number of probable fatalities from sulfur dioxide and carbon and nitrogen oxides were not well assessed. When this information is included, then the list becomes:

Nuclear: lowest risk

Gas and Hydropower: significant risk

Solar: high risk

Oil: very high risk

Coal: highest risk

Another way to look at a comparison between the energy options is to consider such things as thermal pollution in air and water, land used for mining and plants, wastes, and air pollution. This comparison was made by Fowler (1975) for 1000-megawatt coal-fired and nuclear power plants. Table 7–21 is an attempt to incorporate Fowler's data for coal and nuclear with similar estimates for oil, gas, hydropower, and solar; in some places no figure can be estimated.

By using just Table 7–21 one could draw the erroneous conclusion that hydropower is the safest energy option. But when one considers the risks from materials involved in construction, processing, and transportation, the overall risk

TABLE 7–21   Comparison of coal, oil, gas, nuclear, hydropower, and solar photovoltaic 1000-MW plants.

	Coal	Nuclear	Oil	Gas	Hydropower	Solar
Thermal energy (10^6 Btu/s) to be dissipated						
Water	1.1	1.7	1.1	0.5	NA	Small
Air	0.4	0.15	0.4	0.2	NA	Small
Effluent						
Radioactivity (10^3 curies/yr)	[1]5–200	[1]2–200	Small	Small	Very small	Small
Air pollutants (tons/yr)						
SO_2	45,000	1,500	11,200	5,000	None	4,500
NO_x	26,000	900	14,000	6,000	None	2,600
CO	750	25	380	190	None	75
Hydrocarbons	260	9	65	33	None	26
Particulates	3,500	120	850	425	None	350
Wastes (10^3 ft^3/yr)						
Radioactive	[2]	12	[2]	[2]	None	[2]
Ashes	200	7	50	25	None	20
Limestone	300	10	75	37	100	400
Land						
Acres mined	200	13	70	40	100	30
Plant site (acres)	300–400	70	150	100	150	500

[1]The radioactivity released from burning of coal is considerable, but, as it is not classified as a nuclear material, it is not monitored.

[2]Wastes from coal are commonly uraniferous and radiumiferous, as are some wastes from oil, gas, and the coal backup for solar, but the data for these sources are not available.

NA = not available, but not insignificant.

SOURCE: Coal and nuclear data from Fowler (1975) except where noted. Other data are estimates by the author; see text for discussion.

is considerably higher. Further, there is a flood risk associated with hydropower that is not present for other energy options. In 1963, for example, a landslide into the reservoir behind a dam at Belluno, Italy caused water to surge over the dam and killed over 2000 people. It has been estimated that a severe dam accident in California could kill over 100,000 people.

Solar photovoltaic energy is not a panacea. To use solar photovoltaics on a large scale, such as argued for comparative purposes here, requires extensive amounts of iron and steel, aluminum, cement-concrete, and large amounts of chemicals such as arsenic and cadmium. Bernard Cohen (1982) has argued that an expanded use of cadmium in solar photovoltaic could result in an excess of several thousand cancers per year. Large amounts of oils of different kinds are used in some pilot solar photovoltaic systems, and these systems are certainly by no means fail-safe. In 1986 heptane gas ignited at a pilot solar photovoltaic farm near Daggett, California, and the subsequent explosion set fire to more than 240 million gallons of mineral fuel oil used to drive turbines.

Coal adds a considerable risk to both nuclear and solar energy options. The energy used to enrich 235-uranium is coal. This accounts for the entries for SO_2, NO_x, CO, HC (hydrocarbons), and particulates for nuclear (Table 7–21). Similarly, coal is the preferred energy backup for solar, hence coal contributions to solar risk are included there as well.

Oil and gas burning add considerable SO_2 and NO_x, as well as CO and other volatiles, to the atmosphere. In addition, there is also risk from spills of petroleum during ocean shipping, and especially during loading and unloading. One study (Garrels, MacKenzie, and Hunt, 1975) found that some 2 billion kilograms of petroleum is lost to the seas during shipping each year, and this was based on 1971 data. Assuming a 4 percent increase in total shipping and spills, this now amounts to about 3 billion kilograms per year. This mass is greater than all hydrocarbons produced annually by the sea's living organisms. The most obvious effect of these spills is their effect on wildlife, especially birds, and physical eyesores on beaches. Yet the main effect of large hydrocarbon spills is on local marine communities, many of which can be wiped out by such spills, for instance in Prince William Sound, Alaska, after the *Exxon Valdez* spilled 10 million gallons of crude oil in April 1989. Excess hydrocarbon release affects the chemical stimuli of oceanic marine life as well, and carcinogenic petroleum constituents such as benzopyrene tend to concentrate in plankton and in bottom sediments. Although this problem is considered of large magnitude, it has received little media coverage and proportionately little research.

For both oil and gas exploration there is considerable risk from drilling operations, especially for offshore oil. In every year a small, but certainly not insignificant number of workers die from hydrogen sulfide from capped gas wells. Perhaps even more significant is the fact that losses of petroleum to the oceans from offshore production, coastal oil refineries, urban and river runoff, and industrial and municipal wastes add roughly twice as much petroleum hydrocarbons as do oil spills. Sulfur dioxide and nitrogen oxides from oil and gas burning each year may account for several thousand deaths, and carbon monoxide an unknown number. Finally, oil and gas burning also contribute significantly to the greenhouse effect (Chapter 3).

Explosions and fires of oil and gas refinery and storage facilities are not that rare. In 1984, in the Mexico City suburb of San Juan Ixhautepec, over 1 million gallons of liquified gas exploded, killing at least 500 people with another several hundred critically wounded, and the long-range effects of the fire and smoke are not even known. Also in 1984, a leaking gasoline pipeline in Brazil caught fire and killed 500 people as well as burning thousands of homes. Smaller fires are common, sometimes with fatalities and debilitating injuries. Yet the public accepts this as unfortunate but no reason to question the overall safety of oil and gas operations. In 1989, explosion of a gas line in the USSR killed an estimated 400 people.

Further, the EPA estimates that a very large number of gasoline storage tanks (mainly at service stations) in the United States leak, and the subsequent contamination of surrounding waters and soils, well documented in many places, could lead to serious environmental health problems. This is discussed more fully in Chapter 8.

In summary, there is no "free ride" associated with any energy option. Of major options, nuclear energy is the least dangerous in terms of environmental impact and adverse public health effects. Hydropower is second in safety, then gas, then oil; and coal is easily the most dangerous of our energy options.

FURTHER READINGS

Averitt, P. 1967. *Coal resources of the United States, Jan. 1, 1967.* U.S. Geological Survey Bulletin 1275.

Beaumont, E. A., comp. 1982. *Energy minerals.* Tulsa: American Association of Petroleum Geologists, 102 pp.

Brobst, D. A., and Pratt W. A. 1973. Introduction, in *United States mineral resources.* U.S. Geological Survey Professional Paper 820, pp. 1–9.

Brookins, D. G. 1984. *Geochemical aspects of radioactive waste disposal.* New York-Heidelberg: Springer-Verlag.

Brookins, D. G. 1984. *The geological disposal of high level radioactive wastes.* Athens: Theophrastus.

Cohen, B. L., 1982. Applications of ICRP 30, ICRP 23 and radioactive waste assessment techniques to chemical carcinogens. *Jour. Health Physics 42:* 753–757.

Dansereau, P., ed. 1970. *Challenge for survival: Land, air and water for man in megalopolis.* New York: Columbia University Press.

Electric Power Research Institute. 1985. Risk and the human environment, p. 6–14.

Fisher, J. C. 1974. *Energy crises in perspective.* New York: John Wiley & Sons.

Fowler, J. M. 1975. *Energy and the environment.* New York: McGraw-Hill.

Garrels, R. M., Mackenzie, F. T., and Hunt, C. 1975. *Chemical cycles and the global environment: Assessing human influences.* Palo Alto: William Kaufman, 206 pp.

Glasstone, S., 1982. *Energy deskbook.* U.S. Department of Energy DOE/IR/05114-1(DE82013966).

Gray, T. J., and Gashus, D. K. 1972. *Tidal power.* New York: Plenum Press.

Halacy, D. S., Jr. 1977. *Earth, water, wind and sun: Our energy alternatives.* New York: Harper and Row.

Hammond, A. L. Metz, W. D., and Maugh, T. H. 1973. *Energy and the future.* Washington, D.C.: American Association for the Advancement of Science.

Hobson, G. D., and Tiratsoo, E. N. 1981. *Introduction to petroleum geology,* 2d ed. Beaconsfield, England: Scientific Press.

Inhaber, H. 1979. Risk with energy from conventional and nonconventional sources. *Science 203:* 711–718.

League of Women Voters. 1985. *The nuclear waste primer.* Washington, D.C.: League of Women Voters Education Fund.

Marovelli, R. L., and Karbnak, J. M. 1982, The mechanization of mining. *Scientific American* 247: 90–113.

Moore, T., and Dietrich, D. 1987. A special report: Chernobyl and its legacy. *Elec. Power Res. Inst. Bull. 12:* 4–21.

McBride, J. P.,et al. 1987. Radiological impact of airborne effluents of coal and nuclear plants. *Science* 202: 1045–1051.

People's Republic of China. 1980. Health survey in high background radiation areas in China. *Science* 209: 877–880.

National Academy of Science–National Research Council. 1983. *A study of the isolation system for geologic disposal of radioactive wastes.* Washington, D.C.: National Academy of Science–National Research Council Press, 345 pp.

National Academy of Science–National Research Council. 1984. *Review of the scientific and technical criteria for the waste isolation pilot plant (WIPP).* Washington, D.C.: 130 pp.

National Academy of Science–National Research Council. 1987. *Scientific basis for risk assessment and management of uranium mill tailings.* Washington, D.C.

National Academy of Science–National Research Council. 1980. *Trace-element geochemistry of coal resource development related to environmental quality and health.* Washington, D.C., 153 pp.

National Research Council. 1976. *Natural gas from unconventional sources.* Washington, D.C.: National Academy of Sciences.

Odum, H. T., and Odum, E. C. 1976. *Energy basis for man and nature.* New York: McGraw-Hill.

Office of Technology Assessment, 1982. *The regional implications of transported air pollutants: An assessment of acidic deposition and ozone.* Washington, D.C., sections A—M.

Parent, J. D. 1983. *A survey of United States and total world production, proved resources, and remaining recoverable resources of fossil fuels and uranium.* Chicago: Institute of Gas Technology.

Reynolds, W. C. 1974. *Energy: From nature to man.* New York: McGraw-Hill.

Rothman, S., and Litcher, S. R. 1982. The nuclear energy debate: Scientists, the media and the public. *Public Opinion* (Sept.–Oct.): 47–52.

Tiratsoo, E. N. 1973. *Oil fields of the world.* Beaconsfield, England: Scientific Press.

U.S. Department of Energy. 1984. *Coal production, 1984.* U.S. Department of Energy DOE/EIA-0118(84).

U.S. Department of Energy. 1984. *United States uranium mining and milling industry: A comprehensive review.* USDOE/S-00028, 102 pp.

U.S. Department of Energy. 1985. *Annual energy review.* DOE/EIA-0384(85)

U.S. Department of Energy. 1985. *Annual energy outlook 1985.* DOE/EIA-0383(85)

U.S. Department of Energy. 1986. Monthly energy review. DOE/EIA-0035(86/04).

Winston, D. C. 1982. Dead or alive, it's great to be wanted. *Energy Upbeat 3:* 2–8.

World Energy Council. 1980. *World energy resources, 19885–2020; world energy conference.* Guilford, U.K.: IPC Science Technology Press.

Young, L. B. 1973. *Power over people.* London: Oxford University Press.

8

Geochemistry and Human Impact on the Environment

INTRODUCTION

███████████ This chapter will briefly examine several aspects of environmental geochemistry as well as anthropogenic impacts on the environment. Our environment is both rugged and fragile. It is rugged in the sense of the earth machine continually grinding on and on, the plates overriding other plates, the presence of volcanism, erosion, and so on. Yet is is fragile in terms of the thin veneer we call the crust, to which we owe our existence. Even in the oceans, most aquatic life is supported only in the upper 200 meters or so; and on land the surface is where the plant and animal kingdoms fight for survival.

Unlike other life forms, *Homo sapiens* has the ability to profoundly affect the earth's surface. Our efforts in gleaning mineral and energy resources from the earth are not without price. Similarly, where and how we dispose of the vast amounts of waste material we generate each year can adversely affect the earth if we are not careful. Part of the problem lies in attempting to predict how the earth will respond to different materials we expose it to. While we have knowledge of the behavior of many materials such as mercury, copper, certain gases, even nuclear waste, we are in a sense quite ignorant as to how the tens of thousands of artificial compounds we generate each year may behave in the delicate earth environment. The long-term stability of dioxin and PCBs are only two of thousands of examples that can be cited here. Further, our civilization disturbs the earth's surface not only by mining and related activities but much more so by construction, road building, and so on. Our coastal areas in many places are also disturbed by oil spills, dredging operations, barriers, and other examples of our handiwork.

This chapter will attempt to explore some facets of these and other complex problems that affect our existence on this planet.

ENVIRONMENTAL GEOCHEMISTRY

For the last two decades or so there has been a concentrated effort to determine if certain diseases or other health problems can be linked to soil, water, and geochemistry. The National Institute of Health and other organizations have investigated the effects of potential carcinogens in fungicides and insecticides on the ecosystem by looking at virtually all forms of organic life from protozoa to people; however, little had been done until about the mid-1960s to relate rocks or soils to health problems.

In 1971 and 1972 H. L. Cannon and H. C. Hopps published notable treatments of environmental geochemistry and its relation to health and disease.* The latter publication was the result of papers presented before representatives of the medical profession as well as geoscientists. Unfortunately, there were too few medical professionals present for a meaningful exchange of ideas. This is part of the problem. Scientists have compiled an impressive amount of data over the last 15 years. Conferences on trace substances and environmental health are held fairly regularly, but attendance is only fair, the data are not distributed as rapidly or as widely as is desirable, and the average person is usually unaware of the information presented. This is unfortunate because carefully documented data are reported at these meetings and in the publications that result from the meetings.

For example, it has long been apparent in the vicinity of sulfide-ore mines that, prior to regulation, large amounts of potentially carcinogenic materials were being indiscriminately dumped into streams and ponds, or allowed to escape into the atmosphere. The effects of some carcinogens (such as mercury and cadmium) in promoting disease and other health problems have been known for a long time, and measures have been taken to remedy obvious hazards. Not so obvious are the effects of trace substances released into the environment by industrial processes; this includes all ventures concerned with earth resources, not just activities associated with mining. Highway construction, for instance, often causes more land to be excavated than is normally done in most mining efforts. Some trace substances are so tightly locked up in rocks that normal weathering does not release them rapidly. However, if the delicate balance between soil and rock and rain is disturbed, once-immobile elements can be dissolved, transported, and concentrated.

The concentration of a potentially carcinogenic substance is a complex process which includes mobilization in water, uptake by plants, further concentration when plants are ingested by animals—cattle for example—and even further concentration in people, for example when they eat hamburgers. More potential carcinogens may be added if the hamburgers are grilled over charcoal, because organic matter is an excellent chelating and complexing agent for many carcinogenic metals like those concentrated in wood coals. Thus, from soil dis-

---

* H. L. Cannon and H. C. Hopps, *Environmental Geochemistry in Health and Disease.* Geological Society of America Memoir 123, 1971. H. L. Cannon and H. C. Hopps, "Geochemical environment in relation to health and disease," New York Academy of Science *Annals* 199 (1972): 388.

turbed by excavation to human ingestion there may be as much as a hundredfold increase in the concentration of certain potential carcinogens. There are too few quantitative data to properly assess spots in the food chain where excessive concentration of carcinogens may take place. It is not enough to monitor only one spot. All steps in the process—from the original soil or rock, through the transporting media (including airborne materials), and concentration in plants—must be very carefully monitored.

Occasionally the uptake of trace metals by plants can be used to indicate above-normal levels of one or more elements in the soil. Such plants are known as *accumulator plants*. At times they may be used as biogeochemical indicators in searching for ore deposits. A few plants known to concentrate specific elements are

Astragalus (locoweed)—selenium (its occurrence is common in areas over seleniferous coals and commonly over sedimentary uranium deposits)

Clethra (alder)—cobalt

Nyssa (gum tree)—cobalt

Ilex glabra (holly)—zinc

Carya (hickory)—beryllium, rare-earth elements, barium, and scandium

Uptake usually depends on the availability of the elements in the soil.

We distinguish between trace elements added to the environment from weathering of rocks and trace elements added by fertilizing agents, insecticides, and fungicides. Some elements enriched in soil due to chemicals and natural wastes used in agriculture include chromium, iodine, lithium, lead, zinc, cadmium, and copper. Because uptake data are commonly erratic, virtually every combined agricultural chemical used in a given land area should, if possible, be analyzed to see if trace elements are being added to the food system in amounts above or below the recommended dose. Remember also that too little as well as too much of certain elements can be toxic to plants and animals (Table 8–1). Some major elements such as iron and silicon have been omitted from Table 8–1; high amounts of either may be carcinogenic. Also omitted are elements such as barium which, if improperly monitored during milling and purifying, may be carcinogenic.

MERCURY

Mercury deserves some special mention here. It is a scarce element that is easy to recover from ore by simple roasting (see Chapter 4). This characteristic—its volatility—makes it very difficult to control, however. If mercury is locked tightly into its main ore mineral, cinnabar (HgS), it is relatively inert if ingested by animals and may not harm them at all. If mercury is converted to an organic form, either as monomethyl or dimethyl mercury, it is not only hazardous but ex-

TABLE 8–1 Effects of high or low concentrations of trace elements on plants and animals.

Element	Effects
Lithium	Low levels may cause mania in humans.
Fluorine	Low levels cause low growth rates in rats and dental decay in humans. High levels may affect plants and cause bone abnormalities in humans.
Chromium	Toxic in high amounts to some plants; also toxic as Cr^{6+} in animals in some regions.
Nickel	High amounts are not commonly toxic; as inhaled metallic dust and as nickel carbonyl, however, nickel is toxic in humans.
Copper	Low copper content causes impaired growth in some plants and anemia in humans. High amounts are toxic to plants and dangerous to humans (Wilson's disease and possibly increased rate of aging).
Zinc	Low amounts cause stunted growth in plants and all animals, plus other deleterious effects. High amounts are toxic in plants.
Arsenic	Low amounts may cause hair abnormalities in some plants and animals. High amounts are toxic in all animals.
Selenium	Low amounts have been linked to many abnormalities in some plants and animals. High amounts may suppress growth in plants and cattle.
Molybdenum	Low amounts may cause growth retardation in some plants and animals. High amounts can cause a variety of diseases in cattle.
Cadmium	High amounts are thought to be toxic to plants as well as causing hypertension, sterility, and testicular hemorrhage in animals.

tremely easy to concentrate in animal organs. Mercury poisoning during mining and milling has been widespread within the last 100 years, but mercury poisoning from industrial misuse is also known. From 1950 to 1965 at Minamata Bay, Japan, a chemical plant discharge was responsible for 43 deaths out of 116 cases of mercury poisoning. A brief flurry of interest and concern was noted in the press at that time, but most people questioned today would have little knowledge of the Minamata incident.

In the period from February to August 1972 there were 6530 cases of methyl mercury poisoning in Iraq, resulting in 460 deaths.* This was a major disaster, yet even now few in the United States outside professional circles are aware of it. Wheat seed that had been treated with methyl mercury fungicide was planted, and bread made with the wheat was enriched in methyl mercury. Not

*F. Bakir et al., "Methyl mercury poisoning in Iraq," *Science* 181 (1973): 230.

TABLE 8–1 *Continued*	
Element	Effects
Iodine	Low amounts are linked to goiter in animals. High amounts are toxic to plants and humans and impair reproduction in plants and animals.
Lead	Toxic to plants and animals in high amounts. In humans, it affects the central nervous system and causes anemia, urinary disease, and bone damage.
Beryllium	Toxic in high amounts to plants and animals; humans may be more affected by industrially produced beryllium than from natural sources.
Magnesium	Although not a trace element, magnesium deficiency in plants and animals seems to be linked to other diseases and may enhance them.
Manganese	High amounts of manganese, especially when linked with other elements, may cause toxicity in infants, metabolism problems, and intestinal problems.
Tin	High amounts may cause growth abnormalities in laboratory plants and animals; its exact role is not well understood.
Vanadium	Low amounts affect the metabolism of plants and animals; high amounts from industrial exposure are occasionally toxic. No such toxicity is suspected from the normal environment.
Mercury	Toxic in high amounts to plants and animals; most high levels of mercury are directly related to industrial use. In the late nineteenth and early twentieth centuries mercury contamination from mining was widespread.

SOURCE: H. L. Cannon and H. C. Hopps, "Geochemistry and the Environment," in *The Relation of Suspected Trace Elements to Health and Disease*, vol. 1. Washington, D.C.: National Academy of Sciences, 1974.

only was bread intake involved in the epidemic, but mercury was also concentrated by several other means: livestock were fed the contaminated wheat seed and mercury was concentrated in the meat; mercury was found in vegetation other than wheat in the wheat field; game birds and fish fed on the contaminated seed; and some mercury was inhaled and came in contact with skin. All the fatalities occurred in districts outside metropolitan areas. What was even more unfortunate was that warnings in a foreign language appeared on the sacks of seed—but in those rural areas of Iraq, most farmers were not even literate in their native Arabic. This is a good example of poor government policies on the monitoring of such materials. The point is well made, however, that environmental contamination is caused more often by industrial-related (indirect in this case) or distribution factors than by earth resources industries, where monitoring is carried out in a systematic and thorough fashion.

The toxicity of inorganic mercury, in native form or as simple sulfides or oxides, is well known. It is not remarkably toxic, and fairly easy to flush from the body. Mercury vapor can damage lung tissue and cause tremors among workers in areas such as mercury mills or industrial operations.

However, in its organic form, methyl mercury, the substance is extremely toxic, even at very low levels. The mercury at Minamata Bay and in the wheat seed in Iraq was the methyl variety. In some instances methyl mercury , or di-methyl mercury, is discharged directly into the environment. In other cases inorganic mercury and its compounds are released into environments where they concentrate and become methylated. Once methyl mercury has formed, it enters the food chain by microorganisms, fish, plants, and animals. Methyl mercury tends to accumulate, for example, in fish like tuna and swordfish. Study of preserved fish tissue from museums indicates that some of the mercury accumulation may be from natural causes.

Mercury enters the atmosphere due mainly to human activities. Garrels and others (1975) estimated that some 10 million kilograms enters the atmosphere each year from chlor-alkali production (30%), coal-lignite combustion (27%), oil-gas combustion (22%), sulfide ore roasting (20%) and cement manufacturing (1%). Mercury is flushed out of the atmosphere by rainfall, and accumulates on land and sea. Between 1945 and 1985 the level of mercury in the oceans increased 40 percent, while the mercury of rivers increased fourfold. Naturally, the levels found in fish from streams and lakes has also increased markedly over the last 40 years. In Canada, the government has had to forbid fishing from certain lakes due to the possibility of mercury poisoning.

LEAD

Lead contamination in the atmosphere comes primarily (about 80%) from the use of leaded gasoline, but lead contamination also takes place from smelting operations, battery manufacturing, painting, typesetting, and the lead glass industry. Adverse health effects of lead are well known. High levels of lead affect the central nervous system, and may cause mental disorders, urinary tract disorders, anemia, bone damage, and reproductive dysfunctions. The use of leaded drinking vessels in ancient Rome has been suggested as a cause of the very low birthrates in the upper classes. The amount of lead currently cycled in the earth has increased significantly, primarily due to human activities. Garrels and others (1973) estimated that the amount of lead introduced into the earth's cyclical systems from anthropogenic sources rivals that from natural sources, and noted that the lead content of the oceans has roughly doubled in the last 100 years.

The atmospheric lead due to gasoline is strikingly illustrated when one notes that the lead content of soil and plants is high near highways and tapers off with distance from the road. The lead is effectively absorbed into plants and animals which graze on these plants, further concentrating the lead.

FIGURE 8–1 National trend
in lead emissions, 1975–
1983. (Source: U.S.
Environmental Protection
Agency)

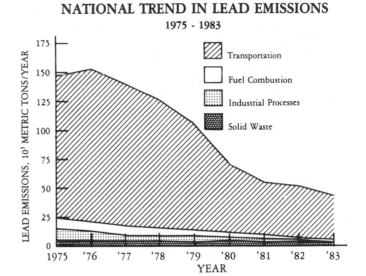

FIGURE 8–1 National trend in lead emissions, 1975–1983. (Source: U.S. Environmental Protection Agency)

Leaded gasoline, which makes up 23 percent of all gasoline sold, is far and away the major source of lead contamination in the United States (Figure 8–1). As shown in Figure 8–2, there is a direct correlation of the amount of lead used in gasoline and the amount of lead found in blood samples from humans. Although the curves shown in Figure 8–2 are decreasing, they have started to level off. This in part is due to the unfortunate practice of fuel switching—people burning leaded gasoline in vehicles designed for the more expensive unleaded fuel. What makes this worse is that lead disables the car's catalytic converter. So not only is this practice responsible for more lead being added to the atmosphere, but the defective catalytic converters also yield four- to eight-fold increases in carbon monoxide and other hydrocarbons and about a threefold increase in nitrogen oxides. The U.S. Environmental Protection Agency argues that leaded fuel may have to be phased out totally; they estimate that at least 16 percent of vehicles designed for unleaded fuels have been switched to leaded gasoline. As more work on the toxicity of lead in humans, and especially children, takes place it is apparent that toxic effects are found at levels of lead in the blood previously considered safe. The EPA in 1985 lowered the threshold for lead from 30 to 25 micrograms of lead per deciliter of blood. By this action, the number of children who may require treatment for lead toxicity was tripled.

ARSENIC

The poisonous nature of arsenic is very well known. Arsenic is added to the environment from smelting of sulfide ores, from fungicides and pesticides, from industrial operations, and from chemical operations in the electronics and pho-

FIGURE 8–2   Lead used in gasoline production and average blood lead levels (NHANES II test), February 1976 to February 1980. (Source: U.S. Environmental Protection Agency)

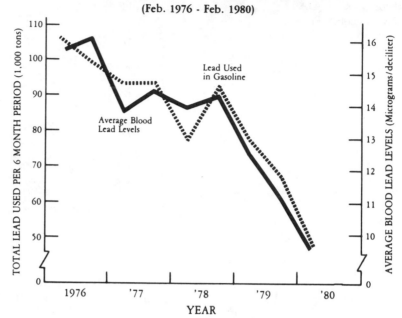

LEAD USED IN GASOLINE PRODUCTION AND AVERAGE NHANES II BLOOD LEAD LEVELS
(Feb. 1976 - Feb. 1980)

tovoltaic industries. Arsenic is a fairly volatile element, and this accounts for its loss from sulfide ores during roasting. In waters, arsenic is oxidized to soluble As^{5+} ions and is easily transported. In the arid playa environments of the western and southwestern United States, arsenic has accumulated to the point where some lands are naturally poisoned. This points to the ease by which arsenic can be concentrated in the environment. Blueberry crops in Maine were for many years treated with arsenic-bearing fungicides, with the result of poisoning of deer and birds. Since arsenic behaves in water much like phosphorus, it started to accumulate in nearshore clams until its use was prohibited.

CADMIUM

Cadmium is a well-known carcinogen, and poses severe present and much greater potential problems to public health. The disease *itai-itai* (ouch-ouch), a very painful decalcification of the bones often accompanied by multiple fractures, was found among the population in the Kinzu River area of Japan; the cause was traced to industrial cadmium pollution of the river. Cadmium is a volatile element, often released during smelting of sulfide ores. It commonly occurs with zinc, and cadmium can be released to the environment from poorly controlled zinc operations and industries using zinc compounds. As mentioned in Chapter

7, a greatly expanded solar energy capacity in the United States, with anticipated use of cadmium salts in photovoltaic cells, could result in many cadmium-induced cancers (Cohen, 1982). Further, the long-term effects of cadmium accumulation in animals are not known, nor is the threshold for cadmium known.

While zinc is an important and essential nutrient, cadmium is not. Thus cadmium's presence in organisms is harmful in two ways: Its high accumulation is toxic and possibly carcinogenic, and it prevents zinc from occupying molecular sites where it is needed as a nutrient. Cadmium poisoning also occurs from food or beverages contaminated from cadmium-plated containers, from cadmium-bearing solders, and cadmium-plated steels. Cigarettes contain about 1 ppm cadmium, well above the crustal average, and may have cumulative effects. Thus the National Academy of Science estimated in 1980 that a rural nonsmoker might ingest some 40 micrograms of cadmium per day, while a smoker in a city might get 190 micrograms per day. Cadmium is present in most foods. It reaches high concentration in oysters and in some other seafood, and in some beef and chicken. The Academy argued that a threshold level in humans might be 170–500 micrograms per day, above which anemia, hypertension, and cancer may result. This figure is not, however, well known, and it is possible that even much smaller amounts of cadmium may be very dangerous if allowed to accumulate over a long time span.

OTHER METALS

Beryllium, selenium, chromium, molybdenum, copper, and nickel are other potentially dangerous metals. All are poisonous in high concentrations, and most, in some form or another, are carcinogenic (Table 8–1). Interestingly, more beryllium is released by coal burning to the environment each year than from all other sources, and coal burning also adds major amounts of selenium and copper. Some of these metals cause adverse health effects by interfering with the action of other elements in animals. Thus molybdenum can interfere with copper and cause copper deficiency. In other cases, high molybdenum may cause bone disease. Chromium as hexavalent Cr is carcinogenic, and carbonyl nickel is both poisonous and carcinogenic.

LAND USE FOR MINING ACTIVITIES

It is interesting to compare the impact of mining with nonmining activities on land availability. Urban growth in the United States is continuing to boom—this comes as no surprise; but how much land is involved? Much discussion tends to be aroused about mining ventures, but relatively little is said about normal urban growth and highway construction. A comparison is useful, especially since land usage plans for mining, for mining-related purposes (tailings and other waste disposal), and for dam construction are by law accompanied by environmental impact statements. Equivalent statements for urban growth or highway construction are not routinely requested.

In the United States, approximately 700 square kilometers of land is stripped by urban construction each year. If we assume that an average of 4500 metric tons of sediment is produced per square kilometer, then some 3,100,000 metric tons of sediment is produced each year—more than three times the load of sediment carried by the Potomac River.

Furthermore, for every kilometer of two-lane highway built, about 35 acres of land is cleared. Thus if more than 800,000 kilometers of highway is built per ten-year period, some 43,000 square kilometers of land is available for erosion and approximately 200 million metric tons of sediment may result. While precise effects of this sediment load are unknown, we do know that urban and highway construction have affected reservoir loads, natural fisheries, and marsh filling. What is not recorded are the types of sediments released. For example, what is the long-range effect of building hundreds of kilometers of highway through rocks relatively rich in carcinogenic elements? Certainly sediments provide a better chance for some elements—especially if they are soluble or easily oxidized—to be removed and possibly concentrated by erosion and transport. The law requires monitoring during mining operations of carcinogenic elements (nickel, arsenic, chromium, beryllium, cadmium, cobalt, iron, lead, selenium, zinc, and titanium), other metals associated with different types of disease (mercury compounds, copper, and vanadium), and radioactive species. No such law exists for urban construction or highway construction.

In mining areas there is a strong potential for release of carcinogenic or potentially carcinogenic elements above threshold level values (TLVs) proposed by the National Institute of Occupational Safety and Health (NIOSH), the Occupational Safety and Health Administration (OSHA), and the Mining Enforcement and Safety Administration (MESA). The TLV refers to the highest level of a substance that a person can be exposed to without suffering harmful effects. But TLVs are poorly defined and even more poorly understood in that knowing the abundance of a potential carcinogen alone is insufficient to establish health effects. Elements are monitored in areas of active mining, but not necessarily in other types of excavations.

Let us now compare land use for all mining with that for specific mining ventures. The National Research Council–National Academy of Sciences has estimated that the total mining, milling, and processing of minerals in the United States results in some 1500 metric tons of materials per year, compared with 3 million metric tons per year from urban construction and more than 19 million metric tons per year from highway construction. Remember also that many of the sediments from mining ventures are used as backfill, stored, or in other ways somewhat isolated from the ecosystem. Similarly, some of the 22 million metric tons of sediment created by urban and highway construction is used for fill and thus is not subjected to severe erosion. Still, we can realistically ask how the effects of monitored mining ventures compare with unmonitored construction for nonmining purposes in the same area. To properly evaluate this we need to know the area's geochemistry before *any* mining or nonmining venture is undertaken; but generally this is unknown. The vigorous testing of the United States (all states but Hawaii) as part of Project NURE (National Uranium Research Evaluation) provided valuable data for uranium and thorium, but only in some areas were multielement analyses proposed. For lack of funds, which could have come from the Department of Health, Education, and Welfare, for instance, a golden opportunity to establish a geochemical background data base for the whole United States was lost.

AIR QUALITY AND AIR POLLUTION

The number of carcinogens made available to the atmosphere by human activities is truly staggering. The EPA lists some 10,000 such substances. Table 8–2 lists a small number of these with known or highly suspected carcinogenic effects. Together, these substances may account for some 1600 to 4000 cancer deaths per year, although this figure must be compared with the 400,000 cancer deaths from all causes per year. The Council on Environmental Quality (1984) estimated that 45 percent of the air pollution cancer deaths are due to organic particulates, about 30 percent to volatile organics, and 25 percent to metals. Admittedly the precise

---

TABLE 8–2   Some common air pollutants.

Organics

Phenol, propylene, propylene oxide, acetaldehyde, acrolein, styrene, 1,3-butadiene, dibenzofurans, chloroform, dioxins, acrylonitrile, methyl chloroform, freon 113, carbon tetrachloride, methylene chloride, hexachlorocyclo pentadiene, phosgene, gasoline vapors, chlorobenzenes, epichlorohydrin, vinylidene chloride, ethylene dichloride, perchloroethylene, trichloroethylene, ethylene oxide, chloroprene, toluene, vinyl chloride, coke oven emissions, benzene, POM (polycyclic organic matter)

Metals

Mercury, lead, arsenic, cadmium, beryllium, nickel, copper, manganese, zinc (oxide), chromium

Chemicals (inorganic)

Hydrogen sulfide, ammonia

Others

Asbestos, radionuclides

SOURCE: Council on Environmental Quality.

documentation of how any cancer is caused is often a guess at best, but nevertheless the available data point strongly to vehicle emissions, industrial emissions, smelting operations, coal combustion, and oil and gas combustion.

The EPA has advocated use of the *bubble concept* to deal with air pollution from industrial sources. This concept assumes there is a metaphorical bubble over an area of known emissions, and that air quality standards must be met for this bubble overall and not at any one or several points within the bubble. This means, for example, that if a particular smokestack is emitting very large quantities of carcinogens but neighboring stacks are not, the former is allowed to pollute unchecked if the overall air standards for the bubble are met. The EPA justifies this stand because (1) the economics are favorable in that not all stacks have to be modified to reduce pollution, and (2) the overall air quality standards are met. If several industries, for example, are covered under any one bubble, then they can work out the system which best accomplishes the overall air requirements. The bubble concept is faulty unless provisions for careful monitoring and liability exist.

WATER QUALITY

The Clean Water Act of 1972 has helped clean up waters in the United States, but many problems remain. Initially, concern was given to so-called traditional contaminants, often of a point-source nature. In more recent years more subtle contaminants have received attention, and contamination from nonpoint sources is now recognized as a major factor affecting water quality in the United States. In this latter category is the contamination from agricultural areas where the run-off from such areas contains fertilizer additives, pesticides, and fungicides.

The Clean Water Act allows streams and other waters to be designated for specific uses. Thus one stream might be selected for water supply reserves, another for lower quality, and still a third for fishing only. Tables 8–3 and 8–4 list the miles of streams and acres of lakes, respectively, by use category.

It is noteworthy that the miles of streams designated as supporting the use for which they were intended has improved somewhat between 1972 and 1982 (Table 8–3), although there has only been a smaller, but not insignificant, increase in the miles of streams not supporting their use as designated. The improvement of stream waters in the United States is due to improved treatment of municipal sewage and improved treatment of industrial waste water. Forty-five states, however, reported problems from nonpoint source pollution, mainly from agricultural sources, but also including road and other construction, land disposal, mining, dams, channels, and saltwater intrusion.

One of the goals of the Clean Water Act has been to make streams both swimmable and fishable. Data for 20 states are shown in Table 8–5, where it is noted that 95 percent of Montana's streams are both fishable and swimmable, while only 43 percent of Delaware's streams meet this goal (although the pollutants in the Delaware River largely originates upstream in other states). There are

TABLE 8–3   Stream water quality 1972–1982.

Use Category	1982 Miles	1982 Percent	1972 Miles	1972 Percent
Supporting uses	488,000	64	272,000	36
Partly supporting uses	167,000	22	46,000	6
Not supporting uses	35,000	5	30,000	4
Unknown or not reported	68,000	9	410,000	54

SOURCE: Association of State and Interstate Water Pollution Control Administrators in cooperation with the U.S. Environmental Protection Agency, *The States' Evaluation of Progress 1972–1982*, (Washington, D.C.: 1984), p. 4.

a wide variety of pollutants recognized in the waters of the United States. Figure 8–3 shows the most widely reported pollutants in the United States, and Table 8–6 notes the number of states reporting elevated levels of metals, pesticides, and other organics. The sources of much of this pollution are indicated in Table 8–7. Nonpoint-source contamination from the agricultural industries is blamed for the eutrophication (buildup of excess nutrients) of lakes in the United States; perhaps 30 percent of the nation's lakes are highly eutrophicated and another 40 percent somewhat polluted. This topic will be discussed later in this chapter.

Contamination of groundwater is a serious problem in the United States. Not only are groundwater supplies being severely overdrawn (see Figure 3–3), but a substantial part of the groundwater still available for use is polluted. The types of contamination of groundwaters are classified into a six-fold series of categories (Table 8–8). There is a tremendous diversity of sources for contamination. A few types are designated to discharge wastes (category I) and a larger number for nonwaste releases (categories II, III). Contamination may occur from other activities (category IV), or from altered flow patterns (category V). Category VI covers those natural sources which are accelerated by human activity.

TABLE 8–4   Lake water quality 1982.

Use Category	1982 Acres	1982 Percent
Supporting uses	13,800,000	84
Partly supporting uses	1,700,000	10
Not supporting uses	400,000	3
Unknown	400,000	3

SOURCE: Association of State and Interstate Water Pollution Control Administrators in cooperation with the U.S. Environmental Protection Agency, *The States' Evaluation of Progress 1972–1982* (Washington, D.C.: 1984), p. 5.

TABLE 8–5   River miles in 20 reported states meeting the fishable/swimmable goal of the Clean Water Act.

State	Total River Miles	Assessed River Miles	Percent Fishable	Percent Swimmable	Percent Swimmable/ Fishable
Arkansas	11,202	11,202	94	53	—
Delaware	—	491	—	—	43
Maine	31,806	2,652	—	—	66
Maryland	9,300	7,440	—	—	92
Massachusetts	10,704	1,630	—	—	47
Minnesota	91,871	2,708	94	39	—
Mississippi	10,274	10,274	—	—	90
Missouri	18,750	18,670	99	21	—
Montana	19,168	17,251	95	96	95
Nebraska	24,000	7,152	74	19	—
New Hampshire	14,544	14,544	—	—	93
New Mexico	3,500	3,500	100	—	—
North Carolina	40,207	37,378	—	—	81
Ohio	43,919	4,949	—	—	62
Oregon	90,000	3,500	—	—	74
Rhode Island	724	724	—	—	81
South Carolina	9,679	2,489	—	—	57
Texas	80,000	16,120	—	—	90
Vermont	4,863	2,325	—	—	93
Virginia	27,240	4,964	81	46	—

SOURCE: U.S. Environmental Protection Agency, *National Water Quality Inventory: 1984 Report to Congress* (Washington, D.C.: 1985).

How groundwater can be contaminated is shown in Figure 8–4. The severity of groundwater contamination is not known with any certainty. The EPA reports numerous chemicals present in groundwater that are known toxins, poisonous, or in some cases carcinogens. A further consequence of groundwater contamination is that it can lead to contamination of surface waters due to recharge. Further, as groundwater is overdrawn, then if the amount of contamination is fairly constant, the amount of contamination in the groundwater increases.

Pollution of the Great Lakes has received a great deal of attention in the last two decades, and steps have been taken to clean them up. The areas of pollution concern around the Great Lakes are shown in Figure 8–5. Lake Michigan is seriously contaminated, especially by polychlorinated biphenyls (PCBs) and other or-

FIGURE 8–3   Pollutants
most widely reported by the
states. (Source: U.S.
Environmental Protection
Agency, *National Water
Quality Inventory: 1984
Report to Congress,* 1985)

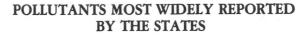

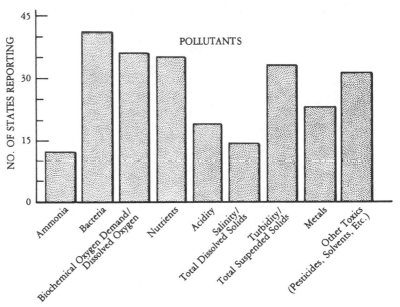

ganics discharged into Waukegan Harbor at Waukegan, Illinois over many years. Trout and other fish still contain highly elevated levels of PCBs and DDT, and fishers are warned not to eat too much fish. There are serious pollution problems in the urban and industrial areas near Calumet River–Indiana Harbor, southern Green Bay, Milwaukee Estuary, and elsewhere. Lake Huron is not as highly polluted as Lakes Michigan, Erie, or Ontario, but the Saginaw Bay area is of concern. Fish from the Saginaw River are high in PCBs and dioxin, and fishers are warned accordingly. Lake Erie, which is the shallowest and smallest in volume of the Great Lakes, is also the most highly polluted. Beaches have been closed due to excess sewage, algae, and dead organic life from the lake, and industrial and other pollution are widespread. Although Lake Erie has been cleaned up considerably, points of concern are found where the Detroit, Raisin, Maumee, Black, Ashtabula, and Cuyahoga Rivers enter the lake. Lake Ontario is less foul but still suffers from eutrophication (primarily from phosphorus buildup) and industrial pollution. The Niagara River, despite a great deal of cleanup, remains the dominant polluter. Lake Superior is the largest, deepest, and cleanest of the Great Lakes; taconite mining residues are no longer dumped into the lake. However, PCB levels in fish are elevated.

Contamination of the Great Lakes by chemicals is well known. Over 800 different chemicals have been identified from the Great Lakes, many of which are mutagenic and carcinogenic based on separate laboratory studies of animals. Most of these chemicals are present in very small amounts, and the long-range effect

TABLE 8–6   Toxics reported by states at elevated levels.

Toxics	States Reporting Elevated Levels
Metals:	
Mercury	21
Copper	16
Zinc	14
Lead	13
Cadmium	11
Chromium	11
Arsenic	6
Nickel	5
Silver	4
Selenium	3
Pesticides:	
Chlordane	15
DDT and its metabolites	8
Dieldrin	7
Other organics:	
PCBs	22
Phthalate esters (for example, di-n-butyl-phthalate)	6
Halogenated aliphatics (for example, carbon tetrachloride)	5
Phenols	5
Monocyclic aromatics (for example, benzene)	4
Dioxin	3

SOURCE: U.S. Environmental Protection Agency, *National Water Quality Inventory: 1984 Report to Congress* (Washington, D.C.: 1985).

of their potential adverse health effects at such small levels is not known. Yet fish concentrate these chemicals to the extent that for several of the lakes warnings about consumption of the fish are given. Yet even now, pollution studies of the Great Lakes reveal over 360 chemicals, over 80 of these in amounts that are not considered safe by the EPA.

But is the quality of the nation's water really improving or not? To help answer this question, the National Stream-Quality Accounting Network (NASQAN) was established by the U.S. Department of the Interior in 1972. This system is designed to keep track of changes in stream quality and quantity. In Table 8–9 we find the tabulated data and recent trends for many metals, organics, and other substances. Most metals show increasing trends, although lead, zinc, and selenium are exceptions (see discussion of lead elsewhere in this chapter).

TABLE 8–7   Sources of pollution.

Pollutant/ Possible Sources	Pollutant Category
Biological oxygen demand	Municipal wastewater treatment plants; industries (particularly pulp and paper mills); combined sewers; natural sources.
Bacteria	Municipal wastewater treatment plants; combined sewers; urban runoff; feedlots; pastures and rangeland; septic systems; natural sources.
Nutrients	Municipal wastewater treatment plants; agriculture; septic systems; silviculture; combined sewers; construction runoff.
Suspended solids/ turbidity	Agriculture; urban runoff; silviculture; construction runoff; mining; industries; combined sewers.
Total dissolved solids	Agriculture; mining; urban runoff; combined sewers.
pH	Atmospheric deposition; mine drainage.
Ammonia	Municipal wastewater treatment plants; combined sewers.
Toxics	Industries; municipal wastewater treatment plants; agriculture; land disposal of wastes; silviculture; urban runoff; spills; combined sewers.

SOURCE: U.S. Environmental Protection Agency, *National Water Quality Inventory: 1984 Report to Congress* (Washington, D.C.: 1985).

Major inorganics also show a greater number of increases than decreases, but bacteria and phytoplankton show more decreases, which reflects the better treatment of municipal wastes. The increase in metals indicates that the steps taken to reduce metal buildup in the nation's waters is not meeting with success. How much of this buildup is due to the huge amount of metals added to the environment by burning of coal is not known, but presumably it is a potentially large factor.

HAZARDOUS CHEMICALS

Chemical wastes were first addressed stringently by the enactment in 1976 of the Resource Conservation and Recovery Act (RCRA) and the Toxic Substances Control Act (TSCA). The RCRA gave the EPA power to begin handling of existing sites of chemical wastes, while the TSCA provided the regulations for new chemicals and new chemical facilities. RCRA and its successor, the Comprehensive Environmental Response, Compensation, and Liability Act of 1980 (CERCLA, which includes Superfund), are discussed later in this section.

Hazardous chemicals are generated, handled, transported, stored, used, and disposed of in virtually an infinite number of ways, even though all such pro-

TABLE 8–8 Sources of groundwater contamination.

Category I—Sources designed to discharge substances
Subsurface percolation (e.g., septic tanks and cesspools)
Injection wells
  Hazardous waste
  Nonhazardous waste (e.g., brine disposal and drainage)
  Nonwaste (e.g., enhanced recovery, artificial recharge, solution mining, and in-situ mining)
Land application
  Wastewater (e.g., spray irrigation)
  Wastewater byproducts (e.g., sludge)
  Hazardous waste
  Nonhazardous waste

Category II—Sources designed to store, treat, and/or dispose of substances; discharge through unplanned release
Landfills
  Industrial hazardous waste
  Industrial nonhazardous waste
  Municipal sanitary
Open dumps, including illegal dumping (waste)
Residential (or local) disposal (waste)
Surface impoundments
  Hazardous waste
  Nonhazardous waste
Waste tailings
Waste piles
  Hazardous waste
  Nonhazardous waste
Materials stockpiles (nonwaste)
Graveyards
Animal burial
Aboveground storage tanks
  Hazardous waste
  Nonhazardous waste
  Nonwaste
Underground storage tanks
  Hazardous waste
  Nonhazardous waste
  Nonwaste
Containers
  Hazardous waste
  Nonhazardous waste
  Nonwaste

Open burning and detonation sites
Radioactive disposal sites

Category III—Sources designed to retain substances during transport or transmission
Pipelines
  Hazardous waste
  Nonhazardous waste
  Nonwaste
Materials transport and transfer operations
  Hazardous waste
  Nonhazardous waste
  Nonwaste

Category IV—Sources discharging substances as consequences of other planned activities
Irrigation practices (e.g., return flow)
Pesticide applications
Fertilizer applications
Animal feeding operations
Deicing salts applications
Urban runoff
Percolation of atmospheric pollutants
Mining and mine drainage
  Surface mine-related
  Underground mine-related

Category V—Sources providing conduit or inducing discharge through altered flow patterns
Production wells
  Oil (and gas) wells
  Geothermal and heat recovery wells
  Water supply wells
Other wells (nonwaste)
  Monitoring wells
  Exploration wells
Construction excavation

Category VI—Naturally occurring sources whose discharge is created and/or exacerbated by human activity
Groundwater–surface water interactions
Natural leaching
Salt-water intrusion/brackish water up-coning (or intrusion of other poor-quality natural water)

SOURCE: Office of Technology Assessment, *Protecting the Nation's Groundwater from Contamination* (Washington, D.C.: OTA-0-233, October 1984).

PATHWAYS OF GROUNDWATER CONTAMINATION

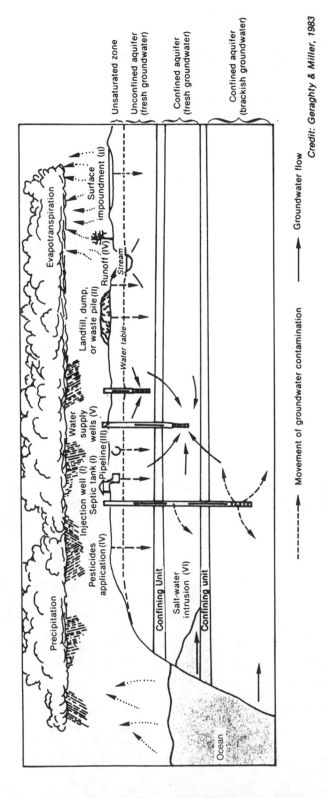

FIGURE 8-4   Pathways of groundwater contamination. Examples are shown for each of the six source categories (I–VI) described in the text. (Source: Office of Technology Assessment, *Protecting the Nation's Groundwater from Contamination: Volume I*, publication OTA-0-233, 1984)

FIGURE 8–5 Areas of pollution concern around the Great Lakes. (Source: U.S. Environmental Protection Agency, *Report on the Environment—Region V, 1983*, 1984)

AREAS OF POLLUTION CONCERN AROUND THE GREAT LAKES

LAKE SUPERIOR
- St. Louis River, Minnesota
- Thunder Bay, Ontario
- Nipigon Bay, Ontario
- Jackfish Bay, Onatario
- Peninsula Harbour, Ontario

LAKE HURON
- Spanish River Mouth, Ontario
- Penetary Bay to Sturgeon Bay, Ontario
- Collingwood Harbour, Ontario

LAKE MICHIGAN
- Manistique River, Michigan
- Menominee River, Michigan
- Sheboygan, Wisconsin
- Muskegon Lake, Michigan
- White Lake Montaque, Michigan
- Fox River/Southern Green Bay, Wisconsin
- Milwaukee Estuary, Wisconsin
- Waukegan Harbor, Illinois
- Grand Calumet River and Indiana Harbor Canal, Indiana

LAKE ERIE
- St. Clair, Michigan / Ontario
- Detroit River, Michigan
- Rouge River, Michigan
- Raisin River, Michigan
- Maumee River, Ohio
- Black River, Ohio
- Cleveland, Ohio
- Ashtabula, Ohio
- Clinton River, Michigan
- Wheatley Harbour, Ontario

LAKE ONTARIO
- Eighteen Mile Creek, New York
- Rochester Embayment, New York
- Oswego, New York
- Toronto Waterfront, Ontario
- Port Hope, Ontario
- Bay of Quinte, Ontario
- Buffalo River, New York
- Niagra River, New York
- Hamilton Harbor, Ontario
- Cornwall-Massena, Ontario-New York

cesses are now regulated by RCRA, CERCLA, and TSCA. Disposal has been accomplished by injection wells, landfills, surface disposal impoundments, land application (treated fill), and in other ways (see Figure 8–6). The EPA inventoried hazardous waste disposal in 1981 and estimated that some 41 million metric tons of hazardous waste was generated in that year. Later estimates indicate the true figure to be much higher, above 265 million metric tons, but even this may be low. In addition to sites for hazardous wastes, there is also concern about the sites for nonhazardous wastes, such as landfills, treated fills, surface impoundments—possibly in excess of 275,000 such sites. While presumed to be for nonhazardous wastes, they do contain some hazardous materials as well, mainly from small industrial operations, household items (oven cleaners, paints and thinners, insecticides, and so on). The magnitude of this hazardous component and how it may affect surrounding waters are unknown.

One very serious aspect of hazardous waste has been the practice of mixing some hazardous materials, such as solvents, with used oils, and then burning the mixture as a blended fuel oil. This material often contains toxic amounts of PCBs, other organics, heavy metals, and probably many other hazardous materials. The EPA estimates that some 1–2 million tons of hazardous wastes and 2.5 million

TABLE 8–9   Summary of trends in selected water-quality constituents and properties as NASQAN stations, 1974 –1981.

Constituents and Properties	Number of Stations with			
	Increasing Trends	No Change	Decreasing Trends	Total Stations
Temperature	39	218	46	303
pH	74	174	56	304
Alkalinity	18	207	79	304
Sulfate	82	182	40	304
Nitrate-nitrite	76	203	25	304
Ammonia	31	221	30	282
Total organic carbon	36	230	13	279
Phosphorus	39	232	30	301
Calcium	23	198	83	304
Magnesium	50	208	46	304
Sodium	103	173	28	304
Potassium	69	193	42	304
Chloride	104	164	36	304
Silica	48	213	41	302
Dissolved solids	68	183	51	302
Suspended sediment	44	204	41	289
Conductivity	69	193	43	305
Turbidity	42	199	18	259
Fecal coliform bacteria	19	216	34	269
Fecal streptococcus bacteria	2	190	78	270
Phytoplankton	22	234	44	300
Dissolved trace metals:				
Arsenic	68	228	11	307
Barium	4	81	1	86
Boron	2	15	3	20
Cadmium	32	264	7	303
Chromium	12	152	2	166
Copper	6	83	6	95
Iron	28	258	21	307
Lead	5	232	76	313
Manganese	30	250	19	299
Mercury	8	194	2	204
Selenium	2	201	21	224
Silver	1	32	0	33
Zinc	19	251	32	302

SOURCE: R. A. Smith, and R. B. Alexander, *A Statistical Summary of Data from the U.S. Geological Survey's National Water Quality Networks,* Open-File Report 83–533, 1983.

FIGURE 8–6   Quantities of hazardous waste disposed in 1981, by disposal process type, in million metric tons. (After U.S. Environmental Protection Agency)

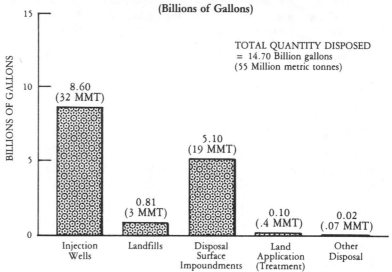

tons of used oils are burned each year as fuels with unknown but potentially very serious health consequences. Further, the residues from these fuels contain even more concentrated amounts of heavy metals and some organics. It is common for these residues to be disposed of at commerical nonhazardous waste facilities. Ways to combat this sloppy use and disposal of hazardous waste and used fuel are under investigation.

Sites for storage of wastes, both legal and illegal, have received a great deal of attention in the United States over the past decade. Love Canal in New York, Times Beach in Missouri, and the Stringfellow Acid Pits in California have all received tremendous press. There is great concern about the release of toxic substances from these sites, by gaseous emission, by surface runoff, by subsurface groundwater flow, and physical disruptions of the sites. Federal legislation to handle newly generated hazardous waste has existed for some years, but it became apparent that new legislation was needed to address the cleanup of uncontrolled waste sites. The first such action was the Resource Conservation and Recovery Act of of 1976, known as RCRA, which was followed by the Comprehensive Environmental Response, Compensation, and Liability Act (CERCLA) of 1980. The CERCLA includes Superfund, the program to deal with emergencies at uncontrolled sites, to clean up the sites, and to deal with related problems. The Office of Technology Assessment (OTA) extensively reviewed the Superfund program in 1985. The agency noted that of the 621,000 sites (Table 8–10) for solid waste disposal, a very large number may require immediate action, and an even larger number will require action in the future. The EPA established a National Priority List (NPL) for waste disposal sites, and identified some 2000 sites requir-

TABLE 8–10   Summary data on solid waste facilities.

Percent of uncontrolled sites that are solid waste facilities:	
Of 1,389 sites with actual or presumed problems of releases of hazardous substances	18%
Of 550 sites on National Priority List	20%
Two most prevalent effects at problem solid waste sites:	
Leachate migration, groundwater pollution: at 89% of sites	
Drinking water contamination: at 49% of sites	
Mean size of problem solid wastes sites	67.4 acres
Median hazard ranking score:*	
Solid waste sites on the NPL	40.8
All NPL sites	42.2
Estimates for national number of solid waste sites:	
Operating sanitary, municipal landfills	14,000
Closed sanitary, municipal landfills	42,000
Operating industrial landfills	75,000
Closed industrial landfills	150,000
Operating surface impoundments	170,000
Closed surface impoundments	170,000
Total	621,000
Estimate of need for future cleanup:	
Low: 5% landfills, 1% impoundments likely to release toxic substances	17,400
High: 10% landfills, 2% impoundments likely to release toxic substances	34,800
Conservative figure used for cleanup by Superfund	5,000

*28.5 required for placement on National Priorities List; current highest site score is 75.6.
SOURCE: Office of Technology Assessment.

ing immediate action. The OTA argued, however, that the number of such sites is closer to 10,000.

There are many problems in dealing with sites requiring cleanup. The first phase of Superfund was oriented at immediate solution, with little emphasis on long-term effects. Hence if a site was leaking chemicals into the water, but the leakage could be stopped by some soil removal or injection of barrier materials, for example, this could stop the immediate problem, but might not be adequate in the long run. The OTA recommended a two-part strategy. Part I is a short period—15 years—where there would be a detailed assessment and evaluation of NPL sites, an immediate response to the worst sites to prevent them from getting worse, permanent remedial cleanup for particularly dangerous sites, and development of institutional capabilities for a long-term program. This part of the program would be much larger than Superfund as it now stands, requiring perhaps $100 billion to deal with the 10,000 sites considered by the OTA as needing some type of remedial cleanup now.

Part II, recommended by the OTA, focuses on more extensive site studies and on permanent cleanup at sites that pose an immediate threat to both public health and the environment. These would draw on the institutional program of Part I and would prevent such things as spending huge amounts of funds before cleanup goals are set or before permanent cleanup technologies are available, and would eliminate inconsistent cleanup policies that have been evident in the first phase of Superfund.

The OTA noted that federal efforts are needed to obtain more information on health and environmental effects, to develop specific national cleanup goals, to provide more support for developing and demonstrating innovative permanent cleanup technologies, to provide more support to the EPA and states to monitor cleanup efforts, to provide support for technical training programs, to improve Superfund goals by more public participation in decisionmaking about initial responses and remedial cleanups, and to provide technical assistance to communities. These are noble goals; they are also expensive. While the EPA lists 2000 NPL sites, the OTA also lists an additional 5000 sites from the open and closed solid waste disposal sites, another 2000 from improved site identification process (that is, those missed by the EPA), and 100 from hazardous waste management sites operating with ineffective groundwater protection standards.

Who will pay for all the cleanup is not clear at all. The EPA anticipates that private parties will pay 40 to 60 percent of the total cleanup amount, and that states will provide an amount to match EPA funds. Currently there is great uncertainty in this matter. Private parties are unwilling to pay for cleanup unless there is some assurance of release from future liability, but the short-term EPA Superfund program is not adequate to address more than near-term treatment, hence they cannot, in a strict sense, enter into such an agreement. Further, funds available from states for such purposes may vary from year to year. The OTA concluded in its 1985 assessment of Superfund that "it is technically and economically impossible to permanently clean up all uncontrolled waste sites in the near term."

The suggested two-part approach by the OTA would treat urgent NPL sites to greatly reduce, but not permanently remove, health hazards from these sites. This would be followed by a more detailed, long-term approach to effectively minimize the potential threat to public health and the environment. The cost for the first part might be roughly $1 million per site, although at sites requiring extensive groundwater cleanup it presumably would be significantly higher. The present Superfund spends about $300,000 per site for immediate removal and about $10 million per site for remedial cleanup, but, as noted by the OTA, neither meets its intended purpose very well. Thus some $10.3 million per site currently spent may not effectively deal with the problem. The OTA argues that while its proposal would cost more initially, more information and technology will be available to address the long-term program and it would be cheaper and more realistic. The OTA also argues that planning now for full study of sites will be expensive, and would require Superfund to increase to at least $100 billion over a few decades, but that such planning, if done now, will prevent severe crisis situations in the future.

This discussion has put the emphasis on CERCLA (Superfund) as opposed to the RCRA. The reason is simply that the RCRA, even though intended to deal with cleanup and monitoring of hazardous waste facilities, is not rigorous enough. For example, at sites where leaks occur, RCRA requires cleanup but does not shut down the facility while the leak is being fixed. RCRA coverage is also limited to 30 years; CERCLA would handle the sites after 30 years. CERCLA regulates more substances and has more stringent drinking water standards than RCRA, hence a site at which groundwater passes RCRA standards might not pass CERCLA standards. RCRA's levels of acceptable groundwater contaminants are not based on long-range health effects. RCRA has no geologic siting criteria or standards to restrict hazardous waste sites to geologically suitable locations. RCRA requirements do not include monitoring in the vadose zone (see Chapter 3), hence early warning of some leaks may be undetected. The RCRA method of implementing cleanup is complex, cumbersome, and time consuming. CERCLA, however, addresses most of the shortfalls of the RCRA program.

Yet there is a growing feeling that Superfund is not working. The 1985 OTA report noted many places where the Superfund approach is inconsistent at best and may be unsatisfactory at worst. First, the sites included on the EPA's high priority NPL list were rated based on a composite score of migration of contaminants by air, by surface water, and by groundwater. An arbitrary cutoff of 28.5 points total was used to restrict the first NPL to about 400 sites. In all three scoring categories, there is a great weight placed on population; that is a contaminated groundwater would have a much higher score in a populated area than in a deserted area. But unpopulated areas of high contamination might reach aquifers which would affect population centers at some distance. The OTA listed numerous sites at which there are severe environmental problems. Case histories for sites in Maryland, Pennsylvania, New York, and California are mentioned, but where the overall score is lower than 28.5, hence the sites are not on the NPL. The very rating system itself, a semiquantitative Hazard Ranking System (HRS), is based on subjective input, and there is a high degree of uncertainty.

But problems are more fundamental than this. The data base from which we must evaluate groundwater contamination and surface water contamination is very sparse. Not only are data few, but the overall system's total toxicity is often so poorly known that attempts at modeling contaminant transport are doomed to failure. Monitoring wells are often too few, too poorly positioned, and too poorly tested to give meaningful results. Background levels of elements, compounds, other materials are often not known; without knowing the background, one cannot detect anomalies. Moreover, the net toxicity of more than one item is very complex. Sometimes two species offset each other's health effects; in other cases they magnify each other. We have a very poor feel for both background toxicity and toxicity of polluted waters.

TABLE 8–11 EPA detection limits for some carcinogens.

Chemical	Highest Permitted EPA Detection Limit (nanograms/liter)	Concentration Projected to Cause One Cancer per 100,000 People (nanograms/liter)	Projected Cancers per 100,000 People*
Aldrin	1,900	0.74	2,600
Dieldrin	2,500	0.71	3,500
1,1,2,2-tetrachloroethane	6,900	1,700	4
3,3'-dichlorobenzidine	16,500	103	160
Heptachlor	1,900	2.78	680
PCBs	36,000	0.79	46,000
Benzo(a)pyrene	2,500	28	90
Benzidine	44,000	1.2	37,000
Chlordane	14	4.6	3
DDT	4,700	0.24	20,000

*Projections based on the consumption of 2 liters a day of the contaminated drinking water over a lifetime. Projections are also based on animal studies that include assumptions on the transfer of results from animals to humans, and extrapolation from high doses to low doses. Despite the uncertainties introduced by these assumptions, these are the projections EPA uses. Column 3 has been calculated by OTA by dividing column 1 by column 2. This calculation converts back to high doses. Uncertainties introduced into column 2 by high-to-low dose extrapolations are thus partially corrected for in deriving column 3. Column 3 contains no correction for uncertainties introduced by applying animal results to humans.

SOURCE: Office of Technology Assessment.

TABLE 8–12   Data on known or probable human carcinogens with primary drinking water standards.

Pollutant	EPA Primary Drinking Water Standard (µg/L)	Concentration Projected to Cause One Cancer per 100,000 People (µg/L)*	Projected Cancers per 100,000 People*
Arsenic	50	0.022	2,300
Barium	1,000	—	—
Cadmium	10	—	—
Lindane	4	0.186	22
Toxaphene	5	0.0071	700

*Projections based on the consumption of 2 liters a day of the contaminated drinking water over a lifetime. Except for arsenic, projections are also based on animal studies that include assumptions on the transfer of results from animals to humans, and extrapolations from high doses to low doses. For arsenic, projections are extrapolated from the effects of high doses in humans. Despite the uncertainties introduced by these assumptions, these are the projections EPA uses. Column 3 has been calculated by OTA by dividing column 1 by column 2. This calculation converts back to high doses. Uncertainties introduced into column 2 by high-to-low dose extrapolations are thus partially corrected for in deriving column 3. Except for the arsenic number, which is based on human data, column 3 retains the uncertainties introduced by applying animal results to humans.

SOURCE: Office of Technology Assessment.

There is also concern that our drinking standards may be misleading as well. As seen in Table 8–11, the permissible level set by the EPA for detection limits is often much higher than the level required to cause one excess cancer per 100,000 people exposed to the carcinogen. Thus, in the case of PCBs, the permissible level is 36,000 nanograms per liter of water, but it takes only 0.79 ng/L to cause one excess cancer per 100,000 people given a lifetime exposure. Thus at the permissible level, which would meet EPA requirements, this might result in an excess 46,000 cancer deaths per 100,000 people. The third column in Table 8–11 shows that in most cases, for the organics listed, a large number of excess cancer deaths might occur. The second column is based on animal studies under laboratory conditions, and the linear hypothesis is used for dose-health effect relationships. Yet even if the figures are off by a large factor, the point is well made that the EPA levels are not realistic, and, since these levels were used in setting NPL scores, it is likely that some seriously contaminated sites are not included on the NPL.

Table 8–12 shows the equivalent data for known or suspected carcinogens for which EPA drinking water standards are set. Here parts of the second and third columns are not filled in because laboratory studies are not conclusive as to how much ingested contaminant can cause cancer. No level is thus set for cadmium, although the EPA has reported that no level of cadmium is without risk. Table 8–13 shows again some of the discrepancy between CERCLA and RCRA groundwater detection levels. The CERCLA values are significantly lower than those of the RCRA; thus a site that passes RCRA levels might still be a candidate for CERCLA action. This picture is made even more complex because the RCRA testing procedures are not good, and the detection limit can be masked and raised by elemental and compound interference. Still further complexity arises from the fact that the EPA has not been able to set levels at which some carcinogenic chemicals can be detected in groundwater; these are shown in Table 8–14.

The magnitude of groundwater contamination is huge. The EPA notes that for 881 of the NPL sites, 526 had recognized and observed releases of hazardous materials into groundwater. Further, over 8 million citizens in the United States are potentially exposed to the groundwater from these sites, and for 350 of these sites the contaminated groundwater is the only source of drinking water for the affected population. And another 6.5 million people may be affected by contaminated surface water from an additional 450 sites. Even the more strict CERCLA standards may be inadequate to protect human health in the generations to come.

AGRICULTURAL POLLUTION

Nonpoint source pollution from agricultural areas is well documented (Council of Environmental Quality, 1984). As noted in Chapter 5, very large amounts of potassium salts, nitrogen salts, phosphates, and sulfur salts (sulfates, pesticides, fungicides) are added to the nation's farmland each year for growing crops. Much of

TABLE 8–13 Some examples of groundwater detection levels of hazardous chemicals which are higher under RCRA than under CERCLA.

Pollutant	CERCLA Detection Levels (ng/L)*	RCRA Detection Levels (ng/L)
Dieldrin	5	2,500†
DDT	10	4,700†
DDE	5	5,600†
DDD	10	2,800†
Heptachlor	5	1,900†
Heptachlor epoxide	5	2,200†
Aldrin	5	1,900†
Antimony	20,000	32,000‡
Arsenic	10,000	53,000‡
Cadmium	1,000	4,000‡
Lead	5,000	42,000‡
Selenium	2,000	75,000‡
Thallium	10,000	40,000‡

*U.S. Environmental Protection Agency, "Statement of Work, Organics Analysis, Contract Laboratory Program" (Washington, DC: EPA, September 1983). U.S. Environmental Protection Agency, "Statement of Work, Inorganics Analysis, Contract Laboratory Program" (Washington, DC: EPA, May 1982).

†U.S. Environmental Protection Agency, *Test Methods for Evaluating Solid Waste, Physical/Chemical Methods,* SW-846, 2d ed. (Washington, DC: Office of Solid Waste, EPA, 1982).

‡Lee M. Thomas, Assistant Administrator, U.S. Environmental Protection Agency, Memorandum to the Administrator proposing additional test methods for reference 17, Oct. 17, 1983.

TABLE 8–14 Some carcinogenic chemicals for which EPA has not yet determined the levels at which they can be detected in groundwater by the methods of reference.

Aflatoxin
4-aminobiphenyl
Aziridine (ethyleneimine)
bis-(chloromethyl)ether
Chloromethyl methyl ether
1,2-dibromo-3-chloropropane (DBCP)
Diethylnitrosamine (n-nitrosodiethylamine)
Diethylstilbesterol*
Dimethylaminoazobenzine
7,12-dimethylbenz(a)anthracene
Dimethylcarbamoyl chloride
1,2-dimethylhydrazine
Ethyl methanesulfonate
Hydrazine
Methylnitrosourea
Nitrosomethylurethane (n-nitroso-n-methylurea)
n-nitosopiperidine
n-nitrosopyrrolidine
Streptozotocin*
2,3,7,8-tetrachlorodibenzo-p-dioxin (TCDD)
Ethylene dibromide (EDB)

*Test methods not yet published by EPA as of Jan. 19, 1984.

SOURCE: U.S. Environmental Protection Agency, *Test Methods for Evaluating Solid Waste, Physical/Chemical Methods*, SW-846, 2d ed. (Washington, DC: Office of Solid Waste, EPA, 1982).

this additive material is consumed by the crops and other growth, but of the remainder part is stored in the soil and the larger part removed by running water and erosion. When these fertilizers enter streams or lakes, they tend to provide an overabundance of nutrients and eutrophication results.

In addition, toxic chemicals are added in large amounts to farmlands. Herbicides, for example, were used in the amount of 76 million pounds on some 70 million acres in 1964 and increased to 433 million pounds on 362 million acres in 1982 (Table 8–15). Their long-range health effects on the environment, including fowl, animal, and plant life, are not well known. Insecticide use grew similarly from 1964 to 1972 (Table 8–16), but has decreased since. DDT was outlawed in the late 1960s, and use of many others is now forbidden or cut back. The toxic residues in fish (Table 8–17) show gradual decreases in some major pesticides for the period 1970–1981, and this is also the case for waterfowl (Table 8–18). Interestingly, in the case of residues from deceased humans (Table 8–19), there was a pronounced decrease in DDT and dieldrin for the period 1970–1983,

TABLE 8–15 Selected herbicides used by farmers on crops, 1964 –1982.

Herbicide	Pounds Used (million)	Acres Treated (million)	Pounds per Acre
All herbicides, total:			
1964	76	NA	NA
1966	112	99	1.1
1971	224	158	1.4
1976	394	197	2.0
1982*	433	362	1.2
Atrazine:			
1964	11	8	1.4
1966	23	15	1.6
1971	57	40	1.4
1976	90	62	1.5
1982*	75	52	1.4
Alachlor:			
1964	NA	NA	NA
1966	NA	NA	NA
1971	15	12	1.3
1976	89	54	1.7
1982*	83	45	1.8
2,4–D:			
1964	30	56	0.5
1966	40	57	0.7
1971	33	55	0.6
1976	38	59	0.7
1982*	21	43	0.5
Trifluralin:			
1964	NA	NA	NA
1966	5	7	0.7
1971	11	17	0.7
1976	28	34	0.8
1982*	35	41	0.9
All other:			
1964	36	NA	NA
1966	44	NA	NA
1971	107	NA	NA
1976	149	NA	NA
1982*	165	166	1.0

NA–Not available
*Preliminary data.
SOURCE: U.S. Environmental Protection Agency.

Insecticide	Pounds Used (million)	Acres Treated (million)	Pounds per Acre
All insecticides total:			
1964	143	NA	NA
1966	138	43	3.3
1971	154	57	2.7
1976	162	75	2.2
1982*	59	62	1.0
Toxaphene:			
1964	34	8	4.3
1966	31	5	5.8
1971	33	6	5.9
1976	31	5	6.3
1982*	3	2	1.5
Methyl parathion:			
1964	10	7	1.5
1966	8	5	1.8
1971	28	12	2.3
1976	23	12	1.9
1982*	9	6	1.5
DDT			
1964	35	12	2.9
1966	29	9	3.3
1971	14	3	4.5
1976	X	X	X
1982*	X	X	X
Carbofuran:			
1964	NA	NA	NA
1966	NA	NA	NA
1971	3	4	0.8
1976	12	11	1.0
1982*	7	7	1.0
Ethyl parathion:			
1964	6	5	1.3
1966	8	6	1.4
1971	9	10	0.9
1976	7	12	0.5
1982*	NA	NA	NA
Aldrin/Dieldrin:			
1964	12	14	0.8
1966	15	15	1.1
1971	8	8	1.0
1976	0.9	0.5	1.9
1982*	X	X	X

NA–Not available.

*Preliminary data.

SOURCE: U.S. Environmental Protection Agency.

TABLE 8-17 Toxic residues in fish, 1970–1981 (parts per million, geometric mean).

Toxic Residue	1970	1971	1972	1973	1974	1976–1977	1978–1979	1980–1981
DDT*	0.98	0.73	0.64	0.44	0.52	0.27	0.25	0.20
PCB	1.20	1.03	1.20	.78	.95	.88	.85	.53
Toxaphene	NA	.01	.13	.17	.17	.35	.29	.27
Dieldrin and aldrin	.08	.07	.07	.05	.09	.05	.05	.04

Freshwater fish samples are collected by Fish and Wildlife Service personnel in all 50 states as part of the National Pesticide Monitoring Program. Two-thirds of the fish sampled are bottom-dwelling species such as carp, suckers, or catfish. The remaining one-third of the fish sampled are predacious species, such as trout, walleye, bass, and bluegill. The whole fish is analyzed, not just the fillet.

NA–Not available. Data for 1970–1974 were collected annually. Data for 1976–1981 were collected on a two-year sampling cycle.

*Includes DDT and its derivatives.

SOURCE: U.S. Environmental Protection Agency.

TABLE 8-18 Toxic residues in waterfowl, by flyway, 1966–1982 (parts per million, mean wet weight).

Year	Pacific Flyway			Central Flyway			Mississippi Flyway			Atlantic Flyway		
	DDE*	Dieldrin	PCBs	DDE*	Dieldrin	PCBs	DDE*	Dieldrin	PCBs	DDE*	Dieldrin	PCBs
1966	0.65	<.01	NA	0.15	<.01	NA	0.25	<.01	NA	0.70	<.01	NA
1969	0.71	0.02	0.20	0.30	0.02	0.20	0.40	0.04	0.44	1.03	0.05	1.29
1972	0.34	0.01	0.11	0.15	0.02	0.10	0.37	0.02	0.66	0.44	0.02	1.24
1976	0.22	0.02	0.16	0.28	0.03	0.15	0.25	0.05	0.23	0.32	0.06	0.52
1979	0.35	0.02	0.07	0.10	0.02	0.06	0.17	0.05	0.11	0.27	0.05	0.45
1982	0.36	NA	0.07	0.11	NA	0.05	0.13	NA	0.14	0.25	NA	0.39

NA–Not available.

Waterfowl residues were sampled in the 48 states as part of the National Pesticides Monitoring Program. Wings were removed and analyzed.

*DDE is a derivative of DDT.

SOURCE: U.S. Environmental Protection Agency.

TABLE 8-19 Toxic residues in humans, 1970–1983 (parts per million, geometric mean, lipid basis).

Year	Number of Specimens	Total DDT*	Dieldrin	Oxychlordane	Heptachlor Epoxide	Trans-Nonachlor	Beta-Benzene Hexachloride	Hexachloro-benzene
1970	1,386	7.95	0.16	NA	0.09	NA	0.37	NA
1971	1,560	8.06	.22	NA	.09	NA	.35	NA
1972	1,886	6.97	.18	0.10	.07	NA	.19	NA
1973	1,092	5.96	.17	.12	.09	NA	.25	NA
1974	901	5.15	.14	.12	.08	0.11	.21	0.03
1975	779	4.76	.12	.11	.08	.06	.19	.04
1976	683	4.34	.09	.11	.08	.13	.18	.04
1977	790	3.15	.09	.10	.07	.10	.14	.04
1978	826	3.52	.09	.11	.07	.12	.14	.04
1979	795	3.10	.08	.10	.07	.12	.15	.04
1980	NA	NA	NA	NA	NA	NA	NA	NA
1981	407	2.38	.05	.09	.11	.13	.10	.04
1982	NA	NA	NA	NA	NA	NA	NA	NA
1983	383	1.63	.06	.10	.09	.12	.08	.03

Data are for fiscal years. Pesticide residues for selected organochlorine compounds were measured in adipose tissue in 48 states as part of the National Human Adipose Tissue Survey. Specimens were collected each year by medical pathologists from selected cities. Individuals with known or suspected pesticide poisoning were excluded. Data are lipid adjusted, design file specimens only.

*DDT includes its metabolites.

SOURCE: U.S. Environmental Protection Agency.

but no significant change for many other hazardous chemicals. These figures attest to the role of agricultural additives in the form of pesticides and insecticides as nonpoint sources of pollution.

Fish kills by pollution are tabulated in Table 8–20. Although figures are widely variable, the average for the period 1976–1981, 12.6 million, is higher than the average of 11.7 million for the period 1970–1975. Of these kills, a major part is due to agriculture pollution.

ASBESTOS AND THE ASBESTOS SCARE

One of the best examples of *enviroscare* is the concern over asbestos. There is no doubt that certain types of asbestos can be linked to lung cancer, mesothelioma, asbestosis, and other respiratory and possibly other adverse health effects. Yet it is the thesis of Malcolm Ross of the U.S. Geological Survey, who has examined this subject in detail, that an overwhelming amount of the asbestos used worldwide is relatively harmless. Let us examine this problem in some detail.

There are three major types of asbestos. There is the common white variety known as chrysotile, which is the variety mined in the United States, Canada, and most other places in the world. In addition, there is the brown variety, amosite, mined in South Africa primarily; and the blue asbestos, crocidolite, mined in both South Africa and Australia. In the United States, 96 percent of the asbestos consumed is the white variety, and 2 percent each of the brown and blue varieties.

Ross (1984) noted that asbestos minerals are found in four main types of occurrences: Alpine ultramafic rocks from which chrysotile is mined, perhaps as much as 85 percent of all asbestos ever mined; stratiform ultramafic intrusions, from which chrysotile is also mined; serpentinized limestone, again a source of chrysotile; and banded ironstones, from which amosite and crocidolite are mined. Tremolite and anthophyllite are asbestos-like minerals mined from Alpine ultramafics, and some tremolite comes from layered intrusive rocks as well. In many instances, miners of asbestos minerals have been exposed to only one kind of asbestos, hence they are well suited to test the adverse health effects of continued exposure to the different types of asbestos. Ross (1984) carefully assessed the studies done through 1983 with some startling results, which will be elaborated on below.

Exactly how asbestos causes disease is not clear. Obviously very small (diameter less than 1.5 μm) particles with an acicular or needlelike nature (longer than about 10 μm) tend to remain in the lower respiratory tract. When these particles are so lodged, they cause buildup of collagen and reticulin fibers that, in turn, impair oxygen intake. In severe cases, lung cancer, mesothelioma, or asbestosis may occur (Table 8–21). Let us examine each of these diseases in turn.

Lung cancer from asbestos exposure is known, but the magnitude is not. Studies in this area tend to be very haphazard when it comes to predicting probable fatalities from asbestos-related lung cancer. Ross (1984) mentioned several:

TABLE 8–20 Fish kills caused by pollution, 1961–1981.

Year	Number of States Responding	Number of Reports	Reports Listing Number of Fish Killed	Total Number of Fish Killed	Average Size of Kill	Largest Kill Reported
1961	45	413	265	14,910,000	6,535	5,387,000
1962	37	421	246	44,001,000	5,710	3,180,000
1963	38	442	304	6,937,000	7,775	2,000,000
1964	40	590	470	22,914,000	5,490	7,887,000
1965	44	625	520	12,140,000	4,310	3,000,000
1966	46	532	453	9,614,000	5,620	1,000,000
1967	40	454	364	11,291,000	6,460	6,549,000
1968	42	542	469	15,815,000	6,015	4,029,000
1969	45	594	492	41,166,000	5,860	25,527,000
1970	45	635	563	22,290,000	6,412	3,240,000
1971	46	860	759	73,670,000	6,154	5,500,000
1972	50	760	697	17,717,000	4,639	2,922,000
1973	50	749	703	37,814,000	5,527	10,000,000
1974	50	721	648	119,052,000	6,532	47,112,000
1975	50	624	543	16,111,210	3,879	10,000,000
1976	50	667	601	13,611,049	4,509	4,800,000
1977	41	503	449	16,538,936	4,386	8,592,000
1978	42	679	621	74,712,651	3,487	33,000,000
1979	47	770	673	8,105,371	4,292	1,000,000
1980	43	801	747	29,949,097	4,046	5,000,000
1981	43	836	760	50,192,644	3,354	25,000,000

SOURCE: U.S. Environmental Protection Agency.

TABLE 8–21   Data on
asbestos-related diseases.

Type of Disease	Years to First Significant Increase in Death Rate	Years to Peaking
Lung cancer	10–14	30 to 35
Mesothelioma	20	Does not peak
Asbestosis	15–20	40 to 45

SOURCE: Ross (1984).

Joseph Califano, then Secretary of Health, Education and Welfare, stated in a speech that 17 percent of all cancer deaths in the United States for the next 30–35 years would be associated with asbestos exposure, which means some 67,000 fatalities per year (Ross said that the reports used by Califano were so grossly in error that *no* conclusions about risk could actually be estimated). Selikoff in 1980 estimated 20,000 deaths per year from asbestos-related disease for the next 40 years, and in 1981 he stated that 10,000 would die each year from some asbestos-related diseases. Finally, Hogan and Hoel (1981) estimated that 3 percent of all U.S. cancer deaths are due to asbestos, or about 12,000. Ross pointed out that these widely varying estimates ranging from 10,000 to 67,000 are mediocre estimates at best. He presented a detailed discussion of this problem, which will be given below.

Mesothelioma is a rare cancer, but one which is associated with exposure to the blue variety of asbestos, crocidolite. Where miners and millers have worked extensively with crocidolite, in South Africa and western Australia, there is a high incidence of mesothelioma, as well as a high incidence of lung cancer. There is also strong evidence that the ratios of lung cancer and asbestosis to mesothelioma for workers exposed to crocidolite are stable. This means that one can estimate the number of lung cancer or asbestosis deaths from the known mesothelioma deaths. Since mesothelioma is both very rare and the one cancer that shows a strong correlation with exposure to crocidolite, then incidence of mesothelioma can be used to estimate fatalities from lung cancer and asbestosis. Because some mesothelioma can be due to causes other than asbestos, such predictions will yield high estimates of lung cancer and asbestosis.

Ross (1984) conducted such an analysis, and compared his predictions with U.S. health data. These are compared in Table 8–22.

The agreement between the estimates and the recorded deaths is quite good. More important, these data, for the period 1967–1977 (early averages), show that annual fatalities from asbestos-related diseases are under 600. The figures in Table 8–22 support the contention that indeed crocidolite is the prime cause of cancer-related deaths. Were, for example, lung cancer but not mesothelioma caused by chrysotile, then the U.S. vital statistics would show a very high figure for lung cancer but a low figure for mesothelioma, yet the data show almost exact agreement with the mesothelioma study by McDonald and McDonald (1981).

TABLE 8–22   Estimated fatalities from asbestos.

Cause of Death	Ross (1984)	McDonald and McDonald (1980)
Mesothelioma	92	103
Lung cancer	199	224
Gastrointestinal tract cancer	21	24
Other cancer	42	47
Asbestosis	88	99
Noninfectious respiratory disease	80	90
Total asbestos-related deaths	522	587

What the data show in Table 8–22 is that actual lung cancer deaths per year in the United States due to asbestos are very low (compare 150,000 deaths from lung cancer due to smoking, or 6,000 to 30,000 due to indoor radon). There is one exception: workers in asbestos mines and mills who smoke indeed have a disproportionately high number of lung cancers.

Further evidence about the relative non–toxicity of white asbestos, chrysotile, has been obtained from a study of miners' wives in Thetford Mines, Quebec, and other small asbestos mining towns in Canada. For many years there were few controls on the amount of asbestos that could be inhaled, and dust controls on the mining and milling operations were few. Thousands of wives in the study received extremely high exposure to asbestos. Yet they showed a below-normal incidence of lung cancer and asbestosis (as well as mesothelioma) (Ross, 1984). Were chrysotile truly carcinogenic, this study would have disclosed it. The study did not. Further, in the United States, most of the incidence of lung cancer, mesothelioma, and asbestosis noted was due to poor working conditions for shipyard workers during the Second World War, hence the figures for total fatalities per year should or may already have started to decrease. Certainly there are no scientific data to back up the statements mentioned earlier in this chapter.

But enviroscare continues to take its toll. For example, in the iron-mining areas of northern Minnesota, the waste rock contains a great deal of amphibole (several varieties). Because amphibole is acicular, and asbestos scare led to a ban on dumping the waste from the iron mining operations into Lake Superior. While for other reasons this makes sense, it did not make sense to require the mining companies not only to come up with very costly excavations for the waste rock, but to put covers on the waste piles. This $500 million expense will probably have a zero effect on public health and disease. In other areas of the United States (Globe, Arizona is a good example), many millions of dollars are being spent to dig up land suspected or known to have minor amounts of chrysotile asbestos. The earth is then transported to another site, dumped, and covered by elaborate schemes. Again, the cost is extreme and positive effects on public health and disease is zero. One could safely argue that because these very operations would expose workers to the normal hazards of such work, the actual effect of this technique on public health is adverse.

There is also a strong push not only to clean up all existing buildings that contain asbestos, but to outlaw asbestos for all uses. Since 96 percent of all asbestos in this country is chrysotile, one can question the merit in these projects. Further, for some uses, such as in brake linings, there are no good substitutes. Thus, are we better off with inferior brake linings, which will very definitely have an adverse effect on public safety, or with chrysotile asbestos with essentially zero risk? In addition, asbestos, which is strong, flexible, and flame resistant, is used as coatings for prevention of fire. Substitutes in the same price range are not as good. Thus is it wise to replace chrysotile asbestos with inferior materials which may result in fires? Asbestos, and especially the common chrysotile variety, is a safe, inexpensive, and effective material with many, many applications for which there is not only no good substitute, but for which no substitute is needed.

As mentioned above in connection with iron ore waste, the EPA has decided, without a good scientific basis, that all fibrous minerals are "like asbestos" and therefore must be treated as asbestos. It makes no difference to the EPA that most of these minerals, such as amphiboles, have never been related to any adverse public health or disease problem. This has had a pronounced effect on some mining and construction industries. Virtually all rocks contain some amphibole. In areas where rocks have very high amphibole content there is no correlation with increased incidence of lung cancer or other disease. In fact, amphiboles appear to be very innocuous minerals. Fibrous or acicular minerals occur nearly everywhere, from water supplies, to soils, sediments, sand and gravel deposits, most rocks, drilling muds and cements, and in ceramics.

Ross (1984) pointed out some other interesting facts. In 1976 there were nearly 900 deaths from anthracosilicosis (black lung) among former coal miners, and nearly 1000 deaths from silica dust (silicosis, silicotuberculosis, other respiratory diseases). Yet money spent on research in these areas of known and significant danger to public health is little, especially when compared to the billions spent on asbestos-related projects due to the enviroscare syndrome. Of interest also is the known deaths from mesothelioma in the United States; the total for the period 1967 through 1976 is 26. If we assume that the ratio of fatalities from either asbestosis or lung cancer to mesothelioma is fixed, and available scientific evidence argues that this is indeed the case, then actual deaths due to asbestos-related causes may be less than 100 for a 10-year period.

Finally, it is interesting to note just how much of a risk asbestos is assigned by EPA scientists, and *not* EPA managers. In the 1987 work *Unfinished Business: A Comparative Assessment of Environmental Problems—Overview Report*, EPA scientists ranked many sources of risk. They included asbestos with formaldehyde, tetrachloroethylene, and methylene chloride. For these four sources combined, a maximum figure of 250 cancer possibilities per year is suggested; that is, possibly 60 per individual item. The EPA scientists noted that while the actual risk from asbestos is thus quite low, it is in terms of public perception extremely high. Since EPA management responds to public pressure, then the overall EPA ranking of the asbestos hazard is the highest of all the topics included. And, although the actual hazard of indoor radon is many, many times greater than that for asbestos, its ranking by the public is much lower, hence EPA efforts are also much less. Thus the lives possibly saved by removal of all asbestos will be close to zero, yet the lives lost from normal incidences of workplace accidents and transportation involved with such a cleanup will be much greater.

FINAL COMMENTS ON ENVIRONMENTAL GEOCHEMISTRY

The problems of environmental geochemistry are so large that, indeed, they warrant volumes devoted just to this topic. But that is not the goal of this book. The topics in this chapter were chosen to illustrate some of our more pressing environmental problems. Coal and nuclear topics, as well as petroleum and gas, were covered in Chapter 7, but they could just as well have been covered here. Basically, this chapter focuses to a large extent on what we don't know, although many regulatory agencies profess to address these problems.

We do not know the environmental consequences of most construction and road building. We do not know the long-term effects of continued pollution of our croplands, waterways, forests, lakes, and living areas. Neither can we relate subtle patterns in health and disease to rock-soil-water geochemistry. Yet all of these can be answered. The U.S. approach to dealing with chemicals in the environment is a scattershot approach: a bit of legislation here, an expensive and often rushed cleanup there, and so on. A sensible policy based on hard data to support such actions is nonexistent. RCRA and CERCLA (Superfund) are obvious examples of the shotgun approach; the actual implementation falls far short of the concept. Instead of pouring many millions of dollars into temporary cleanup, perhaps it would be better to attempt a longer range type of cleanup that would assure environmental safety well into the future. At the same time, a realistic and not an enviroscare basis for making cleanup decisions as well as finding the true extent of the potential harm of the toxins in question is in order.

The United States and the entire world need sensible direction in the area of environmental issues. Without such direction, really devastating results of environmental mishandling can occur. Poland is a prime example.* Roughly one third of Poland's entire population, about 13 of 39 million, is likely to be affected by pollution-caused cancer. Poland's waters are so polluted that well over 50 percent are not fit even for industrial use. Agricultural runoff, untreated sewage, petroleum and coal wastes, chemicals, and metals all contribute to the ruination of these waters. In Chapter 7 it was pointed out that sulfur dioxide due to acid rain from coal-burning power plants is a serious threat to our lives. In the United States coal plants add about 7 tons per square mile of SO_2 to the atmosphere. While this seems large, in Poland this figure is about 34 tons per square mile! At

---

* Lasota, J. P., 1987, Darkness at noon: Time is running out for Poland's environment, *The Sciences* (July/Aug.): 22–30.

the Polish release rate, this would more than double our SO_2-induced fatalities to about 111,000 per year from coal. Further, this staggeringly high figure comes from just burning anthracite, but Poland's energy needs will be soon met by burning lignite, with much higher sulfur content, and thus a terrible air pollution picture becomes worse. Already, 18,000 deaths per year are blamed on SO_2 in Poland. The Polish Academy of Science predicts that 20 percent of the nation's flora and 15 percent of its fauna will die in the next five years due to this pollution. Unsafe levels of lead, mercury, cadmium, and zinc are routinely noted throughout the country. While in all other industrialized countries longevity of males is increasing, in Poland it has decreased to the 1952 level, due to environmental poisoning. Mental retardation and other diseases linked to metals and coal contaminants are on a marked increase as well. The Poland situation may not get better in the foreseeable future. The Polish government needs the production to help its woeful economy, and pollution control measures, such as they are, are routinely ignored (the Polish government has even subsidized some flagrant polluters to pay their fines), and there is also a lack of information internally and externally about Poland's environmental chaos.

The point to make from the Polish experience is obvious. Without a sound and enforced universal environmental policy, and without freedom of exchange of information, and without education of the public, and without fundamental, basic research on the consequences of pollution (including all aspects of background geochemistry to health effects of various chemicals and other substances), then environmental ruin may result. In the United States and in many other market economy countries throughout the world, there is a ready exchange of information and a great deal of public awareness, but even these enlightened countries draw low marks for basic research, lack of policy, and for using enviroscare instead of education to make environmental decisions. In the United States the most obvious example is coal versus nuclear. We endorse coal, a well-known carcinogenic source of energy, and throw hurdles at nuclear. The EPA's stand against asbestos use is another example. And whereas the United States rightly regulates the nuclear industry, there is no equivalent group to oversee coal or the chemical industries, from which most of our pollution is derived! Not that the United States will turn into another Poland, but the very serious long-range effects of the pollution now known can only continue until chemicals and coal are monitored from mining or manufacturing through use to final disposal.

The earth and its atmosphere are very delicate. Mankind has proved to be the greatest polluter, and with increased population, increased consumption of goods and therefore demand, and stagnant education, the world can only expect more and worse pollution. The United States could and should take the lead in promoting a safe environment, then work with other nations to promote these steps worldwide, but this leadership does not seem to be unfolding. However, the next century will bring all of the warnings and menaces discussed here into stark reality, and only then—and perhaps only in response to national tragedy—might recuperative steps be taken.

FURTHER READINGS

Brodeur, P. 1980. *The asbestos hazard*. New York: New York Academy of Sciences.

Cohen, B. L. 1982. Application of ICRP 30, ICRP 23, and radioactive waste risk assessment techniques to chemical carcinogens. *Health Physics 42*: 751–757.

Cohen, B. L., and Lee, I. S. 1979. A catalog of risks. *Health Physics 36*: 707–720.

Dickson, K. L., Maki, A. W., and Cairns, J., Jr. 1982. *Modeling the fate of chemicals in the aquatic environment*. Ann Arbor: Ann Arbor Science Publishers.

Duce, R. A., ed. 1974. *Pollutant transfer to the marine environment*. National Science Foundation/Department of Energy Pollutant Transfer Workshop proceedings.

Enthoven, A. C., and Freeman, A. M., eds. 1973. *Pollution, resources, and the environment*. New York: W. W. Norton.

Fagan, J. J. 1974. *The earth environment*. Englewood Cliffs, N.J.: Prentice-Hall.

Ferguson, J. E. 1982. *Inorganic chemistry and the earth*. Oxford: Pergamon Press, 400 p.

Garrels, R. M., Mackenzie, F. T., and Hunt, C. 1975. *Chemical cycles and the global environment: Assessing human influences*. Palo Alto: William Kaufman.

Gorman, J. 1979. *Hazards to your health: The problem of environmental disease*. New York: New York Academy of Science.

Highland, J. K., ed. 1982. *Hazardous waste disposal—Assessing the problem*. Ann Arbor: Ann Arbor Science Publications.

Hines, L. G. 1973. *Environmental issues: Population, pollution, and economics*. New York: W. W. Norton.

Iverson, O. H. 1986. Cancer risks in relation to other health hazards in a modern society. *Env. Internat. 12*: 499–503.

Kittrick, J. A., Fanning, D. S., and Hossner, L. R. 1982. *Acid sulfate weathering*. Madison: Soil Science Society of America, 234 pp.

Kneip, T. J., and Lioy, P. J., eds. 1980. *Aerosols: Anthropogenic and natural, sources and transport*. New York: New York Academy of Science.

Kraybill, H. F., Dawe, C. J., Harshbarger, J. C., and Tardiff, R. G., eds. 1977. *Aquatic pollutants and biologic effects with emphasis on neoplasia*. New York: New York Academy of Sciences.

Lehman, J. P., ed. 1983. *Hazardous waste disposal*. New York: Plenum.

Moghissi, A. A. 1986. Health risks of passive smoking. *Env. Internat. 11*: 1.

National Academy of Science. 1977. *Drinking water and health*. Washington, D.C.

National Academy of Science–National Research Council. 1974. *Geochemistry and the environment, vol. 1, the relation of selected trace elements to health and disease*. Washington, D.C.

National Academy of Science–National Research Council. 1977. *Geochemistry and the environment, vol. 2, The relation of other selected trace elements to health and disease*. Washington, D.C.

National Academy of Science–National Research Council. 1978. *Distribution of trace elements related to the occurrence of certain cancer, cardiovascular diseases, and urolithiasis*. Washington, D.C.

National Academy of Science–National Research Council. 1979. *Geochemistry of water in relation to cardiovascular disease*. Washington, D.C.

National Academy of Science–National Research Council. 1980. *Trace-element geochemistry of coal resource development related to environmental quality and health*. Washington, D.C.

National Academy of Science–National Research Council. 1984. *Groundwater contamination. Washington, D.C.*

Nriagu, J. O., ed. 1980. *Cadmium in the environment, Part I.* Toronto: John Wiley & Sons.

Pellizzari, E. D., et al. 1986. Comparison of indoor and outdoor residential levels of volatile organic chemicals in five U.S. geographical areas. *Env. Internat. 12*: 619–623.

Repace, J. L., and Lowrey, A. H. 1985. A quantitative estimate of nonsmokers lung cancer risk from passive smoking. *Env. Internat. 11*: 3–22.

Ross, M. A. 1984. A survey of asbestos-related diseases in trades and mining occupations and in factory and mining communities as a means of predicting health risks of nonoccupational exposure to fibrous minerals. *Spec. Tech. Pub. 834*: 51–104. Philadelphia: American Society for Testing and Materials.

Saffiotti, U., and Wagoner, J. K., eds. 1976. *Occupational carcinogenesis.* New York: New York Academy of Sciences.

Salomons, W., and Forstner, U., eds. 1984. *Metals in the hydrocycle.* New York: Springer-Verlag, 349 pp.

Schneider, M. J. 1979. *Persistent poisons, chemical pollutants in the environment.* New York: New York Academy of Sciences.

Singer, S. F., ed. 1970. *Global effects of environmental pollution.* Symposium of the American Association for the Advancement of Science, December 1968. New York: Springer-Verlag.

Tank, R. W., ed. 1973. *Focus on environmental geology.* New York: Oxford University Press.

Turk, A., Turk, J, and Wittes, J. T. 1972. *Ecology pollution environment.* Philadelphia: W. B. Saunders.

Upton, A. C., 1982. The biological effects of low-level ionizing radiation. *Scientific American 246*: 41–49.

U.S. Environmental Protection Agency. 1984. *National survey of hazardous waste generators and treatment, storage and disposal facilities regulated under RCRA in 1981.* Washington, D.C.: U.S. Government Printing Office.

U.S. Office of Technology Assessment. 1985. *Reproductive health hazards in the workplace.* Washington, D.C.

U.S. Office of Technology Assessment. 1985. *Superfund strategy.* Washington, D.C.

9
Conclusions and Speculations

INTRODUCTION

In the preceding eight chapters the reader has been exposed to a wide variety of topics dealing directly and indirectly with energy and mineral resources. This final chapter will recapitulate briefly some of the more salient points raised earlier, make conclusions where appropriate, and speculate on future aspects of resources and energy.

WATER AND AGRICULTURE

The most pressing resource problem facing mankind is the future availability of water. In addition, provisions for the agricultural industry are likely to be in short supply in the next century. World population in 1987 passed the 5 billion mark; at present rates of growth the world will see 10 billion inhabitants by about the year 2020, and a staggering 20 billion by 2050 or so. Further, although the average doubling period for the world population is 33 years, in several undeveloped countries such as Mexico, Kenya, and Bangladesh, this doubling period is closer to 20 years.

Water in the twenty-first century will be very unevenly distributed. As pointed out in Chapter 3, an unchecked greenhouse effect would have very severe ramifications for the United States. The breadbasket states will face unprecedented drought, with a consequence of much less food being produced than now. That in turn will have a serious impact on those nations that depend on the United States for food supplies. Further, worldwide, there is likely to be an ever-increasing amount of contamination of surface and ground waters and estuaries and lakes, even further reducing the availability of water. Despite these ominous warnings, coal continues to be burned with little attempt at reducing its emissions to the atmosphere.

The greenhouse effect will cause worldwide warming and subsequent melting of polar ice, thus causing an increase in sea level as well. Highly industrialized nations such as the United States, the United Kingdom, Japan, and others will be able to cope by construction of barriers and by other means. But what about undeveloped countries such as Bangladesh or low-lying Pacific islands? The toll from storms ravaging low-lying areas given a higher sea level will be devastating.

Water conservation is a must, and, again, this is a worldwide problem. It is imperative that the agricultural and industrial sectors implement conservation strategies, both for water quantity and quality. No longer can unchecked water consumption, waste, or water pollution be tolerated. More efficient ways to withdraw water, to impound and store water, and to make better use of water must be researched and put into practice.

Yet even these steps are not enough. We are the major polluter of the delicate earth environment. While natural volcanic eruptions may be more impressive, it is civilization's relentless race for industrialization and material comfort that is at the root of most pollution. This does not have to be the case. Many governments do not have, and even more important, many do not practice sound environmental policy on contamination. This contamination is perhaps more obvious on waters than on soils and forests, yet unchecked acid rain, acid mine drainage, and releases of wastes from the chemical and agricultural and coal industries march ever on. Even enforcing existing regulations in countries such as the United States would be a healthy step, but this is often tokenism—environmental abuse punished by a meager fine, arguments for excessively long times to implement clean-up procedures, and so on. And this situation is even worse in other countries (see discussion on Poland in Chapter 8).

To ensure a substantial supply of good cropland means a combination of sound farming practices, effective soil and water management, and a realistic government policy and plan put into effect. The world must also be prepared for shortages of agricultural chemicals and equipment in the next century. While the doubling period for world population is 33 years, that for farm equipment and chemicals used by the agricultural industries is only 10 to 15 years. This means that reserves of these vital resources may be in very short supply in the early part of the next century.

The United States has adequate supplies of potash, phosphorus, and, of course, nitrogen. Only sulfur, once thought to be overabundant, will be in short supply. For the rest of the world the picture is uneven and bleak. World potash deposits in evaporites are of limited amount and sporadically distributed. So too are many phosphate deposits. Only nitrogen, which is extracted from the atmosphere, is an assured supply. Sulfur will be hard taxed to meet demand in most industrialized countries in the near future. Part of the problem, as pointed out by Wolfgang Sassin and by Wolfgang Hafele (see references at the end of the chapter), is distribution. Even with adequate supply, making these materials available to the world community demands international cooperation of the fullest degree. Yet political and ideological barriers prevent a smooth transfer of supplies, technology, and knowledge. Many nations responded unselfishly to the famine in

Ethiopia in the mid-1980s, yet even on this small scale much of the food intended for relief purposes rotted on the docks, and funds earmarked for relief were funneled elsewhere. If even a small effort like this is only partially successful, then what arc thc odds for large-scale operations?

The nations of the world must undertake meaningful steps now to check the greenhouse effect, to drastically reduce and eventually eliminate pollution of water and soil, to implement strict conservation and management for waters and soils, to research ways to make more water available, to promote world fair-trade markets for agricultural chemicals, and to do all this in a framework of international cooperation. Even if all this is done, we should be prepared for tragedies.

METALS

There is an adequate supply of some metals, both for the United States and for the world, but this is certainly not true of all metals.

The backbone metal of any industrialized country is iron. Banded iron formations are abundant worldwide, although many are not fully exploited for a variety of reasons including market conditions, surplus, enviroscare-driven concerns, and international competition. Reserves in Canada and the United States are immense, as are reserves in South Africa, Australia, the Soviet Union, Fennoscandia, and elsewhere. In the United States, iron and steel industries are facing difficult times due to the world glut, prices, and competition, as well as high labor costs, antiquated mills, competition from minimills, and other factors.

Aluminum reserves worldwide are marginally sufficient to meet demand. As more and more countries become industrialized, their per capita consumption of aluminum will increase markedly, thus forcing more competition for this valuable metal. The United States is especially vulnerable to aluminum as local resources are impure, of low grade, costly to process, and small.

Copper surpluses have appeared on the world market in the 1980s and are likely to remain. Many foreign producers of copper are government-subsidized operations, and thus they keep producing despite the glut. This forces prices down even further and adds to the surplus. For reasons not clear, the United States actually encourages this practice, at least from some South American and African countries, although it is at a significant cost and impact on domestic operations. The United States has adequate copper supplies for several decades, but the present outlook is not favorable for a healthy copper industry.

United States supplies of lead and zinc are favorable while supplies of nickel and cobalt are not. Worldwide, zinc is very abundant, and nickel is ample (although increased demand from developing countries could change this). Lead reserves worldwide are marginally adequate to the end of the century, but may be inadequate afterwards. Environmental concerns about lead will continue to decrease demand in some areas (gasoline, paints). Cobalt is scarce worldwide, and many major industrialized powers, including the USA and USSR, are in competition for cobalt from Zaire, which contains the largest known continental reserves. Both superpowers also agree to purchase copper from Zaire in order to ensure adequate supplies of vital cobalt, thus adding to the oversupply of copper.

Should seafloor nodules and seamount sulfide crusts be harvested, then a surplus of manganese, cobalt, nickel, and still more copper will result. Yet the problems with this option, from technological to international, are very large, and it is not certain when actual economic recovery of these resources will occur. Clearly the internal bickering over the Law of the Sea in the United Nations will not be resolved quickly, and it is likely that some major power or a consortium of international leaders will take initial steps to reap this resource. The ramifications of this will be watched with worldwide interest.

Molybdenum is overly abundant in the United States, yet the domestic market is hurt by imports of by-product molybdenum from copper operations in Chile and elsewhere. World molybdenum demand is somewhat low.

Chromium supplies could become very short in the next century if turmoil in southern African countries escalates into warfare. Supplies of chromium outside Africa are not sufficient to meet world demand.

Manganese supplies for the world are probably adequate into the next century, and may be very abundant if seafloor nodules are successfully harvested. For the United States, in the absence of seafloor mining, almost 100 percent reliance on imports will continue.

World and domestic sources of titanium are large, although, because of economic factors, the United States imports nearly 100 percent of this commodity. Magnesium is extracted from seawater in most industrialized countries and thus poses no problems.

The future of the precious metals seems moderately secure. The price of gold, which is a good monitor of the world's political scene, should remain elevated well into the foreseeable future. There is escalating, not decreasing, political tension in many areas of the world and, regrettably, little indication of things getting better. Consequently, gold prices will stay high, and gold mining and production will continue to be profitable. So, too, will the platinum-group metals and silver.

Metals such as mercury, gallium, antimony, cadmium, arsenic, and others are more difficult to predict. If there is a real push for a large solar industry worldwide, then cadmium, arsenic, and gallium will be in demand. Arsenic will also continue to be in demand for agricultural purposes in many countries. Mercury, because of its unique properties (from medicinals to explosives), will always be in demand. Antimony will be more subject to local price fluctuations, and will suffer accordingly.

Tin and tungsten use will remain high. The world excess of tin may last for some time, but the United States should take pains to stockpile an adequate supply of tin for most of the world's reserves are located in southeastern Asia. Tungsten, too, is in short supply in the United States and is a vital metal for the aerospace industry.

Niobium will continue to be in demand as long as it is contemplated for fusion nuclear reactors, and the rare-earth elements should increase in demand as more uses for these fascinating metals, from superconductors to ultrahard glass table tops, are found. Beryllium demand should increase since it is not profitably extractable from volcanic rocks.

ENERGY

Chapter 7 dealt with many facets of U.S. and world energy, including availability, exploitation, production, management, risks, wastes, and health and environmental impact. There are only two major options for electric generation the next century, coal and nuclear, as world oil and gas supplies are (despite local abundances at the present time) dwindling. Other energy sources such as hydropower, wind power, solar, geothermal, and biomass and others play only locally important roles in the overall energy picture. So, too, does nuclear fusion. Optimistically, fusion energy will not be a proven source of energy until well into the next century. So let us compare coal and nuclear, recapitulating material presented in Chapter 7 and referred to in the open, refereed literature.

Coal is the most appalling polluter of the world environment. The greenhouse effect can be directly equated as proportional to the amount of coal burned. Coal wastes, from burning waste piles to acid mine drainage problems, ruin and pollute land and water on an ever-increasing scale. Acid rain ruins a very large amount of otherwise fertile forest lands each year. Fatalities due to coal mining, processing, and burning, from SO_2, CO, NO_x, and other volatiles may be in the 70,000 to 80,000 range per year. Further, the long-range effect of particulates and toxic to carcinogenic gases released by coal burning to the environment are unknown, but presumed to have more adverse than positive effects. Given our dwindling oil and gas reserves, and the uncertain status of electric-powered vehicles on a large scale, it may be more prudent to conserve coal as a source of coal liquids for the transportation sector.

The risks from the major energy options have been covered in some detail in Chapter 7. Nuclear energy is by far the safest. Natural gas is a distant second, hydropower a lagging third, then a big jump to petroleum power, another huge jump and then the riskiest, coal, is encountered. Nevertheless coal, the world's biggest polluter, killer, and spreader of carcinogens and debilitating diseases, gets nothing but praise while fear of making the environment "even weakly radioactive" is blown out of proportion. The studies done in China and elsewhere have convincingly shown that above-background radiation (though still an order of magnitude below the conservative safety limit for radiation exposure levels set by the government) has resulted in no identifiable health risks.

In the quest for energy self-sufficiency, conservation has been often promoted. Yet conservation measures have in large part failed due to lack of models, lack of proper incentives, and misconceptions perpetuated by the media and by the government. Perhaps conservation will work only when it is made mandatory for the industry, transportation, and commercial sectors.

Of all the major energy options, that which does the least damage to the environment, is safest in terms of public health, and is readily available with assured supplies for many, many decades of energy production is nuclear. This industry has an unparalleled good safety record, not only in the United States but worldwide as well. Even with the major Chernobyl disaster, this holds true. In part this is due to the small amounts of natural materials involved, as well as to the fact that the industry is regulated more than any other, and in part it is testimony to science's successful efforts to use atomic processes for peaceful purposes. Uranium supplies for conventional light-water reactors are adequate well into the next century worldwide. Should the breeder reactor, now in use in France and being developed for commercial use by the United Kingdom and the Soviet Union, be accepted in the world community, then the resources are automatically available for thousands of years.

Much is made of the potential danger of radioactive waste disposal, yet nature has been convincingly demonstrating that such wastes can be and indeed have been safely disposed of in rocks (the Oklo natural reactor, discussed in Chapter 7). Relative to coal wastes and chemical wastes, neither of which is very carefully regulated, the volume of radioactive wastes is small, the technology to deal with them is known, and the only barriers to initiating safe disposal of radioactive wastes in rocks are political.

MAJOR CHEMICALS AND INDUSTRIAL MATERIALS

The U.S. and world supplies of sodium sulfates and sodium carbonates appear to be adequate for many years, especially in the United States. The large tonnages in Wyoming alone should meet domestic demand for many decades. This is also true of borates, mainly from California, and halite, chlorine, bromine, lithium, and other commodities recovered from evaporites. Fluorine supplies in the United States are low, and imports will be necessary to meet domestic demand. Worldwide fluorine supplies are marginally adequate. Barium usage will depend on the extent of drilling and certain kinds of milling both domestically and worldwide. Should there be a return to an expanded drilling program, and this certainly has to come to the petroleum industry, then domestic barium supplies will be inadequate to meet the demand. World barium supplies appear to be adequate for at least 20 years.

Domestic supplies of some industrial materials may be in short supply in the next century. Easily recoverable sand and gravel supplies have been so exploited in parts of the midwestern and eastern United States that alternate sources must be found. Dredging of offshore sands off the northeastern United States will compensate for much of this. Of the building stones, limestone will remain healthy, sandstone less so, and granite use for buildings will diminish, due mainly to the expense in quarrying. Specialty stone such as marble and slate will stay at a somewhat constant level of consumption. Processed raw materials will continue to increase in consumption. The cement industry will show slow growth, while processed clays will show a greater increase in growth. Reserves of good-quality limestone and other ingredients for the cement industry are abundant, as are good-quality deposits of kaolin, ball clay, fire clay, bentonite, fuller's earth, and miscellaneous clay.

The world and the United States have very ample supplies of abrasives, such as corundum, garnet, and emery. United States supplies of diatomite dominate the world market.

Industrial diamonds pose a problem for the United States. Exploration in the United States has failed to uncover finds of economic significance, and thus imports of a somewhat decreasing supply will result. If the drilling industries return to full-scale operation, the demand for industrial diamonds could strain available world supplies. Artificial diamonds have already impacted on the situation, and these diamonds are major contributors to the use of bort. Since most artificial diamonds are small, they have not affected the demand for the larger industrial diamonds used in drill bits and cutting tools.

Asbestos supplies are adequate in North America to easily meet demand, but the environmental concern, more properly called enviroscare, has made the future of asbestos uncertain. Despite the fact that the common white variety chrysotile, which accounts for 93 percent of domestic consumption, has never been linked to lung cancer directly, the EPA and other agencies continue to call for a curtailment of this important commodity. Blue asbestos, crocidolite, on the other hand, is very definitely linked to mesothelioma and lung cancer, and brown asbestos, amosite, is suspect. The blue variety is mined in South Africa and Australia, and the brown variety only in South Africa. Curtailment of white asbestos use could pose serious problems for U.S. consumers. Should it be outlawed in brake linings, for example, and replaced by something less effective that would put drivers at risk?

CHEMICALS AND OUR DELICATE ENVIRONMENT

The world and the United States are faced with an onslaught of potential environmental pollution of immense magnitude. What is not needed is endless and often pointless enviroscare rhetoric, but rather hard and factually based decisions that will benefit all aspects of our environment. We need sensible policy on mineral and energy resource exploitation, on milling and production, on use, on post-operational treatments of facilities, and on disposal of wastes. One obvious example is that the coal, oil and gas, and chemical industries should be subject to at least as much monitoring and inspection and regulation as the nuclear industries. Since these are our major polluters, why not start making them responsible for their actions?

It is also apparent that we must make the effort to establish background geochemistry for the United States, and to attempt to quantitatively relate this geochemistry to regional patterns of health and disease. This kind of study is badly needed, it is feasible and doesn't bother special interest groups, and there is no good reason to procrastinate further. The same, of course, is true for funding of conservation research.

Analytical methods must improve in terms of sensitivity and speed, and ways to expedite testing of the hundreds of thousands of chemicals as to their short-term and potential long-term effects on health and the environment must be rigorously determined. Our whole pattern of approach must be changed here, for the current ways of establishing such things as "safety levels" are archaic at best.

The United States is capable of taking a leadership role, for example in the United Nations, in advocating sound environmental policy on a worldwide basis. The sad example of Poland (see Chapter 8) is a grim reminder of what unchecked environmental pollution can do, and many other nations are also very backward in their recognition of and treatment of pollution. One nation fighting a good environmental battle, while surrounded by polluters, will fail in its efforts. The Federal Republic of Germany attempts sound environmental steps, yet its efforts in atmospheric pollution are thwarted by pollution from the German Democratic Republic, Poland, and other countries, and its greatest river, the Rhine, has been affected by chemical spills upstream in Switzerland. Only patience, education, and global awareness of the seriousness of the problem can allow a meaningful solution to be reached.

We are, as a nation and as a world, bound to continue toward the goal of a better way of life. We demand raw materials, minerals, food, water, and energy. We must also demand that we reap the benefits of the earth without destroying it.

FURTHER READINGS

Hafele, Wolfgang. 1980. A global and long-range picture of energy developments. *Science* 209: 174–183.

Revelle, P., and Revelle, C. 1988. *The environment.* Boston, Jones and Bartlett, 751 pp.

Sassin, Wolfgang. 1980. Energy. *Sci. American* 243(3): 118–151.

APPENDICES

APPENDICES

APPENDIX 1 The average composition of continental crustal rocks.*

Symbol	Element	Weight Percent	Symbol	Element	Weight Percent
O	Oxygen	46.4	Pr	Praseodymium	0.00065
Si	Silicon	28.15	Dy	Dysprosium	0.00052
Al	Aluminum	8.23	Yb	Ytterbium	0.0003
Fe	Iron	5.63	Hf	Hafnium	0.0003
Ca	Calcium	4.15	Cs	Cesium	0.0003
Na	Sodium	2.36	Er	Erbium	0.00028
Mg	Magnesium	2.33	Be	Beryllium	0.00028
K	Potassium	2.09	U	Uranium	0.00027
Ti	Titanium	0.57	Br	Bromine	0.00025
H	Hydrogen	0.14	Ta	Tantalum	0.0002
P	Phosphorus	0.105	Sn	Tin	0.0002
Mn	Manganese	0.095	As	Arsenic	0.00018
F	Fluorine	0.0625	Ge	Germanium	0.00015
Ba	Barium	0.0425	W	Tungsten	0.00015
Sr	Strontium	0.0375	Mo	Molybdenum	0.00015
S	Sulphur	0.026	Ho	Holmium	0.00015
C	Carbon	0.020	Eu	Europium	0.00012
Zr	Zirconium	0.0165	Tb	Terbium	0.00011
V	Vanadium	0.0135	Lu	Lutetium	0.00008
Cl	Chlorine	0.013	I	Iodine	0.00005
Cr	Chromium	0.010	Tl	Thallium	0.000045
Rb	Rubidium	0.009	Tm	Thulium	0.000025
Ni	Nickel	0.0075	Sb	Antimony	0.00002
Zn	Zinc	0.0070	Cd	Cadmium	0.00002
Ce	Cerium	0.0067	Bi	Bismuth	0.000017
Cu	Copper	0.0055	In	Indium	0.00001
Y	Yttrium	0.0033	Hg	Mercury	0.000008
Nd	Neodymium	0.0028	Ag	Silver	0.000007
La	Lanthanum	0.0025	Se	Selenium	0.000005
Co	Cobalt	0.0025	A(r)	Argon	0.000004
Sc	Scandium	0.0022	Pd	Palladium	0.000001
N	Nitrogen	0.0020	Pt	Platinum	0.000001
Li	Lithium	0.0020	Te	Tellurium	0.000001
Nb	Niobium	0.0020	Ru	Ruthenium	0.000001
Ga	Gallium	0.0015	Rh	Rhodium	0.0000005
Pb	Lead	0.00125	Os	Osmium	0.0000005
B	Boron	0.0010	Au	Gold	0.0000004
Th	Thorium	0.00096	He	Helium	0.0000003
Sm	Samarium	0.00073	Re	Rhenium	0.0000001
Gd	Gadolinium	0.00073	Ir	Iridium	0.0000001

*A few very rare elements and short-lived radioactive elements are omitted.

Periodic table of the elements

Light Metals

Transitional Elements — Heavy Metals

Nonmetals

Period	IA	IIA	IIIB	IVB	VB	VIB	VIIB	VIIIB	VIIIB	VIIIB	IB	IIB	IIIA	IVA	VA	VIA	VIIA	VIIIA
1	1 H 1.0080																	2 He 4.003
2	3 Li 6.939	4 Be 9.012											5 B 10.81	6 C 12.011	7 N 14.007	8 O 15.9994	9 F 18.998	10 Ne 20.183
3	11 Na 22.990	12 Mg 24.31											13 Al 26.98	14 Si 28.09	15 P 30.974	16 S 32.064	17 Cl 35.453	18 Ar 39.948
4	19 K 39.102	20 Ca 40.08	21 Sc 44.96	22 Ti 47.90	23 V 50.94	24 Cr 52.00	25 Mn 54.94	26 Fe 55.85	27 Co 58.93	28 Ni 58.71	29 Cu 63.54	30 Zn 65.37	31 Ga 69.72	32 Ge 72.59	33 As 74.92	34 Se 78.96	35 Br 79.909	36 Kr 83.80
5	37 Rb 85.47	38 Sr 87.62	39 Y 88.91	40 Zr 91.22	41 Nb 92.91	42 Mo 95.94	43 Tc (99)	44 Ru 101.1	45 Rh 102.90	46 Pd 106.4	47 Ag 107.870	48 Cd 112.40	49 In 114.82	50 Sn 118.69	51 Sb 121.75	52 Te 127.60	53 I 126.90	54 Xe 131.30
6	55 Cs 132.91	56 Ba 137.34	57 TO 71	72 Hf 178.49	73 Ta 180.95	74 W 183.85	75 Re 186.2	76 Os 190.2	77 Ir 192.2	78 Pt 195.09	79 Au 197.0	80 Hg 200.59	81 Tl 204.37	82 Pb 207.19	83 Bi 208.98	84 Po (210)	85 At (210)	86 Rn (222)
7	87 Fr (223)	88 Ra 226.05	89 TO 103															

Rare Earth Elements

Lanthanide series

57 La 138.91	58 Ce 140.12	59 Pr 140.91	60 Nd 144.24	61 Pm (147)	62 Sm 150.35	63 Eu 151.96	64 Gd 157.25	65 Tb 158.92	66 Dy 162.50	67 Ho 164.93	68 Er 167.26	69 Tm 168.93	70 Yb 173.04	71 Lu 174.97

Actinide series

89 Ac (227)	90 Th 232.04	91 Pa (231)	92 U 238.03	93 Np (237)	94 Pu (242)	95 Am (243)	96 Cm (247)	97 Bk (249)	98 Cf (251)	99 Es (254)	100 Fm (253)	101 Md (256)	102 No (254)	103 Lw (257)

APPENDIX 2   Periodic table of the elements.

APPENDIX 3   Comparison of metric and English units.

UNITS		
1 kilometer (km)	=	1000 meters (m)
1 meter (m)	=	100 centimeters (cm)
1 centimeter (cm)	=	0.39 inches (in)
1 mile (mi)	=	5280 feet (ft)
1 foot (ft)	=	12 inches (in)
1 inch (in)	=	2.54 centimeters (cm)
1 square mile (mi^2)	=	640 acres (a)
1 kilogram (kg)	=	1000 grams (g)
1 pound (lb)	=	16 ounces (oz)
1 fathom	=	6 feet (ft)

CONVERSIONS		
When you want to convert:	**Multiply by:**	**To find:**
Length		
inches	2.54	centimeters
centimeters	0.39	inches
feet	0.30	meters
meters	3.28	feet
yards	0.91	meters
meters	1.09	yards
miles	1.61	kilometers
kilometers	0.62	miles
Area		
square inches	6.45	square centimeters
square centimeters	0.15	square inches
square feet	0.09	square meters
square meters	10.76	square feet
square miles	2.59	square kilometers
square kilometers	0.39	square miles
Volume		
cubic inches	16.38	cubic centimeters
cubic centimeters	0.06	cubic inches
cubic feet	0.028	cubic meters
cubic meters	35.3	cubic feet
cubic miles	4.17	cubic kilometers
cubic kilometers	0.24	cubic miles
liters	1.06	quarts
liters	0.26	gallons
cubic meters	264.2	gallons
gallons	3.78	liters
barrel (oil)	42.0	gallons

CONVERSIONS (continued)		
When you want to convert:	Multiply by:	To find:
Masses and Weights		
ounces	20.33	grams
grams	0.035	ounces
pounds	0.45	kilograms
kilograms	2.205	pounds
metric tons	10^3	kilograms
long tons	2240.	pounds
short tons	2000.	pounds
metric tons	0.984	long tons
metric tons	1.102	short tons
Energy and Power		
joules	0.239	calories
calories	3.9685×10^{-3}	British thermal unit (BTU)
kilowatt hours	10^3	watt hours
kilowatt hours	3.6×10^6	joules
kilowatt hours	3413.	BTU
watts	3.4129	BTU per hour
watts	1.341×10^{-3}	horsepower
watts	1.	joule per second
watts	14.34	calories per minute

Temperature

When you want to convert degrees Fahrenheit (°F) to degrees Celsius (°C), subtract 32 degrees and divide by 1.8.

When you want to convert degrees Celsius (°C) to degrees Fahrenheit (°F), multiply by 1.8 and add 32 degrees.

When you want to convert degrees Celsius (°C) to kelvins (K), delete the degree symbol and add 273.

When you want to convert kelvins (K) to degrees Celsius (°C), add the degree symbol and subtract 273.

AVERAGE EQUIVALENTS
1 barrel oil weighs approximately 136.4 kilograms.
1 barrel oil is equivalent to approximately 0.22 metric ton coal.
1 barrel oil yields approximately 6.0×10^9 joules of energy.
1 metric ton of coal yields approximately 27.2×10^9 joules of energy.
1 barrel of cement weighs 170.5 kilograms.

APPENDIX 4 Rock classification.

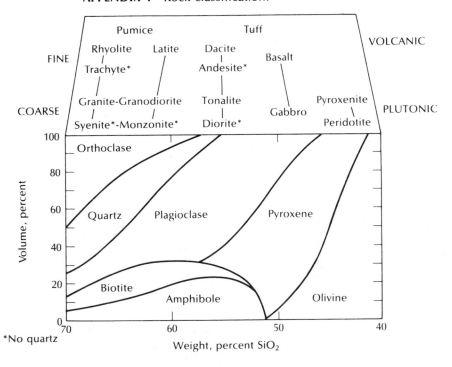

APPENDIX 5 Geologic time scale.

Era	Period	Series	Age Estimates Commonly Used for Boundaries (in million years)
Cenozoic	Quaternary	Holocene	
		Pleistocene	
			1.8
	Tertiary	Pliocene	5.0
		Miocene	22.5
		Oligocene	37.5
		Eocene	53.5
		Paleocene	65.0
Mesozoic	Cretaceous	Upper (Late) Lower (Early)	
			136
	Jurassic	Upper (Late) Middle (Middle) Lower (Early)	
			190–195
	Triassic	Upper (Late) Middle (Middle) Lower (Early)	
			225
Paleozoic	Permian	Upper (Late) Lower (Early)	
			280
	Pennsylvanian	Upper (Late) Middle (Middle) Lower (Early)	
			320
	Mississippian	Upper (Late) Lower (Early)	
			345
	Devonian	Upper (Late) Middle (Middle) Lower (Early)	
			395
	Silurian	Upper (Late) Middle (Middle) Lower (Early)	
			430–440
	Ordovician	Upper (Late) Middle (Middle) Lower (Early)	
			ca. 500
	Cambrian	Upper (Late) Middle (Middle) Lower (Early)	
			570

APPENDIX 5 (continued)

Time subdivisions of the Precambrian:

Precambrian Z—base of Cambrian to 800 m.y.*
Precambrian Y—800 m.y. to 1600 m.y.
Precambrian X—1600 m.y. to 2500 m.y.
Precambrian W—older than 2500 m.y.

*m.y. = million years

NOTE: Subdivisions in use by the U.S. Geological Survey.

Important dates in the last 300 years.

1621	Coal-fired blast furnace used for iron smelting in England.
1627	First record of oil in the United States near the present town of Cuba, N. Y.
1629	The steam-jet turbine described.
1634–74	Coal discovered in northern Illinois.
1648	A gas-turbine device was demonstrated in England (a rotor driven by hot rising gases in a cylinder).
1743	The water turbine demonstrated.
1745	First blast furnace installed in England.
1747	Atmospheric electricity discovered by Benjamin Franklin.
1748	First coal mine opened near Richmond, Va.
1763	Anthracite discovered in the United States.
1769	Anthracite coal used in Pennsylvania forges.
1775	Natural gas discovered in the Ohio Valley.
1780	Iron-making established in all thirteen states.
	Native copper mined at Santa Rita, N. M.
1785	Charles de Coulomb formulated the law of electron charges (France).
1789	Electrolysis demonstrated in Holland.
1791	Principle of bimetallic batteries discovered by Luigi Galvani.
	Gas turbine patented in England by John Barber.
1800	Blast furnace operated by a steam engine.
	Alessandro Volta invented electric battery.
	Electroplating demonstrated in Germany.
1803	Steam locomotive built in England.
1804	First steam-driven electric locomotive operated in Wales.
1809	First steamboat operated between New York and Philadelphia in thirteen days.
1807	Sir Humphry Davy discovered sodium by electrolysis.
1813	First rock drill invented.
1816	Streets lighted with manufactured gas for the first time in the United States in Baltimore, Md.
1817	Coke used instead of charcoal for smelting at Plumsock, Pa.
1820	Electromagnet discovered by William Sturgeon of England, and Hans Christian Oersted of Denmark.
1821	Michael Faraday demonstrated electric motor in England.
	First natural-gas well drilled in Fredonia, N. Y.
1824	Basic principle of thermodynamics discovered by Sadi Carnot in France.

1830	Compressed air used for tunneling in England.
	Steam hoist developed.
	Reaper and steam-powered threshing machine invented by Cyrus McCormick. Mechanized agricultural age begins.
1831	First commercial steam turbine built by William Avery in the United States.
	First electric generator built by Michael Faraday.
	Electric dynamo invented in France.
1832	Water turbine invented in France.
1833	Michael Faraday suggested that current is carried in an electrolyte by ions.
	Anthracite coal used in metal production in the United States.
1835	Coke used to produce good gray-forge iron.
1837	Electric motor patented in Vermont.
1839	First experimental electric locomotive operated between Washington and Bladensburg, Maryland.
1840	Because of discoveries by Galvani, Volta, Ampere, Ohm, Faraday, Franklin, Henry, and others, this is generally regarded as the beginning of the age of electricity.
1841	An improved method of drawing wire and spinning cables for bridges and hoisting invented by John A. Roebling.
1842	Coking coal discovered in Pennsylvania.
1843	Steam-shovel excavator operated in England.
	Artificial fertilizers (superphosphates) sold in England.
1844	Artificial hydraulic cement manufactured.
1845	Blast furnace first used raw bituminous coal.
	Iron ore discovered in Lake Superior district.
	Upper Michigan copper production began. Production from this area totaled 129.3 million pounds by 1895.
1846–49	Gold discovered in the Black Hills of South Dakota but not revealed by discoverers.
1848	First oil refinery established in England.
	California gold discovery.
	Russian mining engineers discover gold on Kenai River, Alaska.
	Iron ore smelted successfully in Lake Superior district.
1849	Microscope used to study physical metallurgy.
	Compressed-air rock drill invented.
	U.S. Department of Interior created.
1855	First practical computer built in Sweden.

1856	Development of synthetic coal-tar dyes.
	Aluminum produced electrolytically by Michael Faraday.
1857	Compressed-air rock drill developed in France.
1858	Discovery of Comstock Lode in Nevada.
1859	Oil discovered at Titusville, Pa. First oil well drilled.
1860	An expedition organized in San Francisco to prospect for gold in the Southwest. Gold discovered near Pinos Altos, N. M.
	First oil refinery built near Titusville, Pa.
	Synthetic rubber made.
	Internal-combustion engine commercially produced in Italy.
1860–70	Prospecting in the Southwest resulted in numerous mining operations and prospects in Arizona and New Mexico.
1860	Leadville, Colorado established.
1863	Natural gas used for industrial purposes in East Liverpool, Ohio.
	Motor car driven by an internal-combustion engine developed by Étienne Lenoir of Belgium.
1864	Lode claims for silver located at Butte, Montana. Output of gold and silver in the area valued at $1.2 million in 1878.
	Open-hearth furnace developed in France.
	Organization of St. Joseph Lead Co. to mine and smelt lead in Missouri.
	Gold-lead-silver mining started near Salt Lake City, Utah.
1865	Development of railroad tank car to transport crude oil.
	Formation of the first company to distribute natural gas at Fredonia, N. Y.
	First oil pipeline laid to transport oil 5 miles from Pithole City to Oil Creek Railroad, Pennsylvania.
1866	Atlantic cable installed.
1867	Invention of dynamite by Alfred Nobel.
1868	Copper shipped from Bingham Canyon, Utah.
	Gold production at Virginia City, Montana, totaled $30 million.
1869	Diamond drill brought from France to Bonne Terre, Mo.
1870s	Recognition of magnetohydrodynamic principle of generating electricity.
1871	Invention of rock drill by Ingersoll Rand.
1875–85	Pennsylvania Geological Society established first principles of modern petroleum engineering and geology.
1876	Homestake mine located in Black Hills, S. D. Organization of the Homestake Mining Co. the following year.
1877	Discovery of copper wealth near Bisbee, Arizona.

1879   Establishment of U.S. Geological Survey.

            Completion of first major oil pipeline 110 miles from Pithole City to Williamsport, Pa.

1879–83   Operation of numerous gold mining claims and prospects in Coeur d'Alene district of Idaho.

1880   Discovery of gold at Cripple Creek, Colorado, by Robert Womack.

            Tunnel bored with a pneumatic tunneler in England.

            Introduction of hydraulic rock drill.

1881   Organization of Anaconda Silver Mining Co. in Butte, Montana area.

1883   Confirmation of the anticlinal theory for accumulation of oil and gas by drilling.

1884   Multiple-stage steam turbine built in England by Sir Charles Parson.

1885   Development of petroleum cracking in the United States.

1886   Discovery of the electrometallurgical process of producing aluminum by Charles M. Hall (United States) and Paul Héroult (France) independently.

1887   Use of steel in pipelines.

1888   Production of manganese steel by Robert Hadfield.

1890s   Development of electrochemical process in the United States.

1891   Installation of first high-pressure long-distance pipeline to transport oil 120 miles from Greentown, Ind., to Chicago, Ill.

1894   Development of first oilfield near Santa Barbara, Calif.

            Export of kerosene to China by Standard Oil Co.

1895   Reorganization of the Anaconda Company to form the Anaconda Copper Mining Co.

1896   Discovery of radioactivity by Henri Becquerel of France.

1900   Marketed production of natural gas totaled 127 billion cubic feet. (Total was 24,700 billion cubic feet by 1975.)

1901   Use of solar energy to power a steam engine.

            Initiation of the first salt-dome production of oil near Beaumont, Texas.

            Motor oil marketed by Mobil Oil Co.

            Use of rotary drilling equipment at Spindletop oilfield in Texas.

1902   Operation of first satisfactory coal-face conveyor in the United States.

            England manufactured first all-metal car body of aluminum.

1903   Organization of Utah Copper Company. (Copper production started in 1907.)

            Installation of a 5,000- kilowatt steam turbine to generate electricity.

1904   Italian engineers drilled a steam well for operation of a small turbine.

Operation of the first steam-powered tractor in California.

1905   Installation of first gasoline pump at Ft. Wayne, Ind.

1906–07   First aircraft manufactured in France and England.

1907   Household detergents manufactured in Germany.

1908   Discovery of oil in Middle East area (Iran).

1909   Installation in Oklahoma of first natural gas processing plant west of Mississippi River.

1910   Establishment of the Bureau of Mines, U.S. Department of Interior.

Onshore drilling for oil in water in Louisiana.

1911   Flotation successfully used in the United States on zinc-lead ore at Butte, Montana.

1912   Demonstration of X-ray analysis of minerals.

1913   Patent of thermal cracking process to increase yield and quality of gasoline from petroleum.

1915   Start of low-cost high-tonnage gold mining at Juneau, Alaska.

1916   Manufacture of first nitrogen fertilizer in England.

1918   Discovery of nation's largest gas field in the Texas Panhandle.

1922   Use of geophysical instruments for oil exploration.

1923   Installation of mechanical coal-loading equipment to replace hand loading.

1925   Production of synthetic petroleum from coal in Europe.

1929   Use of electric logging equipment in oil wells.

1930   Production of wrought iron by continuous processes in Pennsylvania.

1930s   Development of the tungsten-carbide bit in Germany.

1930   Discovery of the largest U.S. oilfield in east Texas.

1932   Fission of the nucleus of the atom achieved by John Cockroft.

1937   Introduction of catalytic cracking in the U.S. petroleum industry.

1938   Commercial gasification of underground coal in Russia.

Development of first offshore field in Louisiana.

1939   Demonstration of nuclear fission in Germany by Otto Hahn.

1942   Production of first man-made chain reaction at the University of Chicago by Enrico Fermi and coworkers.

Preparation of the first pure compound of plutonium at the University of California.

1943   Successful treatment of taconite ore by E. W. Davis in Minnesota.

1945   Explosion of atomic bomb in the United States.

1947   Introduction of the modern scintillation counter to detect and measure radioactivity.

Offshore well out of sight of land drilled off the Louisiana coast.

1949   Largest prospecting boom in the history of the United States triggered by the discovery of uranium ore at Haystack Butte, N. M.

1951   Manufacture of electronic computer.

1953   Gasification of underground coal in Alabama.

1954   The *Nautilus*, first nuclear powered submarine, was built at Groton, Mass. It traveled 69,138 miles on the first fueling.

Establishment of the first atomic power station (5000 kilowatts) near Obninsk, Russia.

1956   Completion of first large-scale (90,000 kilowatts) atomic power station at Calder Hall, England.

1957   Commercial atomic power plant built at Shippingport, Pa.

1960   Power delivered by 12.5-megawatt geothermal steam plant in the United States. (Installed capacity for geothermal power was 600 megawatts by 1975.)

1963   Operation of a 60.9-megawatt liquid-metal fast-breeder reactor (LMFBR) in Michigan.

1967   Operations began at the Peach Bottom plant in Pennsylvania of a high-temperature gas-cooled reactor (HTGR).

1969   *Apollo XI* spacecraft landing of Neil Armstrong and Edwin Aldrin on the moon.

1970   Completion of a 275-mile pipeline to deliver coal from Black Mesa, Arizona, to the Mohave power plant at a cost of $35 million.

1970s   Operation of open-cycle magnetohydrodynamic powerplant with a considerable output by Soviet engineers.

1972   Launching of Landsat–1, formerly called ERTS–1, in the United States to transmit satellite imagery to the earth.

1973   Imposition of Arab Oil Embargo on the United States.

1974   Construction began on a 48-inch 800-mile pipeline from Prudhoe Bay to Valdez, Alaska.

Opening of federal lands to development of geothermal power.

1975   Establishment of the Ocean Mining Administration by the United States.

Index